Control Sensors and Actuators

CLARENCE W. DE SILVA

PRENTICE HALL, Englewood Cliffs, New Jersey 07632

De Silva, Clarence W.
  Control Sensors and actuators.

    Includes bibliographies and index.
    1. Automatic control.   2. Detectors.   3. Actuators.
  I. Title.
  TJ213.D38   1989   629.8   88-28998
  ISBN 0-13-171745-6

*To Charmaine, C.J., and Cheryl—as their senses develop
and as they become increasingly active.*

Editorial/production supervision and
    interior design: David Ershun
Cover design: Ben Santora
Manufacturing buyer: Mary Ann Gloriande

The publisher offers discounts on this book when ordered
in bulk quantities. For more information, write:

  Special Sales/College Marketing
  Prentice Hall
  College Technical and Reference Division
  Englewood Cliffs, NJ 07632

Printed in the United States of America

10 9 8 7 6 5 4 3 2 1

ISBN 0-13-171745-6

PRENTICE-HALL INTERNATIONAL (UK) LIMITED, *London*
PRENTICE-HALL OF AUSTRALIA PTY. LIMITED, *Sydney*
PRENTICE-HALL CANADA INC., *Toronto*
PRENTICE-HALL HISPANOAMERICANA, S.A., *Mexico*
PRENTICE-HALL OF INDIA PRIVATE LIMITED, *New Delhi*
PRENTICE-HALL OF JAPAN, INC., *Tokyo*
SIMON & SCHUSTER ASIA PTE. LTD., *Singapore*
EDITORA PRENTICE-HALL DO BRASIL, LTDA., *Rio de Janeiro*

Contents

Preface

This book is suitable for a course in control system instrumentation or for a second course in feedback control systems. There is adequate material in the book for two fifteen-week courses, one at the senior (fourth-year undergraduate) level and the other at the first-year graduate level. Also, the book can serve as a useful reference tool for practicing engineers in the field of control engineering.

The manuscript for the book evolved from the notes I developed for an undergraduate course entitled Instrumentation and Design of Control Systems and for a graduate course entitled Control System Instrumentation at Carnegie Mellon University. The undergraduate course is a senior elective, but it is a popular course that usually is taken by approximately half of the senior class. The graduate course is offered for electrical and computer engineering, mechanical engineering, and chemical engineering students. The prerequisite for both courses is a conventional undergraduate course in feedback control theory, as well as the consent of the instructor. During the development of the material for this book, a deliberate attempt was made to cover a major part of the syllabuses for two courses—Analog and Digital Control System Synthesis, and Computer Controlled Experimentation—offered in the Department of Mechanical Engineering at the Massachusetts Institute of Technology.

A control system is a dynamic system that contains a controller as an integral part. The purpose of the controller is to generate control signals that will drive the process to be controlled (the plant) in the desired manner. Actuators are needed to perform control actions as well as to drive the plant directly. Sensors and transducers are necessary to measure output signals for feedback control, to measure input signals for feedforward control, to measure process variables for system monitoring, and for a variety of other purposes. Since many different types and levels of signals are present in a control system, signal modification (including signal conditioning and signal conversion) is indeed a crucial function associated with any control system. In particular, signal modification is an important consideration in component interfacing. It is clear that a course in control system instrumentation should deal with sensors and transducers, actuators, signal modification, and controllers. Specifically, the course should address the identification of control system components with respect to functions, operation and interaction, and proper selection and interfacing of these components for various control applications. Parameter selection (including system tuning) is an important step as well. Design is a necessary part of control system instrumentation, for it is design that enables us to build a control system that meets the performance requirements—starting, perhaps, with basic components such as sensors, actuators, controllers, compensators, and signal modification devices.

The approach taken in this book is to treat the basic types of control sensors and actuators in separate chapters, but without losing sight of the fact that various components in a control system have to function as an interdependent and interconnected group in accomplishing the specific control objective. Operating principles, modeling, design considerations, ratings, specifications, and applications of the individual components are discussed. Component integration and design considerations are addressed primarily through examples and problems, which are drawn from such application systems as robotic manipulators, machine tools, ground transit vehicles, aircraft, thermal and fluid process plants, and digital computer components. It is impossible to discuss every available control system component in a textbook of this nature; for example, thick volumes have been written on measurement devices alone. In this book, some types of sensors and actuators are studied in great detail, whereas some others are treated superficially. Once students are exposed to an indepth study of some components, it should be relatively easy for them to extend the same concepts and the same study approach to other components that are functionally or physically similar. Augmenting their traditional role, the problems at the end of each chapter serve as a valuable source of information not found in the main text. In fact, the student is strongly advised to read carefully all problems in addition to the main text.

The book consists of seven chapters. Chapter 1 provides an introduction to the subject of control system instrumentation. Component modeling, rating, and matching aspects are discussed early, in chapter 2, so that the relevance and significance of these considerations can be explored in the subsequent chapters. The next three chapters of the book are devoted to sensors and transducers, and the last two chapters consider actuators. Because of space restrictions, signal modification and microprocessor-based control are not covered in the book, even though these are important topics of control system instrumentation. Although solutions are available for all problems included in the book, answers to only the numerical problems are given at the end of the book, in order to encourage independent thinking.

The planning of the course syllabus is left to the instructor, but proper consideration should be given to the course level (undergraduate or graduate), the department (mechanical, electrical, chemical, aerospace, etc.), and the instructor's own background. My general advice, however, is that all seven chapters be covered, regardless of the course level. A basic treatment of the main topics is appropriate at the undergraduate level, and an in-depth treatment of selected topics from each chapter could be undertaken at the graduate level.

The objective of this book is to provide an introductory course in control system instrumentation, emphasizing sensors and actuators. The book does not claim the virtually impossible undertaking of transforming a student into an expert in control system instrumentation during a fifteen-week period. Rather, the introductory material presented here will serve as a firm foundation for building up expertise in the subject later—perhaps in an industrial setting or in an academic research laboratory—with further knowledge of control hardware and analytical skills (along with the essential hands-on experience) gained during the process. Undoubtedly, for best results, a course in control sensors and actuators should be accompanied by a laboratory component and/or class projects.

Acknowledgments

Many people deserve special mention for making this project possible. Professor David Wormley, head of the Mechanical Engineering Department at MIT—who is an authority on the subject of control systems—and Professor Alistair MacFarlane, head of the Information Engineering Division at the University of Cambridge, have advised and guided me throughout my professional career. A special sense of gratitude is extended to Professor Ian McCausland, professor of electrical engineering at the University of Toronto, who has provided constant support and encouragement during the past fifteen years. Professor Arthur Murphy, presently a Departmental Fellow at DuPont, hired me for his department at Carnegie Mellon University in 1978 and encouraged me to teach a course on control system instrumentation. Writing this book would have been impossible without the support and encouragement given by Professor William Sirignano, presently the dean of engineering at the University of California, Irvine. Those students who attempted some of the problems in the book during the evolutionary period of the manuscript contributed to the book in no uncertain terms. I am grateful to Mr. Bernard Goodwin, executive editor and assistant vice-president of Prentice Hall, for undertaking to publish the manuscript. The two reviewers who provided constructive criticism during the proposal stage and the final manuscript stage have helped in improving the quality of the manuscript. The instrumentation experience I gained at places such as IBM Corporation, Westinghouse Electric Corporation, Bruel and Kjaer, and NASA's Lewis and Langley Research Centers was quite valuable in developing the material for the book. Ms. Mildred Gibb, who painstakingly and carefully generated the final copy of the manuscript on her word processor, deserves special thanks.

But as artificers do not work with perfect accuracy, it comes to pass that mechanics is so distinguished from geometry that what is perfectly accurate is called geometrical; what is less so, is called mechanical. However, the errors are not in the art, but in the artificers.

Sir Isaac Newton, Principia Mathematica,
Cambridge University, May 8, 1686

1

Control, Instrumentation, and Design

1.1 INTRODUCTION

The demand for servomechanisms in military applications during World War II provided much incentive and many resources for the growth of control technology. Early efforts were devoted to the development of analog controllers, which are electronic devices or circuits that generate proper drive signals for a plant (process). Parallel advances were necessary in actuating devices such as motors, solenoids, and valves that drive the plant. For feedback control, further developments in sensors and transducers became essential. With added sophistication in control systems, it was soon apparent that analog control techniques had serious limitations. In particular, linear assumptions were used to develop controllers even for highly nonlinear plants. Furthermore, complex and costly circuitry was often needed to generate even simple control signals. Consequently, most analog controllers were limited to on/off and proportional-integral-derivative (PID) actions, and lead and lag compensation networks were employed to compensate for weaknesses in such simple control actions.

The digital computer, first developed for large number-crunching jobs, was employed as a controller in complex control systems in the 1950s and 1960s. Originally, cost constraints restricted its use primarily to aerospace applications that required the manipulation of large amounts of data (complex models, several hundred signals, and thousands of system parameters) for control and that did not face serious cost restraints. Real-time control requires fast computation, and this speed of computation is determined by the required control bandwidth (or the speed of control) and parameters (e.g., time constants, natural frequencies, and damping constants) of the process that is being controlled. For instance, prelaunch monitoring and control of a space vehicle would require digital data acquisition at very high sampling rates (e.g., 50,000 samples/second). As a result of a favorable decline of computation cost (both hardware and software) in subsequent years, widespread application of digital computers as control devices (i.e., digital control) has become feasible. Dramatic developments in large-scale integration (LSI) technology and microprocessors in the

1970s resulted in very significant drops in digital processing costs, which made digital control a very attractive alternative to analog control. Today, digital control has become an integral part of numerous systems and applications, including machine tools, robotic manipulators, automobiles, aircraft autopilots, nuclear power plants, traffic control systems, and chemical process plants.

Control engineers should be able to identify or select components for a control system, model and analyze individual components or overall systems, and choose parameter values so as to perform the intended functions of the particular system in accordance with some specifications. Component identification, analysis, selection, matching and interfacing, and system tuning (adjusting parameters to obtain the required response) are essential tasks in the instrumentation and design of a control system.

1.2 CONTROL SYSTEM ARCHITECTURE

Let us examine the generalized control system represented by the block diagram in figure 1.1. We have identified several discrete blocks, depending on various functions that take place in a typical control system. Before proceeding, we must keep in mind that in a practical control system, this type of clear demarcation of components might be difficult; one piece of hardware might perform several functions, or more than one distinct unit of equipment might be associated with one function. Nevertheless, figure 1.1 is useful in understanding the architecture of a general control system. This is an analog control system because the associated signals depend on the continuous time variable; no signal sampling or data encoding is involved in the system.

Plant is the system or "process" that we are interested in controlling. By *control,* we mean making the system respond in a desired manner. To be able to accomplish this, we must have access to the *drive system* or *actuator* of the plant. We apply certain *command signals,* or input, to the *controller* and expect the plant to behave in a desirable manner. This is the *open-loop control* situation. In this case, we do not use current information on *system response* to determine the control signals. In *feedback control systems*, the control loop has to be closed; *closed-loop control* means making *measurements* of system response and employing that information to generate control signals so as to correct any output errors. The output measurements are made primarily using analog devices, typically consisting of *sensor-transducer* units.

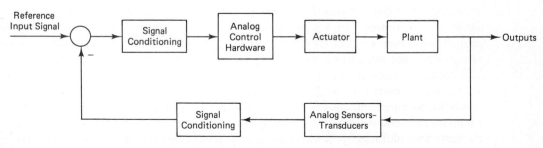

Figure 1.1. Components of a typical analog control system.

An important factor that we must consider in any practical control system is noise, including external disturbances. Noise may represent actual contamination of signals or the presence of other unknowns, uncertainties, and errors, such as parameter variations and modeling errors. Furthermore, weak signals will have to be amplified, and the form of a signal might have to be modified at various points of interaction. In these respects, *signal-conditioning methods* such as *filtering, amplification,* and *modulation* become important.

Identification of the hardware components (perhaps commercially available off-the-shelf items) corresponding to each functional block in figure 1.1 is one of the first steps of instrumentation. For example, in process control applications off-the-shelf analog proportional-integral-derivative (PID) controllers may be used. These controllers for process control applications have knobs or dials for control parameter settings—that is, proportional band or gain, reset rate (in repeats of the proportional action per unit time), and rate time constant. The control bandwidth (frequency range of operation) of these devices is specified. Various control modes—such as on/off, proportional, integral, and derivative, or combinations—are provided by the same control box.

Actuating devices (actuators) include DC motors, AC motors, stepper motors, solenoids, valves, and relays, which are also commercially available to various specifications. Potentiometers, differential transformers, resolvers, synchros, strain gauges, tachometers, piezoelectric devices, thermocouples, thermistors, and resistance temperature detectors (RTDs) are examples of sensors used to measure process response for monitoring performance and possible feedback. Charge amplifiers, lock-in amplifiers, power amplifiers, switching amplifiers, linear amplifiers, tracking filters, low-pass filters, high-pass filters, and notch filters are some of the signal-conditioning devices used in analog control systems. Additional components, such as power supplies and surge-protection units, are often needed in control, but they are not indicated in figure 1.1 because they are only indirectly related to control functions. Relays and other switching devices and modulators and demodulators may also be included.

1.3 DIGITAL CONTROL

Direct digital control (DDC) systems are quite similar to analog control systems. The main difference in a DDC system is that a digital computer takes the place of the analog controller in figure 1.1. Control computers have to be dedicated machines for real-time operation where processing has to be synchronized with plant operation and actuation requirements. This also requires a real-time clock. Apart from these requirements, control computers are basically no different from general-purpose digital computers. They consist of a processor to perform computations and to oversee data transfer, memory for program and data storage during processing, mass storage devices to store information that is not immediately needed, and input/output devices to read in and send out information. Digital control systems might utilize digital instruments and additional processors for actuating, signal-conditioning, or measuring functions, as well. For example, a stepper motor that responds with incremental motion steps when driven by pulse signals can be considered a digital actuator. Furthermore, it usually contains digital logic circuitry in its drive system. Similarly, a two-position solenoid is a digital (binary) actuator. Digital flow control may be accomplished using a digital control valve. A typical digital valve consists of a bank of orifices, each sized in proportion to a place value of a binary word (2^i, $i = 0, 1, 2, \ldots, n$). Each orifice is actuated by a separate rapid-acting on/off solenoid. In this manner, many digital combinations of flow values can be obtained. Direct digital measurement of displacements and velocities can be made using shaft encoders. These are digital transducers that generate coded outputs (e.g., in binary or gray-scale representation) or pulse signals that can be coded using counting circuitry. Such outputs can be read in by the control computer with relative ease. Frequency counters also generate digital signals that can be fed directly into a digital controller. When measured signals are in the analog form, an analog front end is necessary to interface the transducer and the digital controller. Input/output interface boards that can take both analog and digital signals are available with digital controllers.

A block diagram of a direct digital control system is shown in figure 1.2. Note that the functions of this control system are quite similar to those shown in figure 1.1 for an analog control system. The primary difference is the digital controller (processor), which is used to generate the control signals. Therefore, analog measurements and reference signals have to be sampled and encoded prior to digital processing within the controller. Digital processing can be conveniently used for signal conditioning as well. Alternatively, digital signal processing (DSP) chips can function as digital controllers. However, analog signals are *preconditioned,* using analog circuitry prior to digitizing in order to eliminate or minimize problems due to *aliasing distortion* (high-frequency components above half the sampling frequency appearing as low-frequency components) and *leakage* (error due to signal truncation) as well as to improve the signal level and filter out extraneous noise. The drive sys-

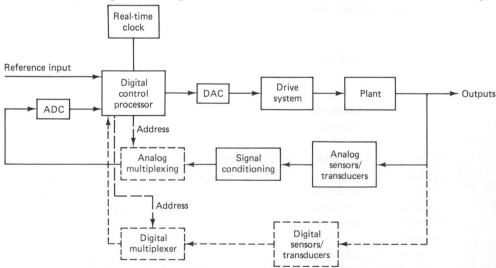

Figure 1.2. Block diagram of a direct digital control system.

tem of a plant typically takes in analog signals. Often, the digital output from controller has to be converted into analog form for this reason. Both *analog-to-digital conversion* (ADC) and *digital-to-analog conversion* (DAC) can be interpreted as signal-conditioning (modification) procedures. If more than one output signal is measured, each signal will have to be conditioned and processed separately. Ideally, this will require separate conditioning and processing hardware for each signal channel. A less expensive (but slower) alternative would be to time-share this expensive equipment by using a *multiplexer*. This device will pick one channel of data from a bank of data channels in a sequential manner and connect it to a common input device. Both analog and digital multiplexers are available. In a digital multiplexer, the input signals come from a bank of digital sensors, and the output signal itself, which would be in digital form, goes directly into the digital controller. High-speed multiplexers (e.g., over 50,000 switchings/second) use electronic switching.

For complex processes with a large number of input/output variables (e.g., a nuclear power plant) and with systems that have various operating requirements (e.g., the space shuttle), centralized direct digital control is quite difficult to implement. Some form of distributed control is appropriate in large systems such as manufacturing cells, factories, and multicomponent process plants. A favorite distributed control architecture is provided by heirarchical control. Here, distribution of control is available both geographically and functionally. An example for a three-level hierarchy is shown in figure 1.3. Management decisions, supervisory control, and coordination between plants are provided by the management (supervisory) computer, which is at the highest level (level 3) of the hierarchy. The next lower level computer generates control settings (or reference inputs) for each control region in the corresponding plant. Set points and reference signals are inputs to the direct digital control (DDC) computers that control each control region. The computers communicate using a suitable information network. Information transfer in both directions (up and down) should be possible for best performance and flexibility. In master–slave distributed control, only downloading of information is available.

1.4 SIGNAL CLASSIFICATION IN CONTROL SYSTEMS

A digital control system can be loosely interpreted as one that uses a digital computer as the controller. It is more appropriate, however, to understand the nature of the signals that are present in a control system when identifying it as a digital control system.

Analog signals are continuous in time. They are typically generated as outputs of a dynamic system. (Note that the dynamic system could be a signal generator or any other device, equipment, or physical system.) Analytically, analog signals are represented as functions of the continuous time variable t.

Sampled data are, in fact, pulse amplitude–modulated signals. In this case, information is carried by the amplitude of each pulse, with the width of the pulses kept constant. For constant sampling rate, the distance between adjacent pulses is also kept constant. In a physical situation, a pulse amplitude–modulated signal is generated through a *sample-and-hold operation,* in which the signal is sampled at

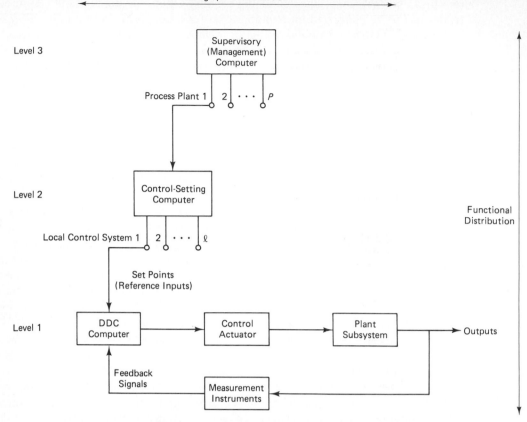

Figure 1.3. A three-level hierarchical control scheme.

the beginning of the *sampling period* and kept constant at that value, irrespective of the true value of the signal over that period. An important advantage of sampling is that expensive equipment can be shared among many signals. Furthermore, sampling is necessary in real-time digital processing to allow for the processing time. Analytically, sampled data consist of a *sequence* of numbers (or a function of integer variable).

   Digital data are coded numerical data. For example, a binary code or ASCII (American Standard Code for Information Interchange) may be used to represent each value in a sequence of digital data. The code itself determines the actual value of a particular unit of digital information. Typically, digital data are generated by *digital processors, digital transducers, counters, encoders,* and other such digital devices.

   Table 1.1 summarizes the identifying characteristics of these three types of data. *Analog systems* generate analog signals only. *Sampled-data systems* depend on analog data as well as sampled data. *Digital systems,* however, utilize all three types of signals, generally at different levels of interaction. Sampled-data systems and digital systems may be modeled using discrete-time models (see table 1.2).

TABLE 1.1 SIGNAL CATEGORIES FOR IDENTIFYING CONTROL SYSTEM TYPES

Signal (data) category	Description
Analog signals (data)	Continuous in time t; typically represents an output of a dynamic system
Sampled data	Pulse amplitude–modulated signals
	Information carried by pulse amplitude
	Typically generated by sample-and-hold process
Digital data	Coded numerical data; the particular code determines the numerical value
	Typically generated by digital processors, digital transducers, and counters

TABLE 1.2 REPRESENTATIVE ANALYTICAL CHARACTERISTICS OF CONTINUOUS-TIME AND DISCRETE-TIME SYSTEMS

System	Analytical model	
	Time domain	Transfer-function domain
Continuous-time systems	Differential equations	Laplace transfer functions or Fourier frequency response functions
Discrete-time systems	Difference equations	Z-transform transfer functions

Example 1.1

The sampling period ΔT for data acquisition is an important parameter in real-time digital control. Discuss the significance of the sampling period.

Solution   On the one hand, ΔT has to be sufficiently large so that required processing and data transfer can be done during that time for each control step. This is crucial in real-time control. On the other hand, ΔT should be small enough to meet control bandwidth and process dynamics requirements. Shannon's sampling theorem states that in a sampled signal, the maximum meaningful frequency is the Nyquist frequency f_c, which is given by half the sampling rate:

$$f_c = \frac{1}{2\Delta T} \tag{1.1}$$

It follows that we should select ΔT such that the significant frequency content of the input/output signals of the particular process stays within the Nyquist frequency. Note that to be able to control the process effectively, all natural frequencies of interest in the plant should be smaller than f_c. Once f_c is chosen in this manner for digital control, it is also important to choose analog components, such as signal-conditioning devices in the control system, to have an operating bandwidth larger than f_c. For example, if the operating bandwidth of a robotic manipulator is specified to be 50 Hz, one must make sure that the associated analog sensors and transducers (resolvers, tachometers, etc.) and signal-conditioning devices (e.g., low-pass filters, charge amplifiers) have an operating bandwidth greater than 50 Hz—preferably about 200 Hz. Furthermore, the sampling period has to be smaller than 10 ms (from equation 1.1), preferably about 2 ms. It is then necessary to make sure that the control computer is capable of doing all the processing needed in each control increment within this time. Otherwise, distribution of control tasks might be needed. Parallel processing is another option. Another alternative is to employ a hardware implementation of the controller. Simplification of control algorithms should also be attempted, but without sacrificing the accuracy requirements. In general, distributed control is better than using a single control computer of larger capacity and faster speed.

1.5 ADVANTAGES OF DIGITAL CONTROL

The current trend toward using dedicated, microprocessor-based, and often decentralized (distributed) digital control systems in industrial applications can be rationalized in terms of the major advantages of digital control. The following are some of the important considerations.

1. Digital control is less susceptible to noise or parameter variation in instrumentation because data can be represented, generated, transmitted, and processed as binary words, with bits possessing two identifiable states.
2. Very high accuracy and speed are possible through digital processing. Hardware implementation is usually faster than software implementation.
3. Digital control can handle repetitive tasks extremely well, through programming.
4. Complex control laws and signal conditioning methods that might be impractical to implement using analog devices can be programmed.
5. High reliability in operation can be achieved by minimizing analog hardware components and through decentralization using dedicated microprocessors for various control tasks (see figure 1.3).
6. Large amounts of data can be stored using compact, high-density data storage methods.
7. Data can be stored or maintained for very long periods of time without drift and without being affected by adverse environmental conditions.
8. Fast data transmission is possible over long distances without introducing dynamic delays, as in analog systems.
9. Digital control has easy and fast data retrieval capabilities.
10. Digital processing uses low operational voltages (e.g., 0–12 V DC).
11. Digital control has low overall cost.

Some of these features should be obvious; the rest should become clear as we proceed through the book.

1.6 FEEDFORWARD CONTROL

Many control systems have inputs that do not participate in feedback control. In other words, these inputs are not compared with feedback (measurement) signals to generate control signals. Some of these inputs might be important variables in the plant (process) itself. Others might be undesirable inputs, such as external disturbances that are unwanted yet unavoidable. Performance of a control system can generally be improved by measuring these (unknown) inputs and somehow using the information to generate control signals. Since the associated measurement and control (and compensation) take place in the forward path of the control system, this method of control is known as *feedforward control*. Note that in feedback control, unknown "outputs" are measured and compared with known (desired) inputs to generate control signals. In feedforward control, unknown "inputs" are measured and that information, along with desired inputs, is used to generate control signals that can reduce errors due to these unknown inputs or variations in them.

A block diagram of a typical control loop that uses feedforward control is shown in figure 1.4. In this system, in addition to feedback control, feedforward control is used to reduce the effects of a disturbance input that enters the plant. The disturbance input is measured and fed into the controller. The controller uses this information to modify the control action so as to compensate for the effect of the disturbance input.

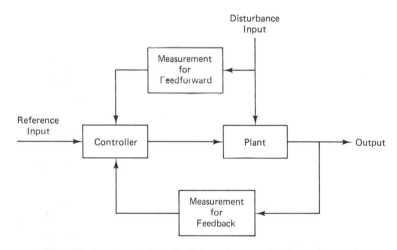

Figure 1.4. A typical feedback loop that uses feedforward control.

As a practical example, consider the natural gas home heating system shown in figure 1.5a. A simplified block diagram of the system is shown in figure 1.5b. In conventional feedback control, the room temperature is measured and its deviation from the desired temperature (set point) is used to adjust the natural gas flow into the furnace. On/off control is used in most such applications. Even if proportional or three-mode (proportional-integral-derivative) control is employed, it is not easy to steadily maintain the room temperature at the desired value if there are large

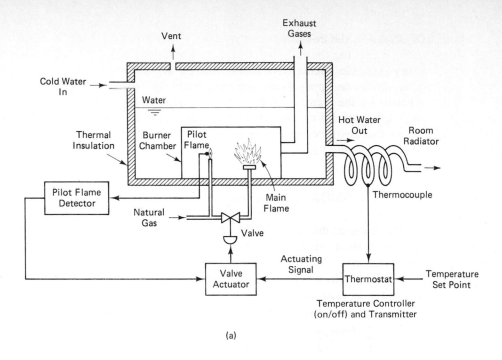

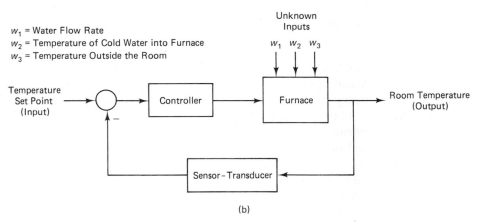

Figure 1.5. (a) A natural gas home heating system. (b) A block diagram representation of the system.

changes in other (unknown) inputs to the system, such as water flow rate through the furnace, temperature of water entering the furnace, and outdoor temperature. Better results can be obtained by measuring these disturbance inputs and using that information in generating the control action. This is feedforward control. Note that in the absence of feedforward control, any changes in the inputs w_1, w_2, and w_3 in figure 1.5 will be detected only through their effect on the feedback signal (room temperature). Hence, the subsequent corrective action can lag behind the cause (changes in w_i) considerably. This delay will lead to large errors and possible instability problems. With feedforward control, information on the disturbance inputs w_i will be available to the controller immediately, thereby speeding up the control action and also improving the response accuracy. Faster action and improved accuracy are two very desirable effects of feedforward control.

In some applications, control inputs are computed using accurate dynamic models for the plants, and the computed inputs are used for control purposes. This is a popular way for controlling robotic manipulators, for example. This method is also known as feedforward control. To avoid confusion, however, it is appropriate to denote this method as *computed-input control*.

1.7 INSTRUMENTATION AND DESIGN

In the previous discussion, we have identified several characteristic constituents of a control system. Specifically, we are interested in

- The *plant*, or the dynamic system to be controlled
- Signal *measurement* for system evaluation (monitoring) and for feedback and feedforward control
- The *drive system* that actuates the plant
- *Signal conditioning* by filtering and amplification and *signal modification* by modulation, demodulation, ADC, DAC, and so forth, into an appropriate form
- The *controller* that generates appropriate drive signals for the plant

Each function or operation within a control system can be associated with one or more physical devices, components, or pieces of equipment, and one hardware unit may accomplish several of the control system functions. By *instrumentation*, in the present context, we mean the identification of these various instruments or hardware components with respect to their functions, operation, and interaction with each other and the proper selection and interfacing of these components for a given application—in short, "instrumenting" a control system.

By *design*, we mean the process of selecting suitable equipment to accomplish various functions in the control system; developing the system architecture; matching and interfacing these devices; and selecting the parameter values, depending on the system characteristics, in order to achieve the desired objectives of the overall control system (i.e., to meet design specifications), preferably in an optimal manner and according to some performance criterion. In the present context, design is included as an instrumentation objective. In particular, there can be many designs that meet a given set of performance requirements.

Identification of key design parameters, modeling of various components, and analysis are often useful in the design process. This book provides fundamentals of sensing and actuation for electromechanical control systems. Emphasis is placed on control systems that perform motion- and force-related dynamic tasks. Sensors and transducers and actuators in this category will be discussed with respect to their performance specification, principles of operation, physical characteristics, modeling and analysis, selection, component interfacing, and determination of parameter values. Both analog and digital devices will be studied. Design examples and case studies drawn from applications such as automated manufacturing and robotics, transit vehicles, dynamic testing, and process control will be discussed throughout the book.

PROBLEMS

1.1. Giving appropriate examples, compare and contrast analog signals, sampled-data signals, and digital signals. What are the relative advantages and disadvantages of these three types of data?

1.2. What are differential equations and what are difference equations? Explain their significanace in the context of control system analysis, discussing the need for their solution as related to a digital control system.

1.3. (a) What is an open-loop control system and what is a feedback control system? Give one example of each case.

(b) A simple mass-spring-damper system (simple oscillator) is excited by an external force $f(t)$. Its displacement response y (see figure P1.3a) is given by the differential equation

$$m\ddot{y} + b\dot{y} + ky = f(t)$$

A block diagram representation of this sytem is shown in figure P1.3b. Is this a feedback control system? Explain and justify your answer.

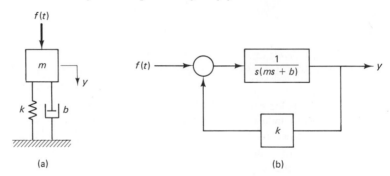

(a)   (b)

Figure P1.3. (a) A mechanical system representing a simple oscillator. (b) A block diagram representation of the simple oscillator.

1.4. You are asked to design a control system to turn on lights in an art gallery at night, provided that there are people inside the gallery. Explain a suitable control system, identifying the open-loop and feedback functions, if any, and describing the control system components.

1.5. Into what classification of control system components (actuators, signal modification devices, controllers, and measuring devices) would you put the following?
(a) Stepping motor   **(e)** DAC
(b) Proportional-plus-intergration circuit   **(f)** Optical incremental encoder
(c) Power amplifier   **(g)** Process computer
(d) ADC   **(h)** FFT analyzer

1.6. In feedforward control (computed-torque control) of robotic manipulators, joint torques (or forces) are computed from joint motion variables (displacements, velocities, and accelerations) using a suitable dynamic model for the manipulator. A 6-df (degree of freedom) model of a particular manipulator, formulated using nonrecursive Lagrangian dynamic equations, required aproximtely 10 s for one cycle of torque computation on a VAX 11/750 minicomputer. If a control bandwidth of at least 10 Hz is needed for typical applications of this robot, discuss whether real-time computed-torque control is possible in this case. Suggest improvements.

1.7. Discuss possible sources of error that can make open-loop control or feedforward control meaningless in some applications. How would you correct the situation?

1.8. A flexible manufacturing cell consists of a set of machine tools, robots, parts transfer units (e.g., conveyors), and gaging stations (e.g., vision-based) managed by a single host computer. Each robot has a separate computer (e.g., Motorola M68000) for trajectory generation and for command setting. Each joint of a manipulator is controlled using a separate microprocessor (e.g., TMS-320), through direct digital control. Identifying control functions in each level, discuss the complete system as a hierarchical control system. A photograph of a flexible manufacturing cell is shown in figure P1.8.

Figure P1.8. A flexible manufacturing cell (courtesy of the Robotics Institute, Carnegie Mellon University).

1.9. Compare analog control and direct digital control for motion control in high-speed applications of industrial manipulators. Give the advantages and disadvantages of each control method for this application.

1.10. Resolution of a device is defined as the smallest useful and detectable increment (change) in the output of the device. Dynamic range is the ratio of output range to resolution, expressed in decibels (dB). A digital control valve has four orifices. It represents a four-digit binary number, and flow through each orifice is proportional to the corresponding position value. The smallest orifice allows a flow of f_o through it. Calculate the resolution and the dynamic range of the valve.

1.11. A soft-drink bottling plant uses an automated bottle-filling system. Desribe the operation of such a system, indicating various components in the control system and their functions. Typical components would include conveyor belt; a motor for the conveyor, with start/stop controls; a measuring cylinder, with inlet valve, exit valve, and level sensors; valve actuators; and an alignment sensor for the bottle and the measuring cylinder.

1.12. Consider the natural gas home heating system shown in figure 1.5. Describe the functions of various components in the system and classify them into controller, actuator, sensor, and signal modification function groups. Explain the operation of the overall system and suggest possible improvements to obtain more stable and accurate temperature control.

1.13. In each of the following examples, indicate at least one (unknown) input that should be measured and used for feedforward control to improve the accuracy of the control system.

(a) A servo system for positioning a mechanical load. The servo motor is a field-controlled DC motor, with position feedback using a potentiometer and velocity feedback using a tachometer.

(b) An electric heating system for a pipeline carrying a liquid. The exit temperature of the liquid is measured using a thermocouple and is used to adjust the power of the heater.

(c) A room heating system. Room temperature is measured and compared with the set point. If it is low, a valve of a steam radiator is opened; if it is high, the valve is shut.

(d) An assembly robot that grips a delicate part to pick it up without damaging the part.

(e) A welding robot that tracks the seam of a part to be welded.

1.14. Consider the system shown by the block diagram in figure P1.14a. Note that

$$G_p(s) = \text{plant transfer function}$$

$$G_c(s) = \text{controller transfer function}$$

$$H(s) = \text{feedback transfer function}$$

$$G_f(s) = \text{feedforward compensation transfer function}$$

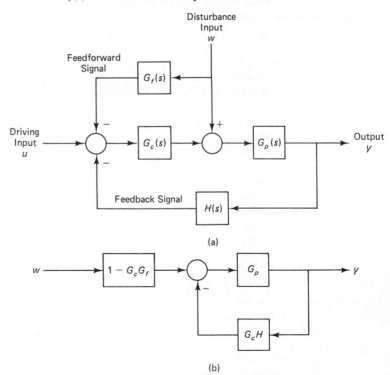

(a)

(b)

Figure P1.14. (a) A block diagram for a system with feedforward control. (b) Reduced form in the absence of the driving input.

The disturbance input w is measured, compensated using G_f, and fed into the controller, along with the driving input u.

(a) Obtain the transfer function relationship between the output y and the driving input u in the absence of the disturbance input w.

(b) Show that in the absence of u, the block diagram can be drawn as in figure P1.14b. Obtain the transfer relationship between y and w in this case.

(c) From (a) and (b), write an expression for y in terms of u and w.

(d) Show that the effect of disturbance is fully compensated if the feedforward compensator is given by

$$G_f(s) = \frac{1}{G_c(s)}$$

1.15. A typical input variable is identified for each of the following examples of dynamic systems. Give at least one output variable for each system.

(a) Human body: neuroelectric pulses

(b) Company: information

(c) Power plant: fuel rate

(d) Automobile: steering wheel movement

(e) Robot: voltage to joint motor

1.16. Measuring devices (sensors-transducers) are useful in measuring outputs of a process for feedback control.

(a) Give other situations in which signal measurement would be important.

(b) List at least five different sensors used in an automobile engine.

1.17. Hierarchical control has been applied in many industries, including steel mills, oil refineries, glass works, and automated manufacturing. Most applications have been limited to two or three levels of hierarchy, however, The lower levels usually consist of tight servo loops, with bandwidths on the order of 1 kHz. The upper levels typically control production planning and scheduling events measured in units of days or weeks.

(a) Estimate event duration at the lowest level and control bandwidth (in hertz) at the highest level for this type of application.

(b) A five-level hierarchy for a flexible manufacturing facility is as follows: The lowest level (level 1) handles servo control of robotic manipulator joints and machine tool degrees of freedom. The second level performs activities such as coordinate transformation in machine tools that are required in generating control commands for various servo loops. The third level converts task commands into motion trajectories (of manipulator end effector, machine tool bit, etc.) expressed in world coordinates. The fourth level converts complex and general task commands into simple task commands. The top level (level 5) performs supervisory control tasks for various machine tools and material-handling devices, including coordination, scheduling, and definition of basic moves. Suppose that this facility is used as a flexible manufacturing cell for turbine blade production. Using diagrams to show tasks at various levels of the hierarchy, describe the operation of this manufacturing cell.

1.18. According to some observers in the process control industry, early brands of analog control hardware had a product life of about twenty years. New hardware controllers can become obsolete in a couple of years, even before their development costs are recovered. As a control instrumentation engineer responsible for developing an off-the-shelf process controller, what features would you incorporate in the controller to correct this problem to a great extent?

1.19. The programmable controller (PC) is a sequential control device that can sequentially and repeatedly activate a series of output devices (e.g., motors, valves, alarms, signal lights) on the basis of the states of a series of input devices (e.g., switches, two-state sensors). Show how a programmable controller and a vision system consisting of a solid-state camera and a simple image processor (say, with an edge-detection algorithm) could be used for sorting fruits on the basis of quality and size for packaging and pricing.

REFERENCES

BARNEY, G. C., *Intelligent Instrumentation*. Prentice-Hall, Englewood Cliffs, N.J., 1985.

BECKWITH, T. G., BUCK, N. L., and MARANGANI, R. D. *Mechanical Measurements,* 3d ed. Addison-Wesley, Reading, Mass., 1982.

DALLY, J. W., RILEY, W. F., and McCONNELL, K. G. *Instrumentation for Engineering Measurements*. Wiley, New York, 1984.

DESILVA, C. W. *Dynamic Testing and Seismic Qualification Practice*. Lexington Books, Lexington, Mass., 1983.

————(Consulting Ed.). *Measurements and Control Journal,* all issues from February 1983 to December 1988.

FRANKLIN, G. F., and POWELL, J. D. *Digital Control of Dynamic Systems*. Addison-Wesley, Reading, Mass., 1980.

GIBSON, J. E., and TUTEUR, F. B. *Control System Components*. McGraw-Hill, New York, 1958.

HORDESKI, M. F. *The Design of Microprocessor, Sensor, and Control Systems*. Reston, Reston, Va., 1985.

JOHNSON, C. D. *Microprocessor-Based Process Control*. Prentice-Hall, Englewood Cliffs, N.J., 1984.

POTVIN, J. *Applied Process Control Instrumentation*. Reston, Reston, Va., 1985.

2

Performance Specification and Component Matching

2.1 INTRODUCTION

In feedback control systems, plant response is measured and compared with a reference input and the error is automatically employed in controlling the plant. It follows that a *measurement system* is an essential component in any feedback control system and forms a vital link between the plant and the controller. Measurements are needed in many engineering applications. The measurement process has to be automated, however, in control systems applications.

A typical measurement system consists of one or more *sensor-transducer* units and associated *signal-conditioning* (and modification) devices (see figure 2.1). Filtering to remove unwanted noise and amplification to strengthen a needed signal are considered signal conditioning. Analog to-digital conversion (ADC), digital-to-analog conversion (DAC), modulation, and demodulation are signal modification methods. Note that signal conditioning can be considered under the general heading of signal modification. Even though data recording is an integral function in a typical data acquisition system, it is not a crucial function in a feedback control system. For this reason, we shall not go into details of data recording devices in this book. In a multiple measurement environment, a *multiplexer* could be employed prior to or following the signal-conditioning process, in order to pick one measured signal at a time from a bank of data channels for subsequent processing. In this manner, one unit of expensive processing hardware can be time-shared between several signals. Sensor-transducer devices are predominantly analog components that generate analog signals, even though *direct digital transducers* are becoming increasingly

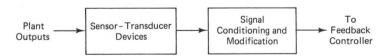

Figure 2.1. Schematic representation of a measurement system.

popular in digital control applications. When analog transducers are employed, *analog-to-digital converters* (ADCs) have to be used to convert analog signals into digital data for digital control. This signal modification process requires sampling of analog signals at discrete time points. Once a value is sampled, it is encoded into a digital representation such as straight binary code, a gray code, binary-coded decimal (BCD) code, or American Standard Code for Information Interchange (ASCII). The changes in an analog signal due to its transient nature should not affect this process of ADC. To guarantee this, a *sample-and-hold operation* is required during each sampling period. For example, the value of an analog signal is detected (sampled) in the beginning of each sampling period and is assumed constant (held) throughout the entire sampling period. This is, in fact, the zero-order hold operation. The operations of multiplexing, sampling, and digitizing have to be properly synchronized under the control of an accurate timing device (a *clock*) for proper operation of the control system. This procedure is shown schematically in figure 2.2.

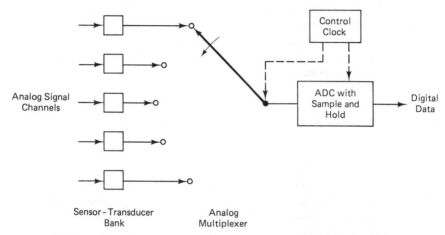

Figure 2.2. Measurement, multiplexing, and analog-to-digital conversion.

All devices that assist in the measurement procedure can be interpreted as components of the measurement system. Selection of available components for a particular application or design of new components should rely heavily on performance specifications for these components. A great majority of instrument ratings provided by manufacturers are in the form of static parameters. In control applications, however, dynamic performance specifications are also very important. In this chapter, we shall study instrument ratings and parameters for performance specification, pertaining to both static and dynamic characteristics of instruments.

When two or more components are interconnected, the behavior of individual components in the overall system can deviate significantly from their behavior when each component operates independently. Matching of components in a multicomponent system, particularly with respect to their impedance characteristics, should be done carefully in order to improve system performance and accuracy. In this chapter, we shall also study basic concepts of impedance and component matching. Although the discussion is primarily limited to components in a measurement system, the ideas are applicable to many other types of components in a control system. Discussions and developments in this chapter are quite general; they do not address specific designs or hardware components. Specific instruments, their operating details, and physical hardware will be discussed in subsequent chapters.

2.2 SENSORS AND TRANSDUCERS

The *output variable* (or response) that is being measured is termed the *measurand*. Examples are acceleration and velocity of a vehicle, temperature and pressure of a process plant, and current through an electric circuit. A measuring device passes through two stages while measuring a signal. First, the measurand is *sensed*. Then, the measured signal is *transduced* (or converted) into a form that is particularly suitable for transmitting, signal conditioning, processing, or driving a controller or actuator. For this reason, output of the transducer stage is often an electrical signal. The measurand is usually an analog signal, because it represents the output of a dynamic system in feedback control applications. Transducer output is discrete in direct digital transducers. This facilitates the direct interface of a transducer with a digital processor. Since the majority of transducers used in control system applications are still analog devices, we shall consider such devices first (in chapters 3 and 4). Digital transducers will be discussed subsequently (in chapter 5).

The sensor and transducer stages of a typical measuring device are represented schematically in figure 2.3a. As an example, consider the operation of a piezoelectric accelerometer (figure 2.3b). In this case, acceleration is the measurand. It is first converted into an inertia force through a mass element and is exerted on a piezoelectric crystal within which a strain (stress) is generated. This is considered the sensing stage. The stress generates a charge inside the crystal, which appears as an electric signal at the output of the accelerometer. This stress-to-charge conversion or stress-to-voltage conversion can be interpreted as the transducer stage.

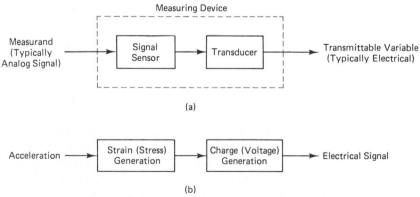

(a)

(b)

Figure 2.3. (a) Schematic representation of a measuring device. (b) Operation of a piezoelectric accelerometer.

A complex measuring device can have more than one sensing stage. More often, the measurand goes through several transducer stages before it is available for control and actuating purposes. Sensor and transducer stages are functional stages, and sometimes it is not easy or even feasible to identify physical elements associated with them. Furthermore, this separation is not very important in using existing devices. Proper separation of sensor and transducer stages (physically as well as functionally) can be crucial, however, when designing new measuring instruments.

In some books, signal-conditioning devices such as electronic amplifiers are also classified as transducers. Since we are treating signal-conditioning and modification devices separately from measuring devices, this unified classification is avoided whenever possible in this book. Instead, the term *transducer* is used primarily in relation to measuring instruments. Following the common practice, however, the terms *sensor* and *transducer* will be used interchangeably to denote measuring instruments.

2.3 TRANSFER FUNCTION MODELS FOR TRANSDUCERS

Pure transducers depend on nondissipative coupling in the transduction stage. *Passive transducers* (sometimes called *self-generating transducers*) depend on their power transfer characteristics for operation. It follows that pure transducers are essentially passive devices. Some examples are *electromagnetic, thermoelectric, radioactive, piezoelectric,* and *photovoltaic* transducers. *Active transducers,* on the other hand, do not depend on power conversion characteristics for their operation. A good example is a *resistive* transducer, such as a potentiometer, that depends on its power dissipation to generate the output signal. Note that an active transducer requires a separate power source (power supply) for operation, whereas a passive transducer derives its power from a measured signal (measurand). In this classification of transducers, we are dealing with power in the immediate transducer stage associated with the measurand, not the power used in subsequent signal conditioning. For example, a piezoelectric charge generation is a passive process. But in order to condition the generated charge, a charge amplifier that uses an auxiliary power source would be needed.

There are advantages and disadvantages in both types of transducers. In particular, since passive transducers derive their energy almost entirely from the measurand, they generally tend to distort (or load) the measured signal to a greater extent than an active transducer would. Precautions can be taken to reduce such loading effects, as will be discussed in a future section. On the other hand, passive transducers are generally simple in design, more reliable, and less costly.

A majority of sensors-transducers can be interpreted as *two-port elements* in which, under steady conditions, energy (or power) transfer into the device takes place at the *input port* and energy (or power) transfer out of the device takes place at the *output port*. Each port of a two-port transducer has a *through variable,* such as force or current, and an *across variable,* such as velocity or voltage, associated with it. Through variables are sometimes called *flux variables,* and across variables are called *potential variables.* Through variables are not always the same as *flow vari-*

ables, which are used exclusively in *bond graph* models. Similarly, across variables are not the same as *effort variables,* which are used in bond graph terminology. For example, force is an effort variable, but it is also a through variable. Similarly, velocity is a flow variable and is also an across variable. The concept of effort and flow chanical impedance, but in analysis, mechanical impedance is not analogous to electrical impedance.

A two-port device can be modeled by the transfer relation

$$\mathbf{G}\begin{bmatrix} v_i \\ f_i \end{bmatrix} = \begin{bmatrix} v_o \\ f_o \end{bmatrix} \tag{2.1}$$

where $\mathbf{G}$ is a 2×2 transfer function matrix, v_i and f_i denote the across and through variables at the input port, and v_o and f_o denote the corresponding variables at the output port (figure 2.4). This representation essentially assumes a *linear model* for transducer, so that the associated transfer functions (elements in matrix $\mathbf{G}$) are defined and valid. Such transducers are known as *ideal transducers.* Note that at a given port, if one variable is considered the input variable to the system, the other automatically becomes the output variable of that system.

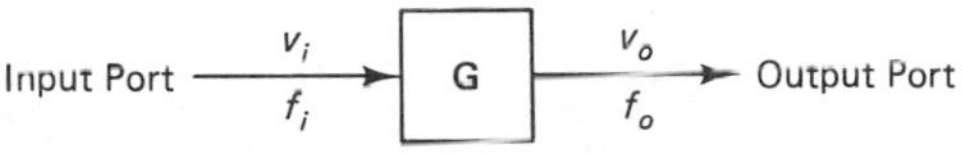

Input Port $\xrightarrow{\begin{array}{c} v_i \\ f_i \end{array}}$ $\boxed{\mathbf{G}}$ $\xrightarrow{\begin{array}{c} v_o \\ f_o \end{array}}$ Output Port

Figure 2.4. Two-port representation of a passive transducer.

Matrix transfer-function models are particularly suitable for transducers whose overall transduction process can be broken down into two or more simpler transducer stages. For example, consider a pressure transducer consisting of a bellows mechanism and a linear variable differential transformer (LVDT). In this device, the pressure signal is converted into a displacement by the pneumatic bellows mechanism. The displacement is converted, in turn, into a voltage signal by the LVDT. (The operation of an LVDT will be discussed in chapter 3.) If the transfer function matrix for each transducer stage is known, the combined model is obtained by simply multiplying the two matrices in the proper order. To illustrate the method further, consider the *generalized series element* (*electrical impedance* or *mechanical mobility*) and the *generalized parallel element* (*electrical admittance* or *mechanical impedance*), which are denoted by Z and Y, respectively. The corresponding circuit representations are shown in figure 2.5. The model for the series-element transducer is

$$\begin{bmatrix} v_i \\ f_i \end{bmatrix} = \begin{bmatrix} 1 & Z \\ 0 & 1 \end{bmatrix}\begin{bmatrix} v_o \\ f_o \end{bmatrix} \quad \text{or} \quad \begin{bmatrix} 1 & -Z \\ 0 & 1 \end{bmatrix}\begin{bmatrix} v_i \\ f_i \end{bmatrix} = \begin{bmatrix} v_o \\ f_o \end{bmatrix} \tag{2.2}$$

and the model for the parallel-element transducer is

$$\begin{bmatrix} v_i \\ f_i \end{bmatrix} = \begin{bmatrix} 1 & 0 \\ Y & 1 \end{bmatrix}\begin{bmatrix} v_o \\ f_o \end{bmatrix} \quad \text{or} \quad \begin{bmatrix} 1 & 0 \\ -Y & 1 \end{bmatrix}\begin{bmatrix} v_i \\ f_i \end{bmatrix} = \begin{bmatrix} v_o \\ f_o \end{bmatrix} \tag{2.3}$$

These relations can be easily verified. The expressions for Z and Y for the three basic (ideal) electrical elements—resistance, inductance, and capacitance—are summarized in table 2.1. Note that s is the Laplace variable. In the frequency domain, s should be replaced by $j\omega$, where ω is the frequency (radians/second) and $j = \sqrt{-1}$.

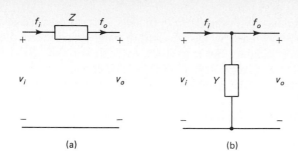

Figure 2.5. (a) Generalized series element (electrical impedance or mechanical mobility). (b) Generalized parallel element (electrical admittance or mechanical impedance).

TABLE 2.1 IMPEDENCE AND ADMITTANCE EXPRESSIONS FOR THE THREE IDEAL ELECTRIC ELEMENTS

Element	Impedance Z	Admittance Y
Resistance R	R	$\dfrac{1}{R}$
Inductance L	Ls	$\dfrac{1}{Ls}$
Capacitance C	$\dfrac{1}{Cs}$	Cs

A disadvantage of using through variables and across variables in the definition of impedance transfer functions is apparent when comparing electrical impedance with mechanical impedance. The definition of mechanical impedance is force/velocity in the frequency domain. This is a ratio of (through variable)/(across variable), whereas electrical impedance, defined as voltage/current in the frequency domain, is a ratio of (across variable)/(through variable). Since both force and voltage are "effort" variables and velocity and current are "flow" variables, it is convenient to use bond graph notation in defining impedance. Specifically,

$$\text{impedance (electrical or mechanical)} = \frac{\text{effort}}{\text{flow}}$$

In other words, impedance measures how much effort is needed to drive a system at unity flow.

Caution must be exercised in analyzing interconnected systems with mechanical impedance, because mechanical impedance cannot be manipulated using the rules for electrical impedance. For example, if two electric components are connected in series, the current (flow variable is the through variable) will be the same for both components, and voltage (effort variable is the across variable) will be additive. Accordingly, impedance of the series-connected electric system is just the sum of the impedances of the individual components. Now consider two mechanical components connected in series. Here the force (effort variable is the through variable) will be the same for both components, and velocity will be additive. Hence, it is mobility, not impedance, that is additive in the case of series-connected mechanical components. It can be concluded that, analytically, mobility behaves like electrical impedance and mechanical impedance behaves like electrical admittance. Hence, in Figure 2.5a, the generalized series element Z could be electrical impedance or mechanical mobility, and in figure 2.5b, the generalized parallel element Y could be electrical admittance or mechanical impedance. Definitions of some mechanical transfer functions are given in table 2.2.

TABLE 2.2 DEFINITIONS OF SOME MECHANICAL TRANSFER FUNCTIONS

Transfer function	Definition (in the frequency domain)
Dynamic stiffness	Force/displacement
Dynamic flexibility, compliance, or receptance	Displacement/force
Impedance	Force/velocity
Mobility	Velocity/force
Dynamic inertia	Force/acceleration
Accelerance	Acceleration/force
Force transmissibility	Magnitude [output force/input force]
Velocity transmissibility	Magnitude [output velocity/input velocity]

Example 2.1

Consider a transducer modeled as in figure 2.6. Obtain a transfer-function relationship.

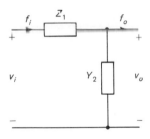

Figure 2.6. An example of a transfer model combination.

Solution This device has a transfer model given by

$$\begin{bmatrix} v_i \\ f_i \end{bmatrix} = \begin{bmatrix} 1 & Z_1 \\ 0 & 1 \end{bmatrix}\begin{bmatrix} 1 & 0 \\ Y_2 & 1 \end{bmatrix}\begin{bmatrix} v_o \\ f_o \end{bmatrix} \quad \text{or} \quad \begin{bmatrix} 1 & 0 \\ -Y_2 & 1 \end{bmatrix}\begin{bmatrix} 1 & -Z_1 \\ 0 & 1 \end{bmatrix}\begin{bmatrix} v_i \\ f_i \end{bmatrix} = \begin{bmatrix} v_o \\ f_o \end{bmatrix}$$

which results in the overall model:

$$\begin{bmatrix} v_i \\ f_i \end{bmatrix} = \begin{bmatrix} 1+Z_1 Y_2 & Z_1 \\ Y_2 & 1 \end{bmatrix}\begin{bmatrix} v_o \\ f_o \end{bmatrix} \quad \text{or} \quad \begin{bmatrix} 1 & -Z_1 \\ -Y_2 & 1+Z_1 Y_2 \end{bmatrix}\begin{bmatrix} v_i \\ f_i \end{bmatrix} = \begin{bmatrix} v_o \\ f_o \end{bmatrix}$$

Notice that when $Y_2 = 0$, equation 2.2 is obtained; and when $Z_1 = 0$, equation 2.3 results.

Example 2.2

The tachometer is a velocity-measuring device (passive) that uses the principle of electromagnetic generation. A DC tachometer is shown schematically in figure 2.7a. The field windings are powered by DC voltage v_f. The across variable at the input port is the measured angular speed ω_i. The corresponding torque T_i is the through variable at the input port. The output voltage v_o of the armature circuit is the across variable at the output port. The corresponding current i_o is the through variable at the output port. Obtain a transfer-function model for this device.

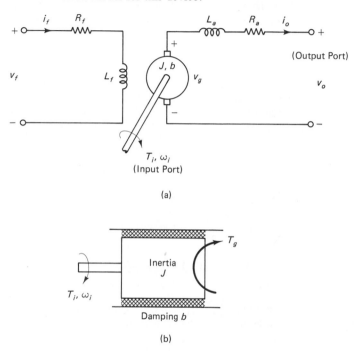

(a)

(b)

Figure 2.7. A DC tachometer example: (a) equivalent circuit; (b) armature free-body diagram.

Solution   The generated voltage v_g at the armature (rotor) is proportional to the magnetic field strength of field windings (which, in turn, is proportional to the field current i_f) and the speed of the armature ω_i. Hence,

$$v_g = K'i_f\omega_i$$

Assuming constant field current, we have

$$v_g = K\omega_i \qquad (i)$$

The rotor magnetic torque T_g that resists the applied torque T_i is proportional to the magnetic field strengths of field windings and armature windings. Consequently,

$$T_g = K'i_fi_o$$

Since i_f is assumed constant, we get

$$T_g = Ki_o \qquad (ii)$$

Note that the same constant K is used in both equations (i) and (ii). This is valid when the same units are used to measure mechanical power and electrical power and no internal dissipation mechanisms are significant in the associated internal coupling. The equation for the armature circuit is

$$v_o = v_g - R_a i_o - L_a \frac{di_o}{dt} \qquad \text{(iii)}$$

where R_a is the armature resistance and L_a is the *leakage inductance* in the armature circuit. With reference to Figure 2.7b, Newton's second law for a tachometer armature having inertia J and damping constant b is expressed as

$$J \frac{d\omega_i}{dt} = T_i - T_g - b\omega_i \qquad \text{(iv)}$$

Now equation (i) is substituted into (iii) in order to eliminate v_g. Similarly, equation (ii) is substituted into (iv) in order to eliminate T_g. Next, the time derivatives are replaced by the Laplace variable s. This results in the two algebraic relations:

$$v_o = K\omega_i - (R_a + sL_a)i_o \qquad \text{(v)}$$

$$(b + sJ)\omega_i = T_i - Ki_o \qquad \text{(vi)}$$

Note that the variables v_o, i_o, ω_i, and T_i in equations (v) and (vi) are actually Laplace transforms (functions of s), not functions of t, as in equations (i) through (iv). Finally, i_o in equation (v) is eliminated using (vi). This gives the matrix transfer function relation

$$\begin{bmatrix} v_o \\ i_o \end{bmatrix} = \begin{bmatrix} K + (R_a + sL_a)(b + sJ)/K & -(R_a + sL_a)/K \\ -(b + sJ)/K & 1/K \end{bmatrix} \begin{bmatrix} \omega_i \\ T_i \end{bmatrix} \qquad \text{(vii)}$$

The corresponding frequency domain relations are obtained by replacing s with $j\omega$, where ω represents the angular frequency (radians/second) in the frequency spectrum of a signal.

Even though transducers are more accurately modeled as two-port elements that have two variables associated with each port, it is useful and often essential, for practical reasons, to relate just one input variable and one output variable so that only one transfer function relating these two variables need be specified. This assumes some form of decoupling in the true model. If this assumption does not hold in the range of operation of the transducer, a measurement error would result. For instance, in the tachometer example, we like to express the output voltage v_o in terms of the measured speed ω_i. In this case, the off-diagonal term $-(R_a + sL_a)/K$ in equation (vii) of example 2.2 has to be neglected. This is valid when the tachometer gain parameter K is large and the armature resistance R_a is negligible, since the leakage inductance L_a is negligible in any case for most practical purposes. Note from equations (i) and (ii) that the tachometer gain K can be increased by increasing the field current i_f. This will not be feasible if the field windings are already saturated, however. Furthermore, K can be increased by increasing K'. Now K' depends on parameters such as number of turns and dimensions of the stator windings and magnetic properties of the stator core. Since there is a limitation on the physical size of the tachometer and the types of materials used in the construction, it follows that K cannot be increased arbitrarily. The instrument designer should take such factors into consideration in developing a design that is optimal in many respects. In practical transducers, the operating range is specified in order to minimize the effect of coupling terms, and the residual errors are accounted for by using correction curves. This approach is more convenient than using a coupled model, which introduces three more transfer functions (in general) into the model.

Another desirable feature for practical transducers is to have a static (nondynamic) input/output relationship so that the output instantly reaches the input value (or the measured variable). In this case, the transducer transfer function is a pure gain. This happens when the transducer time constants are small (i.e., the transducer bandwidth is high). Returning to example 2.2 again, it is clear from equation (vii) that static (frequency-independent) transfer-function relations are obtained when the electrical time constant

$$\tau_e = \frac{L_a}{R_a} \tag{2.4}$$

and the mechanical time constant

$$\tau_m = \frac{J}{b} \tag{2.5}$$

are both negligibly small. The electrical time constant is usually an order of magnitude smaller than the mechanical time constant. Hence, one must first concentrate on the mechanical time constant. Note from equation 2.5 that τ_m can be reduced by decreasing rotor inertia and increasing rotor damping. Unfortunately, rotor inertia depends on rotor dimensions, and this determines the gain parameter K, as we saw earlier. Hence, we face some constraint in reducing K. Next, turning to damping, it is intuitively clear that if we increase b, it will require a larger torque T_i to drive the tachometer, and this will load the system that generates the measurand ω_i, possibly affecting the measurand itself. Hence, increasing b also has to be done cautiously. Now, going back to equation (vii), we note that the dynamic terms in the transfer function between ω_i and v_o decrease as K is increased. So we see that increasing K has two benefits: reduction of coupling and reduction of dynamic effects (i.e., increasing the useful frequency range and bandwidth or speed of response).

2.4 PARAMETERS FOR PERFORMANCE SPECIFICATION

A *perfect measuring device* can be defined as one that possesses the following characteristics:

1. Output instantly reaches the measured value (fast response).
2. Transducer output is sufficiently large (high gain or low output impedance).
3. Output remains at the measured value (without drifting or being affected by environmental effects and other undesirable disturbances and noise) unless the measurand itself changes (stability).

4. The output signal level of the transducer varies in proportion to the signal level of the measurand (static linearity).
5. Connection of measuring device does not distort the measurand itself (loading effects are absent and impedances are matched).
6. Power consumption is small (high input impedance).

All of these properties are based on dynamic characteristics and therefore can be explained in terms of the dynamic behavior of the measuring device. In particular, items 1 through 4 can be specified in terms of the device (response), either in the *time domain* or in the *frequency domain*. Items 2, 5, and 6 can be specified using the *impedance* characteristics of device. In this section, we shall discuss response characteristics, leaving the discussion of impedance characteristics to section 2.5.

Time Domain Specifications

Figure 2.8 shows a typical *step response* in the dominant mode of a device. Note that the curve is normalized with respect to the steady-state value. We have identified several parameters that are useful for the time domain performance specification of the device. Definitions of these parameters are as follows:

Rise time.   This is the time taken to pass the steady-state value of the response for the first time. In overdamped systems, the response is nonoscillatory; consequently, there is no overshoot. So that the definition is valid for all systems, rise time is often defined as the time taken to pass 90 percent of the steady-state value. Rise time is often measured from 10 percent of the steady-state value in order

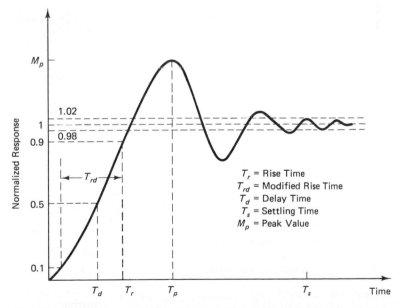

Figure 2.8.   Response parameters for the time domain specification of performance.

to leave out start-up irregularities and time lags that might be present in a system. A modified rise time (T_{rd}) may be defined in this manner (see figure 2.8). An alternative definition of rise time, particularly suitable for nonoscillatory responses, is the reciprocal slope of the step response curve at 50 percent of the steady-state value, multiplied by the steady-state value. In process control terminology, this is in fact the *cycle time*. Note that no matter what definition is used, rise time represents the speed of response of a device: a small rise time indicates a fast response.

Delay time. This is usually defined as the time taken to reach 50 percent of the steady-state value for the first time. This parameter is also a measure of speed of response.

Peak time. This is the time at the first peak. This parameter also represents the speed of response of the device.

Settling time. This is the time taken for the device response to settle down within a certain percentage (e.g., ± 2 percent) of the steady-state value. This parameter is related to the degree of damping present in the device as well as degree of stability.

Percentage overshoot (P.O.). This is defined as

$$\text{P.O.} = 100\,(M_p - 1)\,\% \tag{2.6}$$

using the normalized-to-unity step response curve, where M_p is the peak value. Percentage overshoot is a measure of damping or relative stability in the device.

Steady-state error. This is the deviation of the actual steady-state value from the desired value. Steady-state error may be expressed as a percentage with respect to the (desired) steady-state value. In a measuring device, steady-state error manifests itself as an offset. This is a systematic (deterministic) error that normally can be corrected by recalibration. In servo-controlled devices, steady-state error can be reduced by increasing loop gain or by introducing lag compensation. Steady-state error can be completely eliminated using integral control (*reset*) action.

For the best performance of a measuring device, we wish to have the values of all the foregoing parameters as small as possible. In actual practice, however, it might be difficult to meet all specifications, particularly for conflicting requirements. For instance, T_r can be decreased by increasing the dominant natural frequency ω_n of the device. This, however, increases the P.O. and sometimes the T_s. On the other hand, the P.O. and T_s can be decreased by increasing device damping, but it has the undesirable effect of increasing T_r.

Frequency Domain Specifications

Figure 2.9 shows a representative *frequency transfer function* (often termed frequency response function) of a device. This constitutes *gain* and *phase angle* plots, using frequency as the independent variable. This pair of plots is commonly known

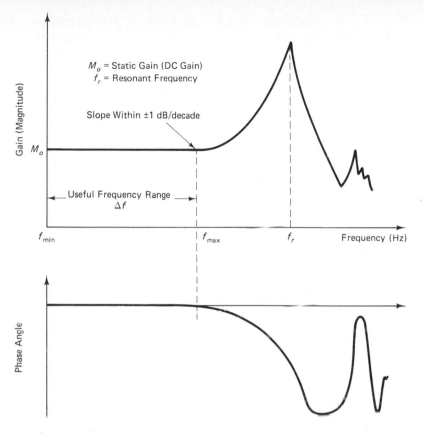

M_o = Static Gain (DC Gain)
f_r = Resonant Frequency

Slope Within ±1 dB/decade

Gain (Magnitude)

M_o

Useful Frequency Range
Δf

f_{min} | f_{max} | f_r | Frequency (Hz)

Phase Angle

Figure 2.9. Response parameters for the frequency domain specification of performance.

as the *Bode diagram,* particularly when the magnitude axis is calibrated in *decibels* (dB) and the frequency axis in a log scale such as *octaves* or *decades.* Experimental determination of these curves can be accomplished either by applying a harmonic excitation and noting *amplitude gain* and *phase lead* in the response signal at steady state or by Fourier analysis of excitation and response signals for either transient or random excitations. Experimental determination of transfer functions is known as *system identification in the frequency domain.* Note that transfer functions provide complete information concerning system response to a sinusoidal excitation. Since any time signal can be decomposed into sinusoidal components through Fourier transform, it is clear that the response of a system to an arbitrary input excitation also can be determined using transfer-function information for that system. In this sense, transfer functions are frequency domain models that can completely describe linear systems. For this reason, one could argue that it is redundant to use both time domain specifications and frequency domain specifications, as they carry the same information. Often, however, both specifications are used simultaneously, because this can provide a better picture of the system performance. Frequency domain parameters are more suitable in representing some characteristics of a system under some types of excitation.

In the frequency domain, several system parameters have special significance for a measuring instrument:

Useful frequency range.   This corresponds to the flat region (static region) in the gain curve and the zero-phase-lead region in the phase curve. It is determined by the dominant resonant frequency f_r of the instrument. The maximum frequency f_{max} in the useful frequency range is several times smaller than f_r for a typical measuring instrument (e.g., $f_{max} = 0.25 f_r$). Useful frequency range may also be determined by specifying the flatness of the static portion of the frequency response curve. For example, since a single pole or a single zero introduces a slope on the order of ∓ 20 dB/decade, a slope within 5 percent of this value (i.e., ± 1 dB/decade) may be considered flat for most practical purposes. Operation in the useful frequency range of a measuring device implies measurement of a signal whose significant frequency content is limited to this band. In that case, faithful measurement and fast response are guaranteed, because measuring device dynamics do not corrupt the measurement.

Instrument bandwidth.   This is a measure of the useful frequency range of an instrument. Furthermore, the larger the bandwidth, the faster the speed of response of the device will be. Unfortunately, the larger the bandwidth, the more susceptible the instrument will be to high-frequency noise as well as stability problems. Filtering will be needed to eliminate unwanted noise. Stability can be improved by dynamic compensation. There are many definitions for bandwidth. Common definitions include the frequency range over which the transfer-function magnitude is flat, the resonant frequency, and the frequency at which the transfer-function magnitude drops to $1/\sqrt{2}$ (or 70.7 percent) of the zero-frequency (or static) level. The last definition corresponds to the *half-power bandwidth,* because a reduction of amplitude level by a factor of $\sqrt{2}$ corresponds to a power drop by a factor of 2.

Control bandwidth.   This is used to specify speed of control. It is an important specification in both analog control and digital control. In digital control, the data sampling rate (in samples/second) has to be at least double the control bandwidth (in hertz) so that the control action can be generated at the full speed. This follows from *Shannon's sampling theorem.* Control bandwidth should be addressed from two points of view. For a system to respond faithfully to a control action (input), the control bandwidth has to be sufficiently small (i.e., input has to be slow enough) in comparison to the dominant (smallest) resonant frequency of the system. This is similar to the bandwidth requirement for measuring devices, mentioned previously. On the other hand, if a certain mode of response in a system is to be insensitive to control action, the control action has to be several times larger than the frequency of that mode. For example, if the bending natural frequency (in the fundamental mode) of a robotic manipulator is 10 Hz, control bandwidth has to be 30 Hz or more so that the robot actuators would not seriously excite that bending mode of the manipulator structure. In digital control, this will require a sampling rate of 60 samples/second or more. In other words, each control cycle in real-time control has to be limited to 1/60 s (approximately 17 ms) or less. Data acquisition and processing, including control computations, have to be done within this time. This calls for fast control processors, possibly hardware implementations, and efficient control algorithms.

Static gain. This is the gain (transfer function magnitude) of a measuring instrument within the useful range (or at low frequencies) of the instrument. It is also termed *DC gain*. A high value for static gain results in a high-sensitivity measuring device, which is a desirable characteristic.

Example 2.3

A mechanical device for measuring angular velocity is shown in figure 2.10. The main element of the tachometer is a rotary viscous damper (damping constant b) consisting of two cylinders. The outer cylinder carries a viscous fluid within which the inner cylinder rotates. The inner cylinder is connected to the shaft whose speed ω_i is to be measured. The outer cylinder is resisted by a linear torsional spring of stiffness k. The rotation θ_o of the outer cylinder is indicated by a pointer on a suitably calibrated scale. Neglecting the inertia of moving parts, perform a bandwidth analysis on this device.

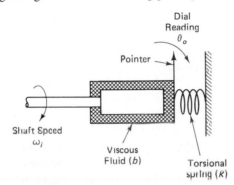

Figure 2.10 A mechanical tachometer.

Solution The damping torque is proportional to the relative velocity of the two cylinders and is resisted by the spring torque. The equation of motion is given by

$$b(\omega_i - \dot{\theta}_o) = k\theta_o$$

or $\qquad\qquad\qquad\qquad\qquad\qquad\qquad\qquad\qquad\qquad\qquad$ (i)

$$b\dot{\theta}_o + k\theta_o = b\omega_i$$

The transfer function is determined by first replacing the time derivative by the Laplace operator s; thus,

$$\frac{\theta_o}{\omega_i} = \frac{b}{[bs + k]} = \frac{b/k}{[(b/k)s + 1]} = \frac{k_g}{[\tau s + 1]} \qquad (ii)$$

Note that the static gain or DC gain (transfer-function magnitude with $s = o$) is

$$k_g = \frac{b}{k} \qquad (iii)$$

and the time constant is

$$\tau = \frac{b}{k} \qquad (iv)$$

We face conflicting design requirements in this case. On the one hand, we like to have a large static gain so that a sufficiently large reading is available. On the other hand, the time constant must be small in order to obtain a quick reading that faithfully follows the measured speed. A compromise must be reached here, depending on the specific design requirements. Alternatively, a signal-conditioning device could be employed to amplify the sensor output.

Also, let us examine the half-power bandwidth of the device. The frequency transfer function is

$$G(j\omega) = \frac{k_g}{[\tau j\omega + 1]} \tag{v}$$

By definition, the half-power bandwidth ω_b is given by

$$\frac{k_g}{|\tau j\omega_b + 1|} = \frac{k_g}{\sqrt{2}}$$

Hence

$$(\tau\omega_b)^2 + 1 = 2$$

Since both τ and ω_b are positive, we have

$$\tau\omega_b = 1$$

or

$$\omega_b = \frac{1}{\tau} \tag{vi}$$

Note that the bandwidth is inversely proportional to the time constant. This confirms our earlier statement that bandwidth is a measure of the speed of response.

Two other system parameters in the frequency domain that play crucial roles in interconnected devices are *input impedance* and *output impedance*. Impedance characteristics will be discussed in section 2.5.

Linearity

A device is considered linear if it can be modeled by linear differential equations, with time t as the independent variable. Nonlinear devices are often analyzed using linear techniques by considering small excursions about an operating point. This linearization is accomplished by introducing incremental variables for inputs and outputs. If one increment can cover the entire operating range of a device with sufficient accuracy, it is an indication that the device is linear. If the input/output relations are nonlinear algebraic equations, it represents a *static nonlinearity*. Such a situation can be handled simply by using nonlinear calibration curves, which linearize the device without introducing nonlinearity errors. If, on the other hand, the input/output relations are nonlinear differential equations, analysis usually becomes quite complex. This situation represents a *dynamic nonlinearity*.

Transfer-function representation of an instrument implicitly assumes linearity. According to industrial terminology, a linear measuring instrument provides a measured value that varies linearly with the value of the measurand. This is consistent with the definition of static linearity. All physical devices are *nonlinear* to some degree. This results from any deviation from the ideal behavior due to causes such as saturation, deviation from Hooke's law in elastic elements, coulomb friction, creep at joints, aerodynamic damping, backlash in gears and other loose components, and component wearout.

Nonlinearities in devices are often manifested as some peculiar characteristics. In particular, the following properties are important in detecting nonlinear behavior in dynamic systems:

Saturation. Nonlinear devices may exhibit saturation (see figure 2.11a). This may result from such causes as magnetic saturation, which is common in transformer devices such as differential transformers (see chapter 3), plasticity in mechanical components, or nonlinear deformation in springs.

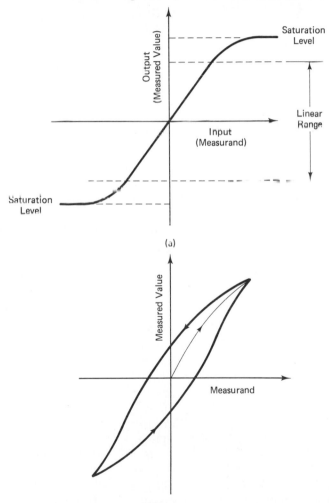

(a)

(b)

Figure 2.11. Common manifestations of nonlinearity in dynamic systems: (a) saturation; (b) hysteresis; (c) the jump phenomenon; (d) limit cycle response.

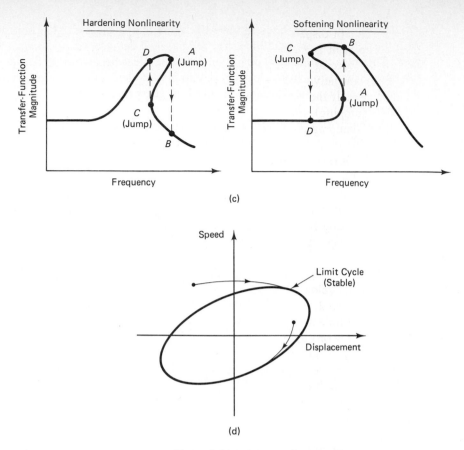

Figure 2.11. (*continued*)

Hysteresis. Nonlinear devices may produce hysteresis. In this case, the input/output curve changes, depending on the direction of motion (as indicated in figure 2.11b), resulting in a hysteresis loop. This is common in loose components such as gears, which have backlash; in components with nonlinear damping, such as coulomb friction; and in magnetic devices with ferromagnetic media and various dissipative mechanisms (e.g., eddy current dissipation). For example, consider a coil wrapped around a ferromagnetic core. If a DC current is passed through the coil, a magnetic field is generated. As the current is increased from zero, the field strength will also increase. Now, if the current is decreased back to zero, the field strength will not return to zero because of residual magnetism in the ferromagnetic core. A negative current has to be applied to demagnetize the core. It follows that the current-field strength curve looks somewhat like figure 2.11b. This is magnetic hysteresis. Note that linear viscous damping also exhibits a hysteresis loop in the force-displacement curve; this is a property of any mechanical component that dissipates energy. (Area within the hysteresis loop gives the energy dissipated in one cycle of motion.) In general, if force depends on both displacement (as in the case of a spring) and velocity (as in the case of a damping element), the value of force for a given value of displacement will change with velocity. In particular, the force when the component is moving in one direction (say, positive velocity) will be different from the force at the same location when the component is moving in the opposite direction (negative velocity), thereby giving a hysteresis loop in the force-displacement plane. If the relationship of displacement and velocity to force is linear (as in viscous damping), the hysteresis effect is linear. If the relationship is nonlinear (as in coulomb damping and aerodynamic damping), however, hysteresis is nonlinear.

The jump phenomenon. Some nonlinear devices exhibit an instability know as the jump phenomenon (or *fold catastrophe*) in the frequency response (transfer) function curve. This is shown in figure 2.11c for both hardening devices and softening devices. With increasing frequency, jump occurs from A to B; and with decreasing frequency, it occurs from C to D. Furthermore, the transfer function itself may change with the level of input excitation in the case of nonlinear devices.

Limit cycles. Nonlinear devices may produce limit cycles. An example is given in figure 2.11d on the phase plane of displacement and velocity. A limit cycle is a closed trajectory in the state space that corresponds to sustained oscillations without decay or growth. Amplitude of these oscillations is independent of the initial location from which the response started. In the case of a stable limit cycle, the response will move onto the limit cycle irrespective of the location in the neighborhood of the limit cycle from which the response was initiated (see figure 2.11d). In the case of an unstable limit cycle, the response will move away from it with the slightest disturbance.

Frequency creation. At steady state, nonlinear devices can create frequencies that are not present in the excitation signals. These frequencies might be harmonics (interger multiples of the excitation frequency), subharmonics (integer fractions of the excitation frequency), or nonharmonics (usually rational fractions of the excitation frequency).

Example 2.4

Consider a nonlinear device modeled by the differential equation

$$\left\{\frac{dy}{dt}\right\}^{1/2} = u(t)$$

in which $u(t)$ is the input and y is the output. Show that this device creates frequency components that are different from the excitation frequencies.

Solution First, note that the steady-state response is given by

$$y = \int_0^t u^2(t)dt + y(0)$$

Now, for an input given by

$$u(t) = a_1 \sin \omega_1 t + a_2 \sin \omega_2 t$$

straightforward integration using properties of trigonometric functions results in the following response:

$$y = (a_1^2 + a_2^2)\frac{t}{2} - \frac{a_1^2}{4\omega_1}\sin 2\omega_1 t - \frac{a_2^2}{4\omega_2}\sin 2\omega_2 t$$

$$+ \frac{a_1 a_2}{2(\omega_1 - \omega_2)}\sin(\omega_1 - \omega_2)t - \frac{a_1 a_2}{2(\omega_1 + \omega_2)}\sin(\omega_1 + \omega_2)t + y(0)$$

Note that the discrete frequency components $2\omega_1$, $2\omega_2$, $(\omega_1 - \omega_2)$, and $(\omega_1 + \omega_2)$ are created. Also, there is a continuous spectrum that is contributed by the linear function of t present in the response.

The fact that nonlinear systems create new frequency components is the basis of well-known *describing function analysis* of nonlinear control systems. In this case, the response of a nonlinear component to a sinusoidal (harmonic) input is represented by a Fourier series, with frequency components that are multiples of the input frequency. Details of the describing function approach can be found in textbooks on nonlinear control theory.

Several methods are available to reduce or eliminate nonlinear behavior in systems. They include calibration (in the static case), use of linearizing elements, such as resistors and amplifiers to neutralize the nonlinear effects, and the use of nonlinear feedback. It is also a good practice to take the following precautions:

1. Avoid operating the device over a wide range of signal levels.
2. Avoid operation over a wide frequency band.
3. Use devices that do not generate large mechanical motions.
4. Minimize coulomb friction.
5. Avoid loose joints and gear coupling (i.e., use *direct drive* mechanisms).

2.5 IMPEDANCE CHARACTERISTICS

When components such as measuring instruments, control boards, process (plant) hardware, and signal-conditioning equipment are interconnected, it is necessary to *match* impedances properly at each interface in order to realize their rated performance level. One adverse effect of improper impedance matching is the *loading effect*. For example, in a measuring system, the measuring instrument can distort the signal that is being measured. The resulting error can far exceed other types of measurement error. Loading errors result from connecting measuring devices with low input impedance to a signal source.

Impedance can be interpreted either in the traditional electrical sense or in the mechanical sense, depending on the signal being measured. For example, a heavy accelerometer can introduce an additional dynamic load that will modify the actual acceleration at the monitoring location. Similarly, a voltmeter can modify the currents (and voltages) in a circuit, and a thermocouple junction can modify the temperature that is being measured. In mechanical and electrical systems, loading errors can appear as phase distortions as well. Digital hardware also can produce loading errors. For example, an analog-to-digital conversion (ADC) board can load the amplifier output from a strain gage bridge circuit, thereby significantly affecting digitized data (see chapter 4).

Another adverse effect of improper impedance consideration is inadequate output signal levels, which make signal processing and transmission very difficult. Many types of transducers (e.g., piezoelectric accelerometers, impedance heads, and microphones) have high output impedances on the order of a thousand megohms. These devices generate low output signals, and they would require conditioning to step up the signal level. *Impedance-matching amplifiers*, which have high input impedances and low output impedances (a few ohms), are used for this purpose (e.g., charge amplifiers are used in conjunction with piezoelectric sensors). A device with a high input impedance has the further advantage that it usually consumes less power (v^2/R is low) for a given input voltage. The fact that a low input impedance device extracts a high level of power from the preceding output device may be interpreted as the reason for loading error.

Cascade Connection of Devices

Consider a standard two-port electrical device. The *output impedance Z_o* of such a device is defined as the ratio of the open-circuit (i.e., no-load) voltage at the output port to the short-circuit current at the output port.

Open-circuit voltage at output is the output voltage present when there is no current flowing at the output port. This is the case if the output port is not connected to a load (impedance). As soon as a load is connected at the output of the device, a current will flow through it, and the output voltage will drop to a value less than that of the open-circuit voltage. To measure open-circuit voltage, the rated input voltage is applied at the input port and maintained constant, and the output voltage is measured using a voltmeter that has a very high (input) impedance. To measure short-circuit current, a very low-impedance ammeter is connected at the output port.

The *input impedance Z_i* is defined as the ratio of the rated input voltage to the corresponding current through the input terminals while the output terminals are maintained as an open circuit.

Note that these definitions are associated with electrical devices. A generalization is possible by interpreting voltage and velocity as *across variables,* and current and force as *through variables*. Then mechanical *mobility* should be used in place of electrical impedance.

Using these definitions, input impedance Z_i and output impedance Z_o can be represented schematically as in figure 2.12a. Note that v_o is the open-circuit output voltage. When a load is connected at the output port, the voltage across the load will be different from v_o. This is caused by the presence of a current through Z_o. In the frequency domain, v_i and v_o are represented by their respective *Fourier spectra*. The corresponding transfer relation can be expressed in terms of the complex frequency response (transfer) function $G(j\omega)$ under open-circuit (no-load) conditions:

$$v_o = Gv_i \qquad (2.7)$$

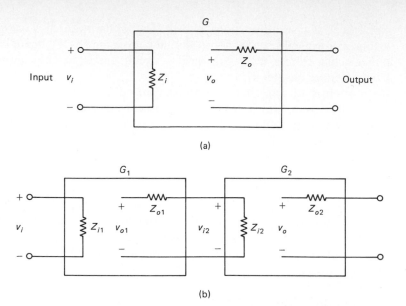

Figure 2.12. (a) Schematic representation of input impedance and output impedance. (b) The influence of cascade connection of devices on the overall impedance characteristics.

Now consider two devices connected in cascade, as shown in figure 2.12b. It can be easily verified that the following relations apply:

$$v_{o1} = G_1 v_i \tag{2.8}$$

$$v_{i2} = \frac{Z_{i2}}{Z_{o1} + Z_{i2}} v_{o1} \tag{2.9}$$

$$v_o = G_2 v_{i2} \tag{2.10}$$

These relations can be combined to give the overall input/output relation:

$$v_o = \frac{Z_{i2}}{Z_{o1} + Z_{i2}} G_2 G_1 v_i \tag{2.11}$$

We see from equation 2.11 that the overall frequency transfer function differs from the ideally expected product $(G_2 G_1)$ by the factor

$$\frac{Z_{i2}}{Z_{o1} + Z_{i2}} = \frac{1}{Z_{o1}/Z_{i2} + 1} \tag{2.12}$$

Note that cascading has "distorted" the frequency response characeristics of the two devices. If $Z_{o1}/Z_{i2} << 1$, this deviation becomes insignificant. From this observation, it can be concluded that when frequency response characteristics (i.e., dynamic characteristics) are important in a cascaded device, cascading should be done such that the output impedance of the first device is much smaller than the input impedance of the second device.

Example 2.5

A lag network used as the compensatory element of a control system is shown in figure 2.13a. Show that its transfer function is given by

$$\frac{v_o}{v_i} = \frac{Z_2}{R_1 + Z_2}$$

where

$$Z_2 = R_2 + \frac{1}{Cs}$$

What is the input impedance and what is the output impedance for this circuit? Also, if two such lag circuits are cascaded as shown in figure 2.13b, what is the overall transfer function? How would you make this transfer function close to the ideal result:

$$\left\{\frac{Z_2}{R_1 + Z_2}\right\}^2$$

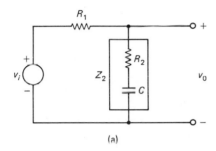

(a)

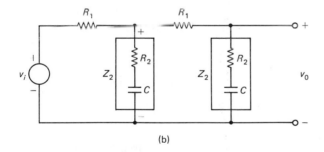

(b)

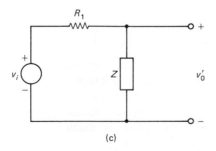

(c)

Figure 2.13. (a) Single circuit module. (b) Cascade connection of two modules. (c) Equivalent circuit for (b).

Solution   To solve this problem, first note that in figure 2.13a, voltage drop across the element $R_2 + 1/(Cs)$ is

$$v_o = \left(R_2 + \frac{1}{Cs}\right) \bigg/ \left\{R_1 + R_2 + \frac{1}{Cs}\right\} v_i$$

$$= Z_2/(R_1 + Z_2)v_i$$

Hence,

$$\frac{v_o}{v_i} = \frac{Z_2}{R_1 + Z_2} \tag{i}$$

Now, input impedance Z_i is derived by

$$\text{input current } i = \frac{v_i}{R_1 + Z_2}$$

$$Z_i = \frac{v_i}{i} = R_1 + Z_2$$

and output impedance Z_o is derived by

$$\text{short-circuit current } i_{sc} = \frac{v_i}{R_1}$$

$$Z_o = \frac{v_o}{i_{sc}} = \frac{Z_2/(R_1 + Z_2)v_i}{v_i/R_1} = \frac{R_1 Z_2}{R_1 + Z_2}$$

Next, consider the equivalent circuit shown in Figure 2.13c. Since Z is formed by connecting Z_2 and $(R_1 + Z_2)$ in parallel, we have

$$\frac{1}{Z} = \frac{1}{Z_2} + \frac{1}{R_1 + Z_2} \tag{ii}$$

Voltage drop across Z is

$$v_o' = \frac{Z}{R_1 + Z} v_i \tag{iii}$$

Now apply the single-circuit module result (i) to the second circuit stage in figure 2.13b; thus,

$$v_o = \frac{Z_2}{R_1 + Z_2} v_o'$$

Substituting equation (iii), we get

$$v_o = \frac{Z_2}{(R_1 + R_2)} \frac{Z}{(R_1 + Z)} v_i$$

The overall transfer function for the cascaded circuit is

$$G = \frac{v_o}{v_i} = \frac{Z_2}{(R_1 + Z_2)} \frac{Z}{(R_1 + Z)} = \frac{Z_2}{(R_1 + R_2)} \frac{1}{(R_1/Z + 1)}$$

Now substituting equation (ii), we get

$$G = \left[\frac{Z_2}{R_1 + Z_2}\right]^2 \frac{1}{1 + R_1 Z_2/(R_1 + Z_2)^2}$$

We observe that the ideal transfer function is approached by making $R_1 Z_2/(R_1 + Z_2)^2$ small compared to unity.

Impedance-Matching Amplifiers

From the analysis given in the preceding section, it is clear that the signal-conditioning circuitry should have a considerably large input impedance in comparison to the output impedance of the sensor-transducer unit in order to reduce loading errors. The problem is quite serious in measuring devices such as piezoelectric sensors, which have very high output impedances. In such cases, the input impedance of the signal-conditioning unit might be inadequate to reduce loading effects; also, the output signal level of these high-impedance sensors is quite low for signal transmission, processing, and control. The solution for this problem is to introduce several stages of amplifier circuitry between the sensor output and the data acquisition unit input. The first stage is typically an *impedance-matching amplifier* that has very high input impedance, very low output impedance, and almost unity gain. The last stage is typically a stable high-gain amplifier stage to step up the signal level. Impedance-matching amplifiers are, in fact, *operational amplifiers* with feedback.

Operational amplifiers. Operational amplifiers (opamps) are voltage amplifiers with very high gain K (typically 10^5 to 10^9), high input impedance Z_i (typically greater than 1MΩ), and low output impedance Z_o (typically smaller than 100 Ω). Thanks to the advances in integrated circuit technology, opamps —originally made with conventional transistors, diodes, and resistors—are now available as miniature units with integrated circuit elements. Because of their small size, the recent trend is to make signal-conditioning hardware an integral part of the sensor-transducer unit.

A schematic diagram for an opamp is shown in figure 2.14a. Supply voltage v_s is essential to power the opamp. This may be omitted, however, in schematic diagrams and equivalent circuits within the scope of present considerations. In the standard design of opamps, there are two input leads, denoted by 1 and 2 in figure 2.14a. If one of the two input leads is grounded, it is a *single-ended amplifier*. If neither lead is grounded, it is a *differential amplifier* that requires two input signals. The latter arrangement rejects noise common to the two inputs (e.g., line noise, thermal noise, magnetic noise) because signal 2 (at the $-$ terminal of the opamp) is subtracted from signal 1 (at the $+$ terminal) and amplified to give the output signal. Figure 2.14a is analogous to figure 2.13a, except that the amplifier gain K has replaced the transfer function $G(j\omega)$. Strictly speaking, K is a transfer function, and it depends on the frequency variable ω of the input signal. Typically, however, the bandwidth of an opamp is on the order of 10kHz; consequently, K may be assumed frequency-independent in that frequency range. This assumption is satisfactory for

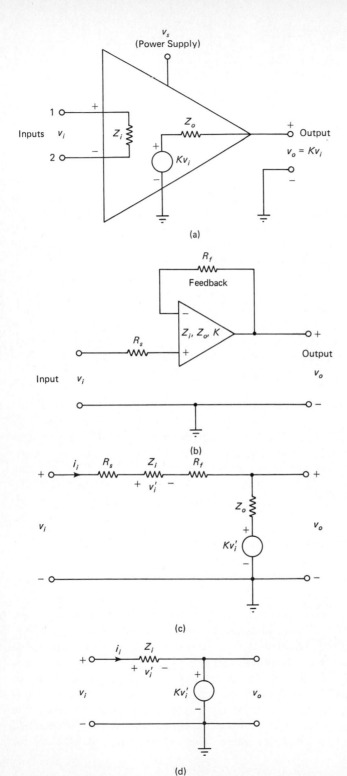

Figure 2.14. Impedance-matching amplifiers: (a) schematic representation of an operational amplifier; (b) schematic representation of a voltage follower; (c) equivalent circuit for the voltage follower; (d) Simplified equivalent circuit for the voltage follower; (e) charge amplifier.

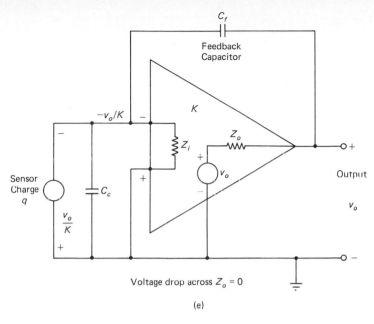

Feedback Capacitor C_f

$-v_o/K$

K

Z_i

Z_o

v_o

Output

v_o

Sensor Charge q

C_c

$\dfrac{v_o}{K}$

Voltage drop across $Z_o = 0$

(e)

Figure 2.14. (*continued*)

most practical applications. Nevertheless, an operational amplifier in its basic form has poor stability characteristics; hence, the amplifier output can drift while the input is maintained steady. Furthermore, its gain is too high for direct voltage amplification of practical signals. For these reasons, additional passive elements, such as feedback resistors, are used in conjunction with opamps in practical applications.

Voltage followers. Voltage followers are impedance-matching amplifiers (or *impedance transformers*) with very high input impedance, very low output impedance, and almost-unity gain. For these reasons, they are suitable for use with high output impedance sensors such as piezoelectric devices. A schematic diagram for a voltage follower is shown in figure 2.14b. It consists of a standard (differential) opamp with a feedback resistor R_f connected between the output lead and the negative input lead. The sensor output, which is the amplifier input v_i, is connected to the positive input lead of the opamp with a series resistor R_s. The amplifier output is v_o, as shown. By combining figures 2.14a and 2.14b, the equivalent circuit for the voltage follower is drawn in figure 2.14c. Since the input impedance Z_i of the opamp is much larger than the other impedances (Z_o, R_s, and R_f) in the circuit, the simplified equivalent circuit shown in figure 2.14d is obtained. Note that v_i' is the voltage drop across Z_i.

To obtain an expression for gain $\tilde{K}$ of the voltage follower, examine figure 2.14d. It is clear that

$$v_i = v_i' + Kv_i' = (1 + K)v_i'$$

and

$$v_o = Kv_i'$$

The gain $\tilde{K}$ is given by

$$\frac{v_o}{v_i} = \frac{K v_i'}{(1 + K)v_i'} = \frac{K}{1 + K}$$

or

$$\tilde{K} = \frac{K}{1 + K} \tag{2.13}$$

which is almost unity for large K.

To determine input impedance $\tilde{Z}_i$ of the voltage follower, first note that the input current i_i is given by

$$i_i = \frac{v_i'}{Z_i} = \frac{v_i}{(1 + K)Z_i}$$

It follows that the input impedance $\tilde{Z}_i$ is given by

$$\frac{v_i}{i_i} = (1 + K)Z_i$$

or

$$\tilde{Z}_i = (1 + K)Z_i \tag{2.14}$$

Since both Z_i and K are very large, it follows that a voltage follower clearly provides a high input impedance. Accordingly, it is able to reduce loading effects of sensors that have high output impedances.

To determine the output impedance $\tilde{Z}_o$ of a voltage follower, note that $v_o = 0$ when the output leads are shorted. Then, from figure 2.14c, the short-circuit output current is found to be (by current summation at the output node)

$$i_{sc} = \frac{v_i'}{Z_i} + \frac{K v_i'}{Z_o}$$

Note that the value of v_i' under short-circuit conditions is different from that under open-circuit conditions. But v_i would not be affected by output shorting. When the output leads are shorted, it is clear from figure 2.14d that $v_i' \sim v_i$. Hence,

$$i_{sc} = \left(\frac{1}{Z_i} + \frac{K}{Z_o}\right) v_i$$

The output impedance is

$$\tilde{Z}_o = \frac{v_o}{i_{sc}} = \left(\frac{K}{1 + K}\right) v_i \Big/ \left[\left(\frac{1}{Z_i} + \frac{K}{Z_o}\right) v_i\right]$$

Now, since $Z_i \gg Z_o/K$, we can neglect $1/Z_i$ in comparison to K/Z_o. Consequently,

$$\tilde{Z}_o = \frac{Z_o}{1 + K} \tag{2.15}$$

Since Z_o is small to begin with and K is very large, it is clear that the output impedance of a voltage follower is very small, as desired. Accordingly, the voltage follower has a unity gain, a very high input impedance, and a very low output impedance; it can be used as an impedance transformer. By connecting a voltage follower to a high-impedance measuring device (sensor-transducer), a low-impedance output signal is obtained. Signal amplification might be necessary before this signal is transmitted or processed, however.

In many data acquisition systems, output impedance of the output amplifier is made equal to the transmission line impedance. When maximum power amplification is desired, *conjugate matching* is recommended. In this case, input impedance and output impedance of the matching amplifier are made equal to the complex conjugates of the source impedance and the load impedance, respectively.

Charge amplifiers. The principle of capacitance feedback is uitilized in charge amplifiers. They are commonly used for conditioning output signals from piezoelectric transducers. A schematic diagram for this device is shown in figure 2.14e. The feedback capacitance is denoted by C_f and the connecting cable capacitance by C_c. The charge amplifier views the sensor as a charge source (q), even though there is an associated voltage. Using the fact that charge = voltage $\times$ capacitance, a charge balance equation can be written:

$$q + \frac{v_o}{K}C_c + \left(v_o + \frac{v_o}{K}\right)C_f = 0$$

From this, we get

$$v_o = -\frac{K}{(K + 1)C_f + C_c}q \tag{2.16}$$

If the feedback capacitance is large in comparison with the cable capacitance, the latter can be neglected. This is desirable in practice. In any event, for large values of gain K, we have the approximate relationship

$$v_o = -\frac{q}{C_f} \tag{2.17}$$

Note that the output voltage is proportional to the charge generated at the sensor and depends only on the feedback parameter C_f. This parameter can be appropriately chosen in order to obtain the required output impedance characteristics. Practical charge amplifiers also have a feedback resistor R_f in parallel with the feedback capacitor C_f. Then the relationship corresponding to equation 2.16 becomes a first-order ordinary differential equation, which in turn determines the time constant of the charge amplifier. This time constant should be high. If it is low, the charge generated by the piezoelectric sensor will leak out quickly, giving erroneous results at low frequencies (see problem 2.11).

Example 2.6

Suppose that the output signal from a sensor with output impedance Z_s is directly connected to an operational amplifier with gain K, input impedance Z_i, and output impedance Z_o. The resulting signal is read directly into a digital controller using an analog-to-digital conversion (ADC) board with an equivalent load impedance Z_L. Pick parameters so as to reduce possible distortion of the digitized signal due to loading.

Solution A schematic representation of this arrangement of data acquisition is shown in figure 2.15. Straightforward analysis provides the following input/output relationship:

$$v_o = K\left(\frac{Z_i}{Z_s + Z_i}\right)\left(\frac{Z_L}{Z_o + Z_L}\right) v_i$$

If the input impedance Z_i of the opamp is very high in comparison with the sensor impedance Z_s, then $Z_i/(Z_s + Z_i)$ will approach unity. Furthermore, if the load impedance Z_L is very high in comparison with the output impedance of the opamp, then $Z_L/(Z_o + Z_L)$ will also approach unity. In that case, the input/output relation reduces to

$$v_o = Kv_i$$

which corresponds to a simple amplification of measured voltage by the gain factor K. In practice, however, the parameter K may drift because of such reasons as bandwidth limitations and stability problems in the opamp. Hence, using an opamp is not the best way to achieve signal amplification.

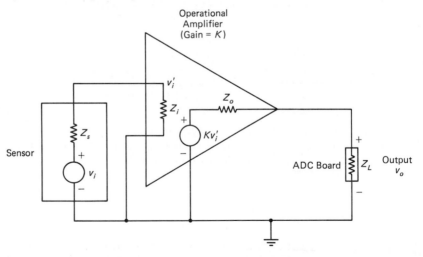

Figure 2.15. A data acquisition system example.

Measurement of Across Variables and Through Variables

Impedance concepts are very useful in selecting (and designing) instruments to measure across variables (voltage, velocity, pressure, temperature) and through variables (current, force, fluid flow rate, heat transfer rate). To develop some general concepts relating to the measurement of across variables and through variables, consider a device (an electronic device such as an amplifier, filter, or control circuit or a mechanical, fluid, or thermal device) that has input impedance Z_i and output impedance Z_o. It is connected to a load of impedance Z_L. This device is shown schematically in figure 2.16a. Variable v_o across the load and variable i_o through the load are to be measured. It is seen that

$$v_o = Gv_i \frac{Z_L}{Z_o + Z_L}$$

or

$$v_o = \frac{G}{[Z_o/Z_L + 1]} v_i \qquad (2.18)$$

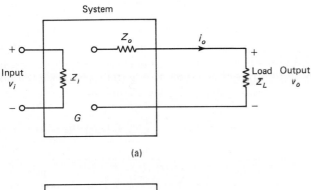

(a)

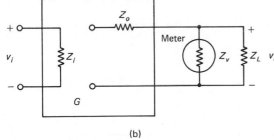

(b)

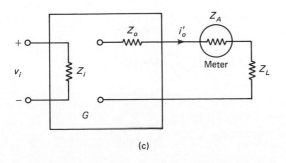

(c)

Figure 2.16. Impedance of measuring instruments: (a) system representation; (b) measurement of an across variable; (c) measurement of a through variable.

and

$$i_o = \frac{G}{Z_o + Z_L} v_i \qquad (2.19)$$

where G is the system (frequency) transfer function. In writing these relationships, we must remember that in the case of mechanical devices, the generalized impedance Z should be interpreted as mechanical mobility (velocity/force), not mechanical impedance (force/velocity). Otherwise, the combination rules used in getting these relationships (i.e., impedances additive in series and admittances additive in parallel) would not be valid.

Suppose that a meter of impedance Z_V is connected across the load to measure v_o, as shown in figure 2.16b. Since Z_V and Z_L are in parallel, their equivalent impedance Z is given by

$$\frac{1}{Z} = \frac{1}{Z_v} + \frac{1}{Z_L} \qquad (2.20)$$

Again, for this relationship to be generally valid, impedance should be interpreted as (across variable)/(through variable), not as (effort variable)/(flow variable). This interpretation, however, contradicts the commonly used definition of mechanical impedance—force/velocity in the frequency domain. Alternatively, Z should be interpreted as "mobility" in mechanical systems. No such ambiguity exists in electrical, fluid, and thermal systems.

Due to loading effects from the meter, across variable v_o changes to v_o', and

$$v_o' = Gv_i \frac{Z}{Z_o + Z} = \frac{Gv_i}{Z_o/Z + 1}$$

In view of equation 2.20, we have

$$v_o' = \frac{Gv_i}{Z_o/Z_v + Z_o/Z_L + 1} \qquad (2.21)$$

By comparing equation 2.21 with equation 2.18, we observe that for high accuracy of measurement (i.e., v_o' nearly equal to v_o), we must have either $Z_o/Z_V \ll 1$ or $Z_o/Z_V \ll Z_o/Z_L$. In other words, we must have $Z_V \gg Z_o$ or $Z_V \gg Z_L$. Hence, in general, a measuring instrument for an across variable must have a high impedance.

Now suppose that a meter of impedance Z_A is connected in series with the load to measure i_o, as shown in figure 2.16c. Because of instrument loading, the through variable i_o changes to i_o', and

$$i_o' = \frac{Gv_i}{Z_o + Z_L + Z_A} \qquad (2.22)$$

By comparing equation 2.22 with equation 2.19, we note that for high accuracy (i.e., i_o' almost equal to i_o), we must have either $Z_A \ll Z_o$ or $Z_A \ll Z_L$. It follows that, in general, an instrument measuring a through variable has to be a low-impedance device.

Ground Loop Noise

In devices that handle low-level signals (e.g., accelerometers and strain gage bridge circuitry), electrical noise can create excessive error. One form of noise is caused by fluctuating magnetic fields due to nearby AC lines. This can be avoided either by taking precautions not to have strong magnetic fields and fluctuating currents near delicate instruments or by using *fiber optic* (optically coupled) signal transmission (see chapter 3). Furthermore, if the two signal leads (positive and negative) are twisted or if shielded cables are used, the induced noise voltages become equal in the two leads, which cancel each other. Another cause of electrical noise is ground loops.

If two interconnected devices are grounded at two separate locations, ground loop noise can enter the signal leads because of the possible potential difference between the two ground points. The reason is that ground itself is not generally a uniform potential medium, and a nonzero (and finite) impedance may exist from point to point within the ground medium. This is, in fact, the case with typical ground media, such as instrument housings and common ground wire. An example is shown schematically in figure 2.17a. In this example, the two leads of a sensor are directly

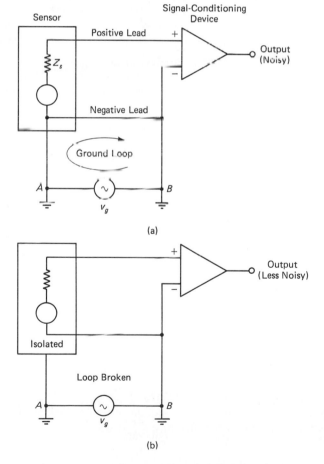

(a)

(b)

Figure 2.17. (a) Illustration of a ground loop. (b) Device isolation to eliminate ground loops (an example of internal isolation).

connected to a signal-conditioning device such as an amplifier. Because of nonuniform ground potentials, the two ground points A and B are subjected to a potential difference v_g. This will create a ground loop with the common negative lead of the two interconnected devices. The solution to this problem is to isolate (i.e., provide an infinite impedance to) either one of the two devices. Figure 2.17b shows internal isolation of the sensor. External isolation, by insulating the casing, is also acceptable. Floating off the power supply ground will also help eliminate ground loops.

2.6 INSTRUMENT RATINGS

Instrument manufacturers do not usually provide complete dynamic information for their products. In most cases, it is unrealistic to expect complete dynamic models (in the time domain or the frequency domain) and associated parameter values for complex instruments. Performance characteristics provided by manufacturers and vendors are primarily static parameters. Known as instrument ratings, these are available as parameter values, tables, charts, calibration curves, and empirical equations. Dynamic characteristics such as transfer functions (e.g., transmissibility curves expressed with respect to excitation frequency) might also be provided for more sophisticated instruments, but the available dynamic information is never complete. Furthermore, definitions of rating parameters used by manufacturers and vendors of instruments are in some cases not the same as analytical definitions used in textbooks on dynamic systems and control. This is particularly true in relation to the term *linearity*. Nevertheless, instrument ratings provided by manufacturers and vendors are very useful in the selection, installation, operation, and maintenance of instruments. In this section, we shall examine some of these performance parameters.

Rating Parameters

Typical rating parameters supplied by instrument manufacturers are

1. Sensitivity
2. Dynamic range
3. Resolution
4. Linearity
5. Zero drift and full-scale drift
6. Useful frequency range
7. Bandwidth
8. Input and output impedances

We have already discussed the meaning and significance of some of these terms with respect to dynamic behavior of instruments. In this section, we shall look at the conventional definitions given by instrument manufacturers and vendors.

Sensitivity of a transducer is measured by the magnitude (peak, rms value, etc.) of the output signal corresponding to a unit input of the measurand. This may be expressed as the ratio of (incremental output)/(incremental input) or, analytically, as the corresponding partial derivative. In the case of vectorial or tensorial signals (e.g., displacement, velocity, acceleration, strain, force), the direction of sensitivity should be specified. Cross-sensitivity is the sensitivity along directions that are orthogonal to the direction of sensitivity; it is expressed as a percentage of direct sensitivity. High sensitivity and low cross-sensitivity are desirable for measuring instruments. Sensitivity to parameter changes and noise has to be small in any device, however. On the other hand, in *adaptive control,* system sensitivity to control parameters has to be sufficiently high. Often, sensitivity and robustness are conflicting requirements.

Dynamic range of an instrument is determined by the allowed lower and upper limits of its input or output (response) so as to maintain a required level of measurement accuracy. This range is usually expressed as a ratio, in *decibels.* In many situations, the lower limit of dynamic range is equal to the resolution of the device. Hence, the dynamic range ratio is usually expressed as (range of operation)/(resolution).

Resolution is the smallest change in a signal that can be detected and accurately indicated by a transducer, a display unit, or any pertinent instrument. It is usually expressed as a percentage of the maximum range of the instrument or as the inverse of the dynamic range ratio. It follows that dynamic range and resolution are very closely related.

Example 2.7

The meaning of dynamic range (and resolution) can easily be extended to cover digital instruments. For example, consider an instrument that has a 12-bit analog-to-digital converter (ADC). Estimate the dynamic range of the instrument.

Solution   In this example, dynamic range is determined (primarily) by the word size of the ADC. Each bit can take the binary value 0 or 1. Since the resolution is given by the smallest possible increment, a change by the least significant bit (LSB),

$$\text{digital resolution} = 1$$

The largest value represented by a 12-bit word corresponds to the case when all twelve bits are unity. This value is decimal $2^{12} - 1$. The smallest value (when all twelve bits are zero) is zero. Hence, using the definition

$$\text{dynamic range} = 20 \log_{10} \left[\frac{\text{range of operation}}{\text{resolution}} \right] \tag{2.23}$$

the dynamic range of the instrument is given by

$$20 \log_{10} \left[\frac{2^{12} - 1}{1} \right] = 72 \text{ dB}$$

Another (perhaps more correct) way of looking at this problem is to consider the resolution to be some value δy, rather than unity, depending on the particular application.

For example, δy may represent an output signal increment of 0.0025 V. Next, we note that a 12-bit word can represent a combination of 2^{12} values (i.e., 4,096 values), the smallest value being y_{min} and the largest value being

$$y_{max} = y_{min} + (2^{12} - 1)\,\delta y$$

Note that y_{min} can be zero, positive, or negative. The smallest increment between values is δy, which is, by definition, the resolution. There are 2^{12} values within y_{min} and y_{max}, the two end values inclusive. Then

$$\text{dynamic range} = \frac{y_{max} - y_{min}}{\delta y} = \frac{(2^{12} - 1)\delta y}{\delta y} = 12^{12} - 1 = 4,095 = 72 \text{ dB}$$

So we end up with the same result for dynamic range, but the interpretation of resolution is different.

Linearity is determined by the calibration curve of an instrument. The curve of output amplitude (peak or rms value) versus input amplitude under static conditions within the dynamic range of an instrument is known as the *static calibration curve*. Its closeness to a straight line measures the degree of linearity. Manufacturers provide this information either as the maximum deviation of the calibration curve from the least squares straight-line fit of the calibration curve or from some other reference straight line. If the least squares fit is used as the reference straight line, the maximum deviation is called *independent linearity* (more correctly, independent nonlinearity, because the larger the deviation, the greater the nonlinearity). Nonlinearity may be expressed as a percentage of either the actual reading at an operating point or the full-scale reading.

Zero drift is defined as the drift from the null reading of the instrument when the measurand is maintained steady for a long period. Note that in this case, the measurand is kept at zero or any other level that corresponds to null reading of the instrument. Similarly, *full-scale drift* is defined with respect to the full-scale reading (the measurand is maintained at the full-scale value). Usual causes of drift include instrument instability (e.g., instability in amplifiers), ambient changes (e.g., changes in temperature, pressure, humidity, and vibration level), changes in power supply (e.g., changes in reference DC voltage or AC line voltage), and parameter changes in an instrument (due to aging, wearout, nonlinearities, etc.). Drift due to parameter changes that are caused by instrument nonlinearities is known as *parametric drift, sensitivity drift,* or *scale-factor drift*. For example, a change in spring stiffness or electrical resistance due to changes in ambient temperature results in a parametric drift. Note that parametric drift depends on the measurand level. Zero drift, however, is assumed to be the same at any measurand level if the other conditions are kept constant. For example, a change in reading caused by thermal expansion of the readout mechanism due to changes in ambient temperature is considered a zero drift. In electronic devices, drift can be reduced by using alternating current (AC) circuitry rather than direct current (DC) circuitry. For example, AC-coupled amplifiers have fewer drift problems than DC amplifiers. Intermittent checking for instrument response level with zero input is a popular way to calibrate for zero drift. In digital devices, for example, this can be done automatically from time to time between sample points, when the input signal can be bypassed without affecting the system operation.

Useful frequency range corresponds to a flat gain curve and a zero phase curve in the frequency response characteristics of an instrument. The maximum frequency in this band is typically less than half (say, one-fifth) of the dominant resonant frequency of the instrument. This is a measure of instrument bandwidth.

Bandwidth of an instrument determines the maximum speed or frequency at which the instrument is capable of operating. High bandwidth implies faster speed of response. Bandwidth is determined by the dominant natural frequency ω_n or the dominant resonant frequency ω_r of the transducer. (Note: For low damping, ω_r is approximately equal to ω_n.) It is inversely proportional to rise time and the dominant time constant. Half-power bandwidth (defined earlier) is also a useful parameter. Instrument bandwidth has to be several times greater than the maximum frequency of interest in the measured signal. Bandwidth of a measuring device is important, particularly when measuring transient signals. Note that bandwidth is directly related to the useful frequency range.

Accuracy and Precision

The instrument ratings mentioned in the preceding section affect the overall *accuracy* of an instrument. Accuracy can be assigned either to a particular reading or to an instrument. Note that instrument accuracy depends not only on the physical hardware of the instrument but also on the operating environment, including arbitrary factors such as the practices of a particular user. Usually, instrument accuracy is given with respect to a standard set of operating conditions (e.g., design conditions that are the normal steady operating conditions or extreme and transient conditions, such as emergency start up and shutdown). *Measurement accuracy* determines the closeness of the measured value to true value. *Instrument accuracy* is related to the worst accuracy obtainable within the dynamic range of the instrument in a specific operating environment. *Measurement error* is defined as

$$\text{error} = (\text{measured value}) - (\text{true value})$$

Correction, which is the negative of error, is defined as

$$\text{correction} = (\text{true value}) - (\text{measured value})$$

Each of these can also be expressed as a percentage of the true value. Accuracy of an instrument may be determined by measuring a parameter whose true value is known, near the extremes of the dynamic range of instrument, under certain operating conditions. For this purpose, standard parameters or signals that can be generated at very high levels of accuracy would be needed. The National Bureau of Standards (NBS) is usually responsible for generation of these standards. Nevertheless, accuracy and error values cannot be determined to 100 percent exactness in typical applications, because the true value is not known to begin with. In a given situation, we can only make estimates for accuracy, by using ratings provided by the instrument manufacturer or by analyzing data from previous measurements and models.

Causes of error include instrument instability, external noise (disturbances), poor calibration, inaccurate information (e.g., poor analytical models, inaccurate control laws and digital control algorithms), parameter changes (e.g., due to environmental changes, aging, and wearout), unknown nonlinearities, and improper use of instrument.

Errors can be classified as *deterministic* (or *systematic*) and *random* (or *stochastic*). Deterministic errors are those caused by well-defined factors, including nonlinearities and offsets in readings. These usually can be accounted for by proper calibration and analysis practices. Error ratings and calibration charts are used to remove systematic errors from instrument readings. Random errors are caused by uncertain factors entering into instrument response. These include device noise, line noise, and effects of unknown random variations in the operating environment. A statistical analysis using sufficiently large amounts of data is necessary to estimate random errors. The results are usually expressed as a mean error, which is the systematic part of random error, and a standard deviation or confidence interval for instrument response. These concepts will be addressed in section 2.7.

Precision is not synonymous with accuracy. Reproducibility (or repeatability) of an instrument reading determines the precision of an instrument. Two or more identical instruments that have the same high offset error might be able to generate responses at high precision, even though these readings are clearly inaccurate. For example, consider a timing device (clock) that very accurately indicates time increments (say, up to the nearest microsecond). If the reference time (starting time) is set incorrectly, the time readings will be in error, even though the clock has very high precision.

Instrument error may be represented by a random variable that has a mean value μ_e and a standard deviation σ_e. If the standard deviation is zero, the variable is deterministic. In that case, the error is said to be deterministic or repeatable. Otherwise, the error is said to be random. The precision of an instrument is determined by the standard deviation of error in the instrument response. Readings of an instrument may have a large mean value of error (e.g., large offset), but if the standard deviation is small, the instrument has high precision. Hence, a quantitative definition for precision would be

$$\text{precision} = (\text{measurement range})/\sigma_e \qquad (2.24)$$

Lack of precision originates from random causes and poor construction practices. It cannot be compensated for by recalibration, just as precision of a clock cannot be improved by resetting the time. On the other hand, accuracy can be improved by recalibration. Repeatable (deterministic) accuracy is inversely proportional to the magnitude of the mean error μ_e.

In selecting instruments for a particular application, in addition to matching instrument ratings with specifications, several additional considerations should be looked into. These incude geometric limitations (size, shape, etc.), environmental conditions (e.g., chemical reactions including corrosion, extreme temperatures, light, dirt accumulation, electromagnetic fields, radioactive environments, and shock and vibration), power requirements, operational simplicity, availability, past record and reputation of the manufacturer and of the particular instrument, and cost and related economic aspects (initial cost, maintenance cost, cost of supplementary components such as signal-conditioning and processing devices, design life and associated frequency of replacement, and cost of disposal and replacement). Often, these considerations become the ultimate deciding factors in the selection process.

2.7 ERROR ANALYSIS

Analysis of error is a very challenging task. Difficulties arise for many reasons, particularly the following:

1. True value is usually unknown.
2. The instrument reading may contain random error that cannot be determined exactly.
3. The error may be a complex (not simple) function of many variables (input variables and state variables or response variables).
4. The instrument may be made up of many components that have complex interrelations (dynamic coupling, multiple degree-of-freedom responses, nonlinearities, etc.), and each component may contribute to the overall error.

The first item is a philosophical issue that would lead to an argument similar to the chicken-and-egg controversy. For instance, if the true value is known, there is no need to measure it; and if the true value is unknown, it is impossible to determine exactly how inaccurate a particular reading is. In fact, this situation can be addressed to some extent by using statistical representations of error, which takes us to the second item listed. The third and fourth items may be addressed by error combination in multivariable systems and by error propagation in complex multicomponent systems. It is not feasible here to provide a full treatment of all these topics. Only an introduction to available analytical techniques will be given, using illustrative examples.

The concepts discussed in this section are useful not only in statistical error analysis but also in the field of *statistical process control* (SPC)—the use of statistical signals to improve performance of a process. Performing statistical analysis of a response signal and drawing its *control chart,* along with an *upper control line* and a *lower control line,* are key procedures in statistical process control.

Statistical Representation

We have noted that, in general, error is a random variable. It is defined as

$$\text{error} = (\text{instrument reading}) - (\text{true value}).$$

Randomness associated with a measurand can be interpreted in two ways. First, since the true value of the measurand is a fixed quantity, randomness can be interpreted as the randomness in error that is usually originating from the random factors in instrument response. Second, looking at the issue in a more practical manner, error analysis can be interpreted as an "estimation problem" in which the objective is to estimate the true value of a measurand from a known set of readings. In this latter point of view, "estimated" true value itself becomes a random variable. No matter what approach is used, however, the same statistical concepts may be used in representing error. First, let us review some important concepts in probability and statistics.

Cumulative probability distribution function.   Consider a random variable X. The probability that the random variable takes a value equal to or less than a specific value x is a function of x. This function, denoted by $F(x)$, is termed *cumulative probability distribution function,* or simply *distribution functon.* Specifically,

$$F(x) = P[X \leq x] \tag{2.25}$$

Note that $F(\infty) = 1$ and $F(-\infty) = 0$, because the value of X is always less than infinity and can never be less than negative infinity. Furthermore, $F(x)$ has to be a monotonically increasing function, as shown in figure 2.18a, because negative probabilities are not defined.

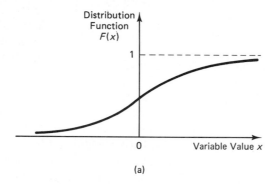

(a)

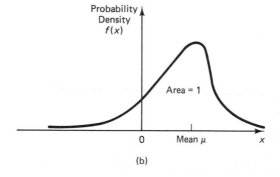

(b)

Figure 2.18.   (a) A cumulative probability distribution function. (b) A probability density function.

Probability density function.   Assuming that random variable X is a continuous variable and, hence, $F(x)$ is a continuous function of x, probability density function $f(x)$ is given by the slope of $F(x)$, as shown in figure 2.18b. Thus,

$$f(x) = \frac{dF(x)}{dx} \tag{2.26}$$

Hence,

$$F(x) = \int_{\infty}^{x} f(x)dx \tag{2.27}$$

Note that the area under the density curve is unity. Furthermore, the probability that the random variable falls within two values is given by the area under the density curve within these two limits. This can be easily shown using the definitions of $F(x)$ and $f(x)$:

$$P[a < X \le b] = F(h) - F(a)$$

$$= \int_{-\infty}^{b} f(x)dx - \int_{-\infty}^{a} f(x)dx = \int_{a}^{b} f(x)dx \tag{2.28}$$

Mean value (expected value).   If a random variable X is measured repeatedly a very large (infinite) number of times, the average of these measurements is the *mean value* μ or *expected value* $E(X)$. It should be easy to see that this may be expressed as the weighted sum of all possible values of the random variable, each value being weighted by the associated probability of its occurrence. Since the probability that X takes the value x is given by $f(x)\,\delta x$, with δx approaching zero, we have

$$\mu = E(X) = \lim_{\delta x \to 0} \sum xf(x)\delta x$$

Since the right-hand-side summation becomes an integral in the limit, we get

$$\mu = E(X) = \int_{-\infty}^{\infty} xf(x)dx \tag{2.29}$$

Root-mean-square (rms) value.   The mean square value of a random variable X is given by

$$E(X^2) = \int_{-\infty}^{\infty} x^2 f(x)dx \tag{2.30}$$

The root-mean-square (rms) value is the square root of the mean square value.

Variance and standard deviation.   Variance of a random variable is the mean square value of the deviation from mean. This is denoted by Var(X) or σ^2 and is given by

$$\text{Var}(X) = \sigma^2 = \int_{-\infty}^{\infty} (x - \mu)^2 f(x)dx \tag{2.31}$$

By expanding equation 2.31, we can show that

$$\sigma^2 = E(X^2) - \mu^2 \tag{2.32}$$

Standard deviation σ is the square root of variance. Note that standard deviation is a measure of statistical "spread" of a random variable. A random variable with smaller σ is less random and its density curve exhibits a sharper peak, as shown in figure 2.19.

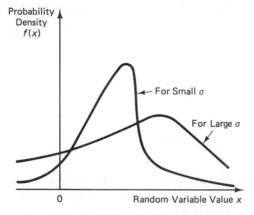

Figure 2.19. Effect of standard deviation on the shape of a probability density curve.

Some thinking should convince you that if the probability density function of random variable X is $f(x)$, then the probability density function of any (well-behaved) function of X is also $f(x)$. In particular, for constants a and b, the probability density function of $(aX + b)$ is also $f(x)$. Note, further, that the mean of $(aX + b)$ is $(a\mu + b)$. Hence, from equation 2.31, it follows that the variance of aX is

$$\text{Var}(aX) = \int_{-\infty}^{\infty} (ax - a\mu)^2 f(x)dx$$

$$= a^2 \int_{-\infty}^{\infty} (x - \mu)^2 f(x)dx$$

Hence,

$$\text{Var}(aX) = a^2 \text{Var}(X) \tag{2.33}$$

Independent random variables. Two random variables, X_1 and X_2, are said to be independent if the event "X_1 assumes a certain value" is completely independent of the event "X_2 assumes a certain value." In other words, the processes that generate the responses X_1 and X_2 are completely independent. Furthermore, probability distributions of X_1 and X_2 will also be completely independent. Hence, it can be shown that for independent random variables X_1 and X_2, the mean value of the product is equal to the product of the mean values. Thus,

$$E(X_1 X_2) = E(X_1)E(X_2) \tag{2.34}$$

for independent random variables X_1 and X_2.

Now, using the definition of variance and equation 2.34, it can be shown that

$$\text{Var}(X_1 + X_2) = \text{Var}(X_1) + \text{Var}(X_2) \tag{2.35}$$

for independent X_1 and X_2.

Sample mean and sample variance. Consider N measurements $\{X_1, X_2, \ldots, X_N\}$ of random variable X. This set of data is termed a *data sample*. It generally is not possible to extract all information about the probability distribution of X from this data sample. We are able, however, to make some useful *estimates*. One would expect that the larger the data sample, the more accurate these statistical estimates would be.

An estimate for the mean value of X would be the *sample mean* $\overline{X}$, which is defined as

$$\overline{X} = \frac{1}{N}\sum_{i=1}^{N} X_i \tag{2.36}$$

An estimate for variance would be the *sample variance* S^2, given by

$$S^2 = \frac{1}{N-1}\sum_{i=1}^{N} (X_i - \overline{X})^2 \tag{2.37}$$

An estimate for standard deviation would be the *sample standard deviation, S*, which is the square root of the sample variance.

One might be puzzled by the denominator $N - 1$ on the right-hand side of equation 2.37. Since we are computing an "average" deviation, the denominator should have been N. But in that case, with just one reading ($N = 1$), we get a finite value for S, which is not correct because one cannot talk about a sample standard deviation when only one measurement is available. Since, according to equation 2.37, S is not defined $(0/0)$ when $N = 1$, this definition of S^2 is more realistic. Another advantage of equation 2.37 is that this equation gives an *unbiased estimate* of variance. This concept will be discussed next. Note that if we use N instead of $N - 1$ in equation 2.37, the computed variance is called *population variance*. Its square root is *population standard deviation*. When $N > 30$, the difference between sample variance and population variance becomes negligible.

Unbiased estimates. Note that each term X_i in the sample data set $\{X_1, X_2, \ldots, X_N\}$ is itself a random variable just like X, because the measured value of X_i contains some randomness and is subjected to chance. In other words, if N measurements were taken at one time and then the same measurements were repeated, the values would be different from the first set, since X was random to begin with. It follows that $\overline{X}$ and S in equations 2.36 and 2.37 are also random variables. Note that the mean value of $\overline{X}$ is

$$E(\overline{X}) = E\left[\frac{1}{N}\sum_{i=1}^{N} X_i\right] = \frac{1}{N}\sum_{i=1}^{N} E(X_i) = \frac{N\mu}{N}$$

Hence,

$$E(\overline{X}) = \mu \tag{2.38}$$

We know that $\overline{X}$ is an estimate for μ. Also, from equation 2.38, we observe that the mean value of $\overline{X}$ is μ. Hence, the sample mean $\overline{X}$ is an *unbiased estimate* of mean value μ. Similarly, from equation 2.37, we can show that the mean value of S^2 is

$$E(S^2) = \sigma^2 \tag{2.39}$$

assuming that X_i are independent measurements. Thus, the sample variance S^2 is an unbiased estimate of variance σ^2. In general, if the mean value of an estimate is equal to the exact value of the parameter that is being estimated, the estimate is said to be unbiased. Otherwise, it is a *biased estimate*.

Example 2.8

An instrument has a response X that is random, with standard deviation σ. A set of N independent measurements $\{X_1, X_2, \ldots, X_N\}$ is made and the sample mean $\bar{X}$ is computed. Show that the standard deviation of $\bar{X}$ is $\sigma/\sqrt{N}$.

Also, a measuring instrument produces a random error whose standard deviation is 1 percent. How many measurements should be averaged in order to reduce the standard deviation of error to less than 0.05 percent?

Solution   To solve the first part of the problem, start with equation 2.36 and use the properties of variance given by equations 2.33 and 2.35:

$$\text{Var}(\bar{X}) = \text{Var}\left[\frac{1}{N}(X_1 + X_2 + \cdots + X_N)\right]$$

$$= \frac{1}{N^2}\text{Var}(X_1 + X_2 + \cdots + X_N)$$

$$= \frac{1}{N^2}[\text{Var}(X_1 + \text{Var}X_2 + \cdots + \text{Var}X_N)]$$

$$= \frac{N\sigma^2}{N^2}$$

Here we used the fact that X_i are indpendent.

Hence,

$$\text{Var}(\bar{X}) = \frac{\sigma^2}{N} \tag{2.40}$$

Accordingly,

$$\text{Std}(\bar{X}) = \frac{\sigma}{\sqrt{N}} \tag{2.41}$$

For the second part of the problem, $\sigma = 1$ percent and $\sigma/\sqrt{N} < 0.05$ percent. Then,

$$\frac{1}{\sqrt{N}} < 0.05$$

or

$$N > 400$$

Thus, we should average more than 400 measurements to obtain the specified accuracy.

Gaussian distribution.   Gaussian distribution, or *normal distribution,* is probably the most extensively used probability distribution in engineering applications. Apart from its ease of use, another justification for its widespread use is provided by the *central limit theorem.* This theorem states that a random variable that is formed by summing a very large number of independent random variables takes Gaussian distribution in the limit. Since many engineering phenomena are consequences of numerous independent random causes, the assumption of normal distribution is justified in many cases. The validity of Gaussian assumption can be checked by plotting data on *probability graph paper* or by using various tests such as the *chi-square test*.

The Gaussian probability density function is given by

$$f(x) = \frac{1}{\sqrt{2\pi}\,\sigma} \exp\left[-\frac{(x - \mu)^2}{2\sigma^2}\right] \tag{2.42}$$

Note that only two parameters, mean μ and standard deviation σ, are necessary to determine a Gaussian distribution completely.

A closed algebraic expression cannot be given for the cumulative probability distribution function $F(x)$ of Gaussian distribution. It should be evaluated by numerical integration. Numerical values for the normal distribution curve are available in tabulated form, with the random variable X being normalized with respect to μ and σ according to

$$Z = \frac{X - \mu}{\sigma} \tag{2.43}$$

Note that the mean value of this normalized variable Z is

$$E(Z) = E[(X - \mu)/\sigma] = [E(X) - \mu]/\sigma$$
$$= (\mu - \mu)/\sigma$$

or

$$E(Z) = 0 \tag{2.44}$$

and the variance of Z is

$$\text{Var}(Z) = \text{Var}[(X - \mu)/\sigma] = \text{Var}(X - \mu)/\sigma^2$$
$$= \text{Var}(X)/\sigma^2 = \sigma^2/\sigma^2$$

or

$$\text{Var}(Z) = 1 \tag{2.45}$$

Furthermore, the probability density function of Z is

$$f(z) = \frac{1}{\sqrt{2\pi}} \exp(-z^2/2) \tag{2.46}$$

What is usually tabulated is the area under the density curve $f(z)$ of the normalized random variable Z for different values of z. A convenient form is presented in table 2.3, where the area under the $f(z)$ curve from 0 to z is tabulated up to four decimal places for different positive values of z up to two decimal places. Since the density curve is symmetric about the mean value (zero for the normalized case), values for negative z do not have to be tabulated. Furthermore, when $z \to \infty$, area A in table 2.3 approaches 0.5. The value for $z = 3.09$ is already 0.4990. Hence, for most practical purposes, area A may be taken as 0.5 for z values greater than 3.0. Since Z is nor-

TABLE 2.3 A TABLE OF GAUSSIAN PROBABILITY DISTRIBUTION

$$f(z) = \frac{1}{\sqrt{2\pi}} \exp(-z^2/2)$$

$$A = \int_0^z f(z)dz$$

Area A

z	.00	.01	.02	.03	.04	.05	.06	.07	.08	.09
0.0	0.0000	0.0040	0.0080	0.0120	0.0160	0.0199	0.0239	0.0279	0.0319	0.0359
0.1	0.0398	0.0438	0.0478	0.0517	0.0557	0.0596	0.0636	0.0675	0.0714	0.0753
0.2	0.0793	0.0832	0.0871	0.0910	0.0948	0.0987	0.1026	0.1064	0.1103	0.1141
0.3	0.1179	0.1217	0.1255	0.1293	0.1331	0.1368	0.1406	0.1443	0.1480	0.1517
0.4	0.1554	0.1591	0.1628	0.1664	0.1700	0.1736	0.1772	0.1808	0.1844	0.1879
0.5	0.1915	0.1950	0.1985	0.2019	0.2054	0.2088	0.2123	0.2157	0.2190	0.2224
0.6	0.2257	0.2291	0.2324	0.2357	0.2389	0.2422	0.2454	0.2486	0.2517	0.2549
0.7	0.2580	0.2611	0.2642	0.2673	0.2704	0.2734	0.2764	0.2794	0.2823	0.2852
0.8	0.2881	0.2910	0.2939	0.2967	0.2995	0.3023	0.3051	0.3078	0.3106	0.3233
0.9	0.3159	0.3186	0.3212	0.3238	0.3264	0.3289	0.3315	0.3340	0.3365	0.3389
1.0	0.3413	0.3438	0.3461	0.3485	0.3508	0.3531	0.3554	0.3577	0.3599	0.3621
1.1	0.3643	0.3665	0.3686	0.3708	0.3729	0.3749	0.3770	0.3790	0.3810	0.3830
1.2	0.3849	0.3869	0.3888	0.3907	0.3925	0.3944	0.3962	0.3980	0.3997	0.4015
1.3	0.4032	0.4049	0.4066	0.4082	0.4099	0.4115	0.4131	0.4147	0.4162	0.4177
1.4	0.4192	0.4207	0.4222	0.4236	0.4251	0.4265	0.4279	0.4292	0.4306	0.4319
1.5	0.4332	0.4345	0.4357	0.4370	0.4382	0.4394	0.4406	0.4418	0.4429	0.4441
1.6	0.4452	0.4463	0.4474	0.4484	0.4495	0.4505	0.4515	0.4525	0.4535	0.4545
1.7	0.4554	0.4564	0.4573	0.4582	0.4591	0.4599	0.4608	0.4616	0.4625	0.4633
1.8	0.4641	0.4649	0.4656	0.4664	0.4671	0.4678	0.4686	0.4693	0.4699	0.4706
1.9	0.4713	0.4719	0.4726	0.4732	0.4738	0.4744	0.4750	0.4758	0.4761	0.4767
2.0	0.4772	0.4778	0.4783	0.4788	0.4793	0.4799	0.4803	0.4808	0.4812	0.4817
2.1	0.4821	0.4826	0.4830	0.4834	0.4838	0.4842	0.4846	0.4850	0.4854	0.4857
2.2	0.4861	0.4864	0.4868	0.4871	0.4875	0.4878	0.4881	0.4884	0.4887	0.4890
2.3	0.4893	0.4896	0.4898	0.4901	0.4904	0.4906	0.4909	0.4911	0.4913	0.4916
2.4	0.4918	0.4920	0.4922	0.4925	0.4927	0.4929	0.4931	0.4932	0.4934	0.4936
2.5	0.4938	0.4940	0.4941	0.4943	0.4945	0.4946	0.4948	0.4949	0.4951	0.4952
2.6	0.4953	0.4955	0.4956	0.4957	0.4959	0.4960	0.4961	0.4962	0.4963	0.4964
2.7	0.4965	0.4966	0.4967	0.4968	0.4969	0.4970	0.4971	0.4972	0.4973	0.4974
2.8	0.4974	0.4975	0.4976	0.4977	0.4977	0.4978	0.4979	0.4979	0.4980	0.4981
2.9	0.4981	0.4982	0.4982	0.4983	0.4984	0.4984	0.4985	0.4985	0.4986	0.4986
3.0	0.4987	0.4987	0.4987	0.4988	0.4988	0.4988	0.4989	0.4989	0.4989	0.4990

malized with respect to σ, $z = 3$ actually corresponds to three times the standard deviation of the original random variable X. It follows that for a Gaussian random variable, most of the values will fall within $\pm 3\sigma$ about the mean value. It can be stated that approximately

- 68 percent of the values will fall within $\pm \sigma$ about μ
- 95 percent of the values will fall within $\pm 2\sigma$ about μ
- 99.7 percent of the values will fall within $\pm 3\sigma$ about μ

These can be easily verified using table 2.3.

Statistical process control.   In statistical process control (SPC), statistical analysis of process responses is used to generate control actions. This method of control is applicable in many situations of process control, including manufacturing quality control, control of chemical process plants, computerized office management systems, inventory control systems, and urban transit control systems. A major step in statistical process control is to compute control limits (or action lines) on the basis of measured data from the process.

Control Limits or Action Lines.   Since a very high percentage of readings from an instrument should lie within $\pm 3\sigma$ about the mean value, according to the normal distribution, these boundaries (-3σ and $+3\sigma$) drawn about the mean value may be considered *control limits* or *action lines* in statistical process control. If any measurements fall outside the action lines, corrective measures such as recalibration, controller adjustment, or redesign should be carried out.

Steps of SPC.   The main steps of statistical process control are as follows:

1. Collect measurements of appropriate response variables of the process.
2. Compute the mean value of the data, the upper control limit, and the lower control limit.
3. Plot the measured data and the two control limits on a control chart.
4. If measurements fall outside the control limits, take corrective action and repeat the control cycle (go to step 1).

If the measurements always fall within the control limits, the process is said to be in statistical control.

Example 2.9

Error in a satellite tracking system was monitored on-line for a period of one hour to determine whether recalibration or gain adjustment of the tracking controller would be necessary. Four measurements of the tracking deviation were taken in a period of five minutes, and twelve such data groups were acquired during the one-hour period. Sample means and sample variances of the twelve groups of data were computed. The results are tabulated as follows:

Period i	1	2	3	4	5	6	7	8	9	10	11	12
Sample mean $\bar{X}_i$	1.34	1.10	1.20	1.15	1.30	1.12	1.26	1.10	1.15	1.32	1.35	1.18
Sample variance S_i^2	0.11	0.02	0.08	0.10	0.09	0.02	0.06	0.05	0.08	0.12	0.03	0.07

Draw a control chart for the error process, with control limits (action lines) at $\bar{X} \pm 3\sigma$. Establish whether the tracking controller is in statistical control or needs adjustment.

Solution The overall mean tracking deviation,

$$\bar{X} = \frac{1}{12}\sum_{i=1}^{12}\bar{X}_i$$

is computed to be $\bar{X} = 1.214$. The average sample variance,

$$\bar{S}^2 = \frac{1}{12}\sum_{i=1}^{12}S_i^2$$

is computed to be $\bar{S}^2 = 0.069$. Since there are four readings within each period, the standard deviaton σ of group mean $\bar{X}_i$ can be estimated, using equation 2.41, as

$$S = \frac{\bar{S}}{\sqrt{4}} = \frac{\sqrt{0.069}}{\sqrt{4}} = 0.131$$

The upper control limit (action line) is at (approximately)

$$x = \bar{X} + 3S = 1.214 + 3 \times 0.131 = 1.607$$

The lower control limit (action line) is at

$$x = \bar{X} - 3S = 0.821$$

These two lines are shown on the control chart in figure 2.20. Since the sample means lie within the two action lines, the process is considered to be in statistical control, and controller adjustments would not be necessary. Note that if better resolution is required in making this decision, individual readings, rather than group means, should be plotted in figure 2.20.

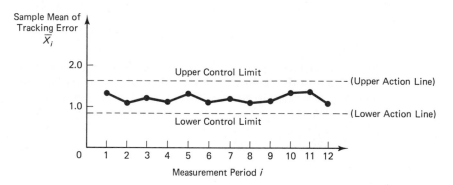

Figure 2.20. Control chart for the satellite tracking error example.

Confidence intervals. The probability that the value of a random variable would fall within a specified interval is called a *confidence level*. As an example, consider a Gaussian random variable X that has mean μ and standard deviation σ. This is denoted by

$$X = N(\mu, \sigma) \tag{2.47}$$

Suppose that N measurements $\{X_1, X_2, \dots, X_N\}$ are made. The sample mean $\bar{X}$ is an unbiased estimate for μ. We also know that the standard deviation of $\bar{X}$ is σ/N. Now consider the normalized random variable

$$Z = \frac{\bar{X} - \mu}{\sigma / \sqrt{N}} \qquad (2.48)$$

This is a Gaussian random variable with zero mean and unity standard deviation. The probability p that the values of Z fall within $\pm z_o$:

$$P(-z_o < Z \leq z_o) = p \qquad (2.49)$$

can be determined from table 2.3 for a specified value of z_o. Now substituting equation 2.48 in 2.49, we get

$$P\left(-z_o < \frac{\bar{X} - \mu}{\sigma / \sqrt{N}} \leq z_o\right) = p$$

or

$$P\left(\bar{X} - \frac{z_o \sigma}{\sqrt{N}} \leq \mu < \bar{X} + \frac{z_o \sigma}{\sqrt{N}}\right) = p \qquad (2.50)$$

Note that the lower limit has the $\leq$ sign and the upper limit has the $<$ sign within the parentheses. These have been used for mathematical precision, but for practical purposes, either $\leq$ or $<$ may be used in each limit. Now, from equation 2.50, it follows that the confidence level is p that the actual mean value μ would fall within $\pm z_o \sigma / \sqrt{N}$ of the estimated (sample) mean value $\bar{X}$.

Example 2.10

The angular resolution of a resolver (a rotary displacement sensor—see chapter 3) was tested sixteen times, independently, and recorded in degrees as follows:

0.11, 0.12, 0.09, 0.10, 0.10, 0.14, 0.08, 0.08

0.13, 0.10, 0.10, 0.12, 0.08, 0.09, 0.11, 0.15

If the standard deviation of the angular resolution of this brand of resolvers is known to be 0.01°, what are the odds that the mean resolution would fall within 5 percent of the sample mean?

Solution To solve this problem, we assume that resolution is normally distributed. The sample mean is computed as

$$\bar{X} = \frac{1}{16}(0.11 + 0.12 + \cdots + 0.11 + 0.15) = 0.10625$$

In view of equation 2.50, we must have

$$\frac{z_o \sigma}{\sqrt{16}} = 5\% \text{ of } \bar{X}$$

Hence,

$$\frac{z_o \times 0.01}{\sqrt{16}} = \frac{5}{100} \times 0.10625$$

or

$$z_o = 2.125$$

Now, from table 2.3,

$$P(-2.125 < Z < 2.125) = 2 \times \frac{(0.4830 + 0.4834)}{2} = 0.9664$$

Sign test and binomial distribution. Sign test is useful in comparing accuracies of two similar instruments. First, measurements should be made on the same measurand (input signal to instrument) using the two devices. Next, the readings of one instrument are subtracted from the corresponding readings of the second instrument, and the results are tabulated. Finally, the probability of getting the number of negative signs (or positive signs) equal to what is present in the tabulated results is computed using *binomial distribution*.

Before discussing binomial distribution, let us introduce some new terminology. First, *factorial r* (denoted by $r!$) of an integer r is defined as the product

$$r! = r \times (r - 1) \times (r - 2) \times \cdots \times 2 \times 1 \tag{2.51}$$

Now, suppose that there are n distinct articles that are distinguishable from one another. The number of ways in which r articles could be picked from the batch of n, giving proper consideration to the order in which the r articles are picked (or arranged), is called the number of *permutations* of r from n. This is denoted by ^nP_r, which is given by

$$^nP_r = n \times (n - 1) \times (n - 2) \times \cdots \times (n - r + 2) \times (n - r + 1)$$

$$= \frac{n!}{(n - r)!} \tag{2.52}$$

This can be easily verified, since the first article can be chosen in n ways and the second article can be chosen from the remaining $(n - 1)$ articles in $(n - 1)$ ways and kept next to the first article, and so on.

If we disregard the order in which the r articles are picked (and arranged), the number of possible choices of r articles is termed the number of *combinations* of r from n. This is denoted by ^nC_r. Now, since each combination can be arranged in $r!$ different ways (if the order of arrangement is considered), we have

$$^nC_r \times r! = {}^nP_r \tag{2.53}$$

Hence, using equation 2.52, we get

$$^nC_r = \frac{n \times (n - 1) \times (n - 2) \times \cdots \times (n - r + 2) \times (n - r + 1)}{r!}$$

$$= \frac{n!}{(n - r)! \, r!} \tag{2.54}$$

With the foregoing notation, we can introduce binomial distribution in the context of sign test. Suppose that n pairs of readings are taken from the two instruments. If the probability that a difference in reading would be positive is p, then the probability that the difference would be negative is $1 - p$. Note that if the systematic error in the two instruments is the same and if the random error is purely random, then $p = 0.5$.

The probability of getting exactly r positive signs among the n entries in the table is

$$p(r) = {}^nC_r p^r(1 - p)^{n-r} \qquad (2.55)$$

To verify equation 2.55, note that this event is similar to picking exactly r items from n items and constraining each picked item to be positive (having probability p) and also constraining the remaining $(n - r)$ items to be negative (having probability $1 - p$). Note that r is a discrete variable that takes values $r = 1, 2, \ldots, n$. Furthermore, it can be easily verified that

$$\sum_{r=1}^{n} p(r) = \sum_{r=1}^{n} {}^nC_r p^r(1 - p)^{n-r} = (p + 1 - p)^n = 1 \qquad (2.56)$$

Hence, $p(r)$, $r = 1, 2, \ldots, n$, is a discrete function that resembles a continuous probability density function $f(x)$. In fact, $p(r)$ given by equation 2.55 represents *binomial probability distribution*. Using equation 2.55, we can perform the sign test. The details of the test are conveniently explained by means of an example.

Example 2.11

To compare the accuracies of two brands of differential transformers (DTs, which are displacement sensors—see chapter 3), the same rotation (in degrees) of a robot arm joint was measured using both brands, DT1 and DT2. The following ten measurement pairs were taken:

DT1	10.3	5.6	20.1	15.2	2.0	7.6	12.1	18.9	22.1	25.2
DT2	9.8	5.8	20.0	16.0	1.9	7.8	12.2	18.7	22.0	25.0

Assuming that both devices are used simultaneously (so that backlash and other types of repeatability errors in manipulators do not enter into our problem), determine whether the two brands are equally accurate at the 70 percent level of significance.

Solution   First, we form the sign table by taking the differences of corresponding measurements:

DT1 − DT2	0.5	−0.2	0.1	−0.8	0.1	−0.2	−0.1	0.2	0.1	0.2

Note that there are six positive signs and four negative signs. If we had tabulated DT2 − DT1, however, we would get four positive signs and six negative signs. Both these cases should be taken into account in the sign test. Furthermore, more than six positive signs or fewer than four positive signs would make the two devices less similar (in accuracy) than what is indicated by the data. Hence, the probability of getting six or more positive signs or four or fewer positive signs should be computed in this example in order to estimate the possible match (in accuracy) of the two devices.

If the error in both transducers is the same, we should have

$$P(\text{positive difference}) = p = 0.5$$

This is the hypothesis that we are going to test. Using equation 2.55, the probability of getting six or more positive signs or four or fewer negative signs is calculated as

$$1 - \text{probability of getting exactly 5 positive signs}$$

$$= 1 - {}^{10}C_5\,(0.5)^5 \times (0.5)^5 = 1 - \frac{10!}{5!\,5!} \times (0.5)^{10}$$

$$= 1 - 0.246 = 0.754$$

Note that the hypothesis of two brands being equally accurate is supported by the test data at a level of significance over 75 percent, which is better than the specified value of 70 percent.

Least squares fit. We have mentioned that instrument *linearity* may be measured by the largest deviation of the input/output data (or calibration curve) from the least squares straight-line fit of data. Since many algebraic expressions become linear when plotted to a logarithmic scale, linear (straight-line) fit is generally more accurate if log-log axes are used. Linear least squares fit can be thought of as an estimation method because it "estimates" the two parameters of an input/output model, the straight line, that fits a given set of data such that the squared error is a minimum. The estimated straight line is also known as the *linear regression line* or *mean calibration curve*.

Consider N pairs of data $\{(X_1, Y_1), (X_2, Y_2), \ldots, (X_N, Y_N)\}$ in which X denotes the *independent variable* (input variable) and Y denotes the *dependent variable* (output variable).

Suppose that the estimated linear regression is given by

$$Y = mX + a \tag{2.57}$$

For the independent variable value X_i, the dependent variable value on the regression line is $(mX_i + a)$, but the actual (measured) value of the dependent variable is Y_i. Hence, the sum of squared error for all data points is

$$e = \sum_{i=1}^{N} (Y_i - mX_i - a)^2 \tag{2.58}$$

We have to minimize e with respect to the two parameters m and a. The required conditions are

$$\frac{\partial e}{\partial m} = 0 \quad \text{and} \quad \frac{\partial e}{\partial a} = 0$$

By carrying out these differentiations in equation 2.58, we get

$$\sum_{i=1}^{N} X_i (Y_i - mX_i - a) = 0$$

and

$$\sum_{i=1}^{N} (Y_i - mX_i - a) = 0$$

Dividing the two equations by N and using the definition of sample mean, we get

$$\frac{1}{N}\sum X_i Y_i - \frac{m}{N}\sum X_i^2 - a\bar{X} = 0 \qquad \text{(i)}$$

$$\bar{Y} - m\bar{X} - a = 0 \qquad \text{(ii)}$$

Solving these two simultaneous equations for m, we obtain

$$m = \left(\frac{1}{N}\sum_{i=1}^{N} X_i Y_i - \bar{X}\bar{Y}\right) \bigg/ \left(\frac{1}{N}\sum_{i=1}^{N} X_i^2 - \bar{X}^2\right) \qquad (2.59)$$

The parameter a does not have to be explicitly expressed, because from equations 2.57 and (ii), we can eliminate a and express the linear regression line as

$$Y - \bar{Y} = m(X - \bar{X}) \qquad (2.60)$$

Note from equation 2.57 that a is the Y-axis intercept (i.e., the value of Y when $X = 0$) and is given by

$$a = \bar{Y} - m\bar{X} \qquad (2.61)$$

Example 2.12

Consider the capacitor circuit shown in figure 2.21. First, the capacitor is charged to voltage v_o using a constant DC voltage source (switch in position 1); then it is discharged through a known resistance R (switch in position 2). Voltage decay during discharge is measured at known time increments. Three separate tests are carried out. The measured data are as follows:

Time t (sec)	0.1	0.2	0.3	0.4	0.5
Voltage v (volts)					
Test 1	7.3	2.8	1.0	0.4	0.1
Test 2	7.4	2.7	1.1	0.3	0.2
Test 3	7.3	2.6	1.0	0.4	0.1

If the resistance is accurately known to be 1,000 Ω, estimate the capacitance C in microfarads (μF) and the source voltage v_o in volts.

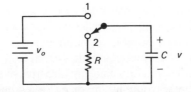

Figure 2.21. A circuit for the least squares estimation of capacitance.

Solution To solve this problem, we assume the well-known expression for the free decay of voltage across a capacitor:

$$v(t) = v_o \exp\left[-t/(RC)\right] \tag{i}$$

Take the natural logarithm of equation (i):

$$\ln v = -\frac{t}{RC} + \ln v_o \tag{ii}$$

With $Y = \ln v$ and $X = t$, equation (ii) represents a straight line with slope

$$m = -\frac{1}{RC} \tag{iii}$$

and the Y-axis intercept

$$a = \ln v_o \tag{iv}$$

Using all the data, the overall sample means can be computed. Thus,

$$\bar{X} = 0.3 \quad \text{and} \quad \bar{Y} = -0.01335$$

$$\frac{1}{N}\sum X_i Y_i = -0.2067 \quad \text{and} \quad \frac{1}{N}\sum X_i^2 = 0.11$$

Now substitute these values in equations 2.59 and 2.61. We get

$$m = -10.13 \quad \text{and} \quad a = 3.02565$$

Next, from equation (iii), with $R = 1,000$, we have

$$C = \frac{1}{10.13 \times 1000}\,\text{F} = 98.72\ \mu\text{F}$$

From equation (iv),

$$v_o = 20.61 \text{ volts}$$

Note that in this problem, the estimation error would be tremendous if we did not use log scaling for the linear fit.

Least squares curve fitting is not limited to linear (i.e., straight-line) fit. The method can be extended to a polynomial fit of any order. For example, in *quadratic fit,* the data are fitted to a second-order (i.e., quadratic) polynomial. In that case, there are three unknown parameters, which would be determined by minimizing the quadratic error.

Error Combination

Error in a response variable of an instrument or in an estimated system parameter would depend on errors present in measured variables and parameter values that are used to determine the unknown variable or parameter. Knowing how component errors are propagated within a multicomponent system and how individual errors in system variables and parameters contribute toward the overall error in a particular response variable or parameter would be important in estimating error limits in complex instruments. For example, if the output power in a gas turbine is computed by measuring torque and speed at the output shaft, error margins in the two measured "response variables" (torque and speed) would be directly combined into the error in the power computation. Similarly, if the natural frequency of a simple suspension system is determined by measuring mass and spring stiffness "parameters" of the suspension, the natural frequency estimate would be directly affected by possible errors in mass and stiffness measurements. Extending this idea further, the overall error in a control system depends on individual error levels in various components (sensors, actuators, controller hardware, filters, amplifiers, etc.) of the system and on the manner in which these components are physically interconnected and physically interrelated. For example, in a robotic manipulator, the accuracy of the actual trajectory of the end effector will depend on the accuracy of sensors and actuators at manipulator joints and on the accuracy of the robot controller. Note that we are dealing with a generalized idea of error propagation that considers errors in system variables (e.g., input and output signals, such as velocities, forces, voltages, currents, temperatures, heat transfer rates, pressures, and fluid flow rates), system parameters (e.g., mass, stiffness, damping, capacitance, inductance, and resistance), and system components (e.g., sensors, actuators, filters, amplifiers, and control circuits).

For the analytical development of a basic result in error combination, we will start with a functional relationship of the form

$$y = f(x_1, x_2, \ldots, x_r) \tag{2.62}$$

Here, x_i are the independent system variables or parameter values whose error is propagated into a dependent variable (or parameter value) y. Determination of this functional relationship is not always simple, and the relationship itself may be in error. Since our intention is to make a reasonable estimate for possible error in y due to the combined effect of errors from x_i, an approximate functional relationship would be adequate in most cases. Let us denote error in a variable by the differential of that variable. Taking the differential of equation 2.62, we get

$$\delta y = \frac{\partial f}{\partial x_1} \delta x_1 + \frac{\partial f}{\partial x_2} \delta x_2 + \cdots + \frac{\partial f}{\partial x_r} \delta x_r \tag{2.63}$$

For those who are not familiar with differential calculus, equation 2.63 should be interpreted as the first-order terms in a *Taylor series expansion* of equation 2.62. Now, rewriting equation 2.63 in the fractional form, we get

$$\frac{\delta y}{y} = \sum_{i=1}^{r} \left[\frac{x_i}{y} \frac{\partial f}{\partial x_i} \frac{\delta x_i}{x_i} \right] \tag{2.64}$$

Here, $\delta y/y$ represents the overall error and $\delta x_i/x_i$ represents the component error, expressed as fractions. We shall consider two types of estimates for overall error.

Absolute error. Since error δx_i could be either positive or negative, an upper bound for the overall error is obtained by summing the absolute value of each right-hand-side term in equation 2.64. This estimate e_{ABS}, which is termed *absolute error*, is given by

$$e_{ABS} = \sum_{i=1}^{r} \left| \frac{x_i}{y} \frac{\partial f}{\partial x_i} \right| e_i \qquad (2.65)$$

Note that component error e_i and absolute error e_{ABS} in equation 2.65 are always positive quantities; when specifying error, however, both positive and negative limits should be indicated or implied. (e.g., $\pm e_{ABS}$, $\pm e_i$).

SRSS error. Equation 2.65 provides a conservative (upper bound) estimate for overall error. Since the estimate itself is not precise, it is often wasteful to introduce such a high conservatism. A nonconservative error estimate that is frequently used in practice is the *square root of sum of squares* (SRSS) error. As the name implies, this is given by

$$e_{SRSS} = \left[\sum_{i=1}^{r} \left(\frac{x_i}{y} \frac{\partial f}{\partial x_i} e_i \right)^2 \right]^{1/2} \qquad (2.66)$$

Note that this is not an upper bound estimate for error and that $e_{SRSS} < e_{ABS}$ when more than one nonzero error contribution is present. The SRSS error relation is particularly suitable when component error is represented by the standard deviation of the associated variable or parameter value and when the corresponding error sources are independent.

We shall conclude this chapter by giving several examples of error combination.

Example 2.13

Using the absolute value method for error combination, determine the fractional error in each item x_i so that the contribution from each item to the overall error e_{ABS} is the same.

Solution   For equal contribution, we must have

$$\left| \frac{x_1}{y} \frac{\partial f}{\partial x_1} \right| e_1 = \left| \frac{x_2}{y} \frac{\partial f}{\partial x_2} \right| e_2 = \cdots = \left| \frac{x_r}{y} \frac{\partial f}{\partial x_r} \right| e_r$$

Hence,

$$r \left| \frac{x_i}{y} \frac{\partial f}{\partial x_i} \right| e_i = e_{ABS}$$

Thus,

$$e_i = e_{ABS} \Big/ \left(r \left| \frac{x_i}{y} \frac{\partial f}{\partial x_i} \right| \right) \qquad (2.67)$$

Example 2.14

The result obtained in example 2.13 is useful in the design of multicomponent systems and in the cost-effective selection of instrumentation for a particular application. Using equation 2.67, arrange the items x_i in their order of significance.

Solution Note that equation 2.67 may be written as

$$e_i = K \Big/ \left| x_i \frac{\partial f}{\partial x_i} \right| \tag{2.68}$$

where K is a quantity that does not vary with x_i. It follows that for equal error contribution from all items, error in x_i should be inversely proportional to $\left| x_i (\partial f / \partial x_i) \right|$. In particular, the item with the largest $\left| x (\partial f / \partial x) \right|$ should be made most accurate. In this manner, allowable relative accuracy for various components can be estimated. Since, in general, the most accurate device is also the most costly one, instrumentation cost can be optimized if components are selected according to the required overall accuracy, using a criterion such as that implied by equation 2.68.

Example 2.15

Tension T at point P in a cable can be computed with the knowledge of cable sag y, cable length s, cable weight w per unit length, and the minimum tension T_o at point O (see figure 2.22). The applicable relationship is

$$1 + \frac{w}{T_o} y = \sqrt{1 + \frac{w^2}{T^2} s^2}$$

For a particular arrangement, it is given that $T_o = 100$ lbf. The following parameter values were measured:

$$w = 1 \text{ lb/ft}, \qquad s = 10 \text{ ft}, \qquad y = 0.412 \text{ ft}$$

Calculate the tension T.

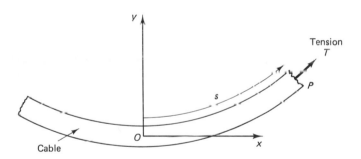

Figure 2.22. Cable tension example of error combination.

In addition, if the measurements y and s each have 1 percent error and the measurement w has 2 percent error in this example, estimate the percentage error in T.

Now suppose that equal contributions to error in T are made by y, s, and w. What are the corresponding percentage error values for y, s, and w so that the overall error in T is equal to the value computed in the previous part of the problem? Which of the three quantities y, s, and w should be measured most accurately, according to the equal contribution criterion?

Solution To make the analysis simpler, let us first square the given relationship:

$$\left(1 + \frac{w}{T_o} y \right)^2 = 1 + \frac{w^2 s^2}{T^2} \tag{i}$$

Substituting numerical values,

$$\left(1 + \frac{1 \times 0.412}{100}\right)^2 = 1 + \frac{1 \times 10^2}{T^2}$$

Hence,

$$T = 110 \text{ lbf}$$

Next, we differentiate equation (i) to get the differential relationship

$$2\left(1 + \frac{w}{T_o}y\right)\left[\frac{y}{T_o}\delta w + \frac{w}{T_o}\delta y\right] = \frac{2ws^2}{T^2}\delta w + \frac{2w^2s}{T^2}\delta s - \frac{2w^2s^2}{T^3}\delta T \quad \text{(ii)}$$

Note that T_o is treated as a constant. The implication is that T_o is known with 100 percent accuracy. On rearranging the terms in equation (ii) and after straightforward algebraic manipulation, we get

$$\frac{\delta T}{T} = (1 - z)\frac{\delta w}{w} + \frac{\delta s}{s} - z\frac{\delta y}{y} \quad \text{(iii)}$$

where

$$z = \frac{T^2 y}{s^2 w T_o}\left\{1 + \frac{wy}{T_o}\right\} \quad \text{(iv)}$$

Using the absolute value method for error combination, we can express the error level in T as

$$e_{\text{ABS}} = |1 - z|e_w + e_s + ze_y \quad \text{(v)}$$

Substituting the given numerical values,

$$z = \frac{110^2 \times 0.412}{10^2 \times 1 \times 100}\left(1 + \frac{1 \times 0.412}{100}\right) = 0.5$$

Hence,

$$e_{\text{ABS}} = 0.5\, e_w + e_s + 0.5\, e_y \quad \text{(vi)}$$

Also, it is given that

$$e_w = 2\%, \qquad e_s = e_y = 1\%$$

Hence,

$$e_{\text{ABS}} = (1 - 0.5) \times 2 + 1 + 0.5 \times 1\% = 2.5\%$$

For equal contribution of error, in view of equation (vi), we have

$$0.5e_w = e_s = 0.5e_y = \frac{2.5}{3}\%$$

Hence,

$$e_w = 1.7\%, \qquad e_s = 0.8\%, \qquad e_y = 1.7\%$$

Note that the variable s should be measured most accurately according to the equal contribution criterion, because the tolerable level of error is the smallest for this variable.

PROBLEMS

2.1. Discuss a type of device that could serve as a clock for a digital control system. Identify the main functions of such a clock in real-time control.

2.2. Explain the operation of two measuring devices of your choice, one to measure voltage and the other to measure temperature. Clearly identify the sensor stage and the transducer stage (or stages) in each of these devices.

2.3. Compare and contrast the following pairs of terms, giving suitable examples:
(a) Measurand and measured value
(b) Active transducers and passive transducers
(c) Through variables and across variables
(d) Effort variables and flow variables
(e) Impedance and admittance

2.4. Obtain transfer relations in the second-order vector-matrix form for the following two-port devices, clearly identifying the through and across variables at the input port and the output port.
(a) Gyroscope (or spinning top)
(b) Cam-follower mechanism
(c) Loudspeaker
(d) Viscous damper
(e) Thermocouple

2.5. What are pure transducers and what are ideal transducers? Discuss the five devices listed in problem 2.4 from this point of view.

2.6. The simple oscillator equation is given by

$$\ddot{y} + 2\zeta\omega_n\dot{y} + \omega_n^2 y = \omega_n^2 u(t)$$

Obtain the response y of this system for a unit step input $u(t)$ under zero initial conditions. In terms of the two parameters ζ and ω_n, obtain expressions for damping ratio, undamped natural frequency, damped natural frequency, rise time, percentage overshoot, and 2 percent settling time. Discuss the significance of each of these parameters with reference to the performance of a sensor-transducer device.

2.7. For the simple oscillator given in problem 2.6, what is the transfer function? Obtain an expression for its resonant frequency in terms of ζ and ω_n. If a sinusoidal excitation of frequency ω_n is applied to this system, what is the amplitude gain and phase lead in its response at steady state?

2.8. What do you consider a perfect measuring device? Suppose that you are asked to develop an analog device for measuring angular position in an application related to control of a kinematic linkage system (a robotic manipulator, for example). What instrument ratings (or specifications) would you consider crucial in this application? Discuss their significance.

2.9. Define electrical impedance and mechanical impedance. Identify a defect in these definitions in relation to the force-current analogy. What improvements would you suggest? What roles do input impedance and output impedance play in relation to the accuracy of a measuring device?

2.10. Discuss and contrast the following terms:
(a) Measurement accuracy
(b) Instrument accuracy

(c) Measurement error

(d) Precision

Also, for an analog sensor-transducer unit of your choice, identify and discuss various sources of error and ways to minimize or account for their influence.

2.11. A schematic diagram for a charge amplifier (with resistive feedback) is shown in figure P2.11. Obtain the differential equation governing the response of the charge amplifier. Identify the time constant of the device and discuss its significance. Would you prefer a charge amplifier to a voltage follower for conditioning signals from a piezoelectric accelerometer? Explain.

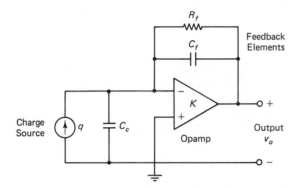

Figure P2.11. Schematic diagram for a charge amplifier.

2.12. List several response characteristics of nonlinear dynamic systems that are not exhibited by linear systems in general. Also, determine the response y of the nonlinear system

$$\left[\frac{dy}{dt}\right]^{1/3} = u(t)$$

when excited by the input $u(t) = a_1 \sin \omega_1 t + a_2 \sin \omega_2 t$. What characteristic of nonlinear systems does this result show?

2.13. What is meant by "loading error" in a signal measurement? Also, suppose that a piezoelectric sensor of output impedance Z_s is connected to a voltage-follower amplifier of input impedance Z_i. The sensor signal is v_i volts and the amplifier output is v_o volts. The amplifier output is connected to a device with very high input impedance. Plot to scale the signal ratio v_o/v_i against the impedance ratio Z_i/Z_s for values of the impedance ratio in the range 0.1 to 10.

2.14. Discuss how the accuracy of a digital controller may be affected by

(a) Stability and bandwidth of amplifier circuitry

(b) Load impedance of the analog-to-digital conversion circuitry.

Also, what methods do you suggest to minimize problems associated with these parameters?

2.15. From the point of view of loading, discuss why an active transducer is generally superior to a passive transducer. Also, discuss why impedance matching amplifiers are generally active devices.

2.16. Suppose that v_i and f_i are the across variable and the through variable at the input port of a two-port device and that v_o and f_o are the corresponding variables at the output port. A valid form for representing transfer characteristics of the device is

$$\begin{bmatrix} v_o \\ f_i \end{bmatrix} = G \begin{bmatrix} v_i \\ f_o \end{bmatrix}$$

What are the advantages and disadvantages of this representation compared to that given by equation 2.1? Also, express transfer relations for a DC tachometer (example 2.2) in the foregoing form.

2.17. Thevenin's theorem states that with respect to the characteristics at an output port, an unknown subsystem consisting of linear passive elements and ideal source elements may be represented by a single across-variable (voltage) source v_{eq} connected in series with a single impedance Z_{eq}. This is illustrated in figures P2.17a and P2.17b. Note that v_{eq} is equal to the open-circuit across variable v_{oc} at the output port, because the current through Z_{eq} is zero. Consider the network shown in figure P2.17c. Determine the equivalent voltage source v_{eq} and the equivalent series impedance Z_{eq}, in the frequency domain, for this circuit.

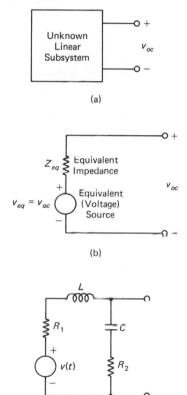

(a)

(b)

(c)

Figure P2.17. Illustration of Thevenin's theorem: (a) unknown linear subsystem; (b) equivalent representation; (c) example.

2.18. Using suitable impedance circuits, explain why a voltmeter should have a high resistance and an ammeter should have a very low resistance. What are some of the design implications of these general requirements for the two types of measuring instruments, particularly with respect to instrument sensitivity, speed of response, and robustness? Use a classical moving-coil meter as the model for your discussion.

2.19. A two-port nonlinear device is shown schematically in figure P2.19. The transfer relations under static equilibrium conditions are given by

$$v_o = F_1(f_o, f_i)$$

$$v_i = F_2(f_o, f_i)$$

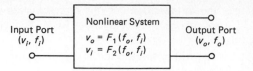

Figure P2.19. Impedance characteristics of a nonlinear system.

where v denotes an across variable, f denotes a through variable, and the subscripts o and i represent the output port and the input port, respectively. Obtain expressions for input impedance and output impedance of the system in the neighborhood of an operating point, under static conditions, in terms of partial derivatives of the functions F_1 and F_2. Explain how these impedances could be determined experimentally.

2.20. The damping constant b of the mounting structure of a machine is determined experimentally. First, the spring stiffness k is determined by applying a static load and measuring the resulting displacement. Next, mass m of the structure is directly measured. Finally, damping ratio ζ is determined using the logarithmic decrement method, by conducting an impact test and measuring the free response of the structure. A model for the structure is shown in figure P2.20. Show that the damping constant is given by

$$b = 2\zeta\sqrt{km}$$

If the allowable levels of error in the measurements of k, m, and ζ are ± 2 percent, ± 1 percent, and ± 6 percent, respectively, estimate a percentage absolute error limit for b.

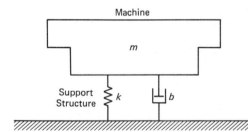

Figure P2.20. A model for the mounting structure of a machine.

2.21. In example 2.13 in the text, suppose that the square root of sum of squares (SRSS) method is used for error combination. What are the corresponding component error limits?

2.22. In example 2.15 in the text, suppose that the percentage error values specified are in fact standard deviations in the measurements of y, s, and w. Estimate the standard deviation in the estimated value of tension T.

2.23. The quality control system in a steel rolling mill uses a proximity sensor to measure the thickness of rolled steel (steel gage) at every two feet along the sheet, and the mill controller adjustments are made on the basis of the last twenty measurements. Specifically, the controller is adjusted unless the probability that the mean thickness lies within ± 1 percent of the sample mean exceeds 0.99.

A typical set of twenty measurements (in millimeters) is as follows:

$$5.10, \quad 5.05, \quad 4.94, \quad 4.98, \quad 5.10, \quad 5.12, \quad 5.07, \quad 4.96, \quad 4.99, \quad 4.95,$$

$$4.99, \quad 4.97, \quad 5.00, \quad 5.08, \quad 5.10, \quad 5.11, \quad 4.99, \quad 4.96, \quad 4.90, \quad 4.10,$$

Check whether adjustments would be made in the gage controller on the basis of these measurements.

2.24. Consider a mechanical component whose response x is governed by the relationship

$$f = f(x, \dot{x})$$

where f denotes applied (input) force and $\dot{x}$ denotes velocity. Three special cases are
(a) Linear spring:

$$f = kx$$

(b) Linear spring with a viscous (linear) damper:

$$f = kx + b\dot{x}$$

(c) Linear spring with coulomb friction:

$$f = kx + f_c \, \text{sgn}(\dot{x})$$

Suppose that a harmonic excitation of the form

$$f = f_o \sin \omega t$$

is applied in each case. Sketch the force-displacement curves for the three cases at steady state. Which components exhibit hysteresis? Which components are nonlinear? Discuss your answers.

2.25. A tactile (distributed touch) sensor of a robotic manipulator gripper consists of a matrix of piezoelectric sensor elements placed 2 mm apart. Each element generates an electric charge when it is strained by an external load. Sensor elements are multiplexed at very high speed in order to avoid charge leakage and to read all data channels using a single high-performance charge amplifier. Load distribution on the surface of the tactile sensor is determined from the charge amplifier readings, since the multiplexing sequence is known. Each sensor element can read a maximum load of 50 N and can detect load changes on the order of 0.01 N.
(a) What is the spatial resolution of the tactile sensor?
(b) What is the load resolution (in N/m^2) of the tactile sensor?
(c) What is the dynamic range?

2.26. Four sets of measurements were taken on the same response variable of a process using four different sensors. The true value of the response was known to be constant. Suppose that the four sets of data are as shown in figure P2.26a–d. Classify these data sets, and hence the corresponding sensors, with respect to precision and deterministic (repeatable) accuracy.

2.27. Dynamics and control of inherently unstable systems, such as rockets, can be studied experimentally using simple scaled-down physical models of the prototype systems. One such study is the classic inverted pendulum problem. An experimental setup for the inverted pendulum is shown in figure P2.27. The inverted pendulum is supported on a trolley that is driven on a tabletop in a straight line, using a chain-and-sprocket transmission operated by a DC motor. The motor is turned by commands from a microprocessor that is interfaced with the control circuitry of the motor. The angular position of the pendulum rod is measured using a resolver and is transmitted to the microprocessor. A strategy of statistical process control is used to balance the pendulum rod. Specifically, control limits are established from an initial set of measurement samples of the pendulum angle. Subsequently, if the angle exceeds one control limit, the trolley is accelerated in the opposite direction, using an automatic command to the motor. The control limits are also updated regularly. Suppose that the following

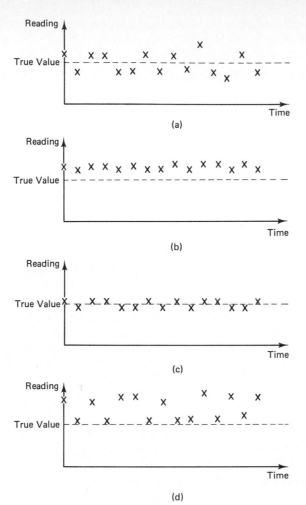

Figure P2.26. Four sets of measurements on the same response variable using different sensors.

twenty readings of the pendulum angle were measured (in degrees) after the system had operated for a few minutes:

$$0.5, \quad -0.5, \quad 0.4, \quad -0.3, \quad 0.3, \quad 0.1, \quad -0.3, \quad 0.3, \quad 4.0, \quad 0.0,$$
$$0.4, \quad -0.4, \quad 0.5, \quad -0.5, \quad -5.0, \quad 0.4, \quad -0.4, \quad 0.3, \quad -0.3, \quad -0.1$$

Establish whether the system was in statistical control during the period in which the readings were taken. Comment on this method of control.

2.28. (a) What is a two-port element? Discuss how dynamic coupling in a two-port device affects its accuracy when the device is used as a measuring device.

(b) What is the significance of bandwidth in a measuring device? Discuss methods to improve bandwidth.

(c) Using a two-port model for a DC tachometer, discuss methods of decreasing dynamic coupling and improving the useful frequency range.

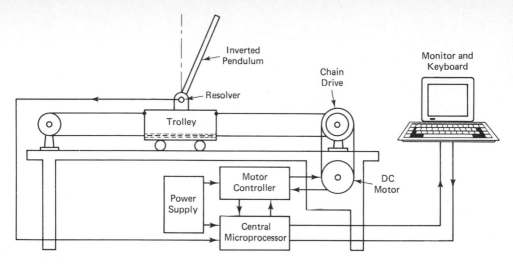

Figure P2.27. A microprocessor-controlled inverted pendulum—an application of statistical process control.

2.29. (a) Explain why mechanical loading error due to tachometer inertia can be significantly higher when measuring transient speeds than when measuring constant speeds.

(b) A DC tachometer has an equivalent resistance $R_a = 20\ \Omega$ in its rotor windings. In a position plus velocity servo system, the tachometer signal is connected to a feedback control circuit with equivalent resistance 2 kΩ. Estimate the percentage error due to electrical loading of the tachometer at steady state.

(c) If the conditions were not steady, how would the electrical loading be affected in this application?

2.30. A single-degree-of-freedom model of a mechanical manipulator is shown in figure P2.30a. The joint motor has rotor inertia J_m. It drives an inertial load that has moment of inertia J_ℓ through a speed reducer of gear ratio $1:r$ (Note: $r < 1$). The control scheme used in this system is the so-called feedforward control (strictly, *computed-torque control*) method. Specifically, the motor torque T_m that is required to accelerate or decelerate the load is computed using a suitable dynamic model and a desired motion trajectory for the manipulator, and the motor windings are excited so as to generate that torque. A typical trajectory would consist of a constant angular acceleration segment followed by a constant angular velocity segment, and finally a constant deceleration segment, as shown in figure P2.30b.

(a) Neglecting friction (particularly bearing friction) and inertia of the speed reducer, show that a dynamic model for torque computation during accelerating and decelerating segments of the motion trajectory would be

$$T_m = (J_m + r^2 J_\ell)\ddot{\theta}_\ell / r$$

where $\ddot{\theta}_\ell$ is the angular acceleration of the load, hereafter denoted by α_ℓ. Show that the overall system can be modeled as a single inertia rotating at the motor speed. Using this result, discuss the effect of gearing on a mechanical drive.

(b) Given that $r = 0.1$, $J_m = 0.1$ kg m^2, $J_\ell = 1.0$ kg m^2, and $\alpha_\ell = 5.0$ rad/s^2, estimate the allowable error for these four quantities so that the combined error in the

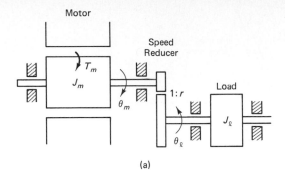

Motor

Speed
Reducer

T_m

J_m

$1:r$

Load

θ_m

J_ℓ

θ_ℓ

(a)

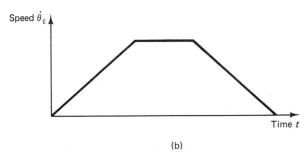

Speed $\dot{\theta}_\ell$

Time t

(b)

Figure P2.30. (a) A single-degree-of-freedom model of a mechanical manipulator. (b) A typical reference (desired) speed trajectory for computed-torque control.

computed torque is limited to ± 1 percent and so that each of the four quantities contributes equally toward this error in computed T_m. Use the absolute value method for error combination.

(c) Arrange the four quantities r, J_m, J_ℓ, and α_ℓ in the descending order of required accuracy for the numerical values given in the problem.

(d) Suppose that $J_m = r^2 J_\ell$. Discuss the effect of error in r on the error in T_m.

2.31. A useful rating parameter for a mechanical tool is *dexterity*. Though not complete, an appropriate analytical definition for dexterity of a device is

$$\text{dexterity} = \frac{\text{number of degrees of freedom}}{\text{motion resolution}}$$

where the number of degrees of freedom is equal to the number of independent variables that is required to completely define an arbitrary position increment of the tool (i.e., for an arbitrary change in its kinematic configuration).

(a) Explain the physical significance of dexterity and give an example of a device for which the specification of dexterity would be very important.

(b) The power rating of a tool may be defined as the product of maximum force that can be applied by it in a controlled manner and the corresponding maximum speed. Discuss why the power rating of a manipulating device is usually related to the dexterity of the device. Sketch a typical curve of power versus dexterity.

2.32. Resolution of a feedback sensor (or resolution of a response measurement used in feedback) has a direct effect on the accuracy that is achievable in a control system. This is true because the controller cannot correct a deviation of the response from the desired value (set point) unless the response sensor can detect that change. It follows that the resolution of a feedback sensor will govern the minimum (best) possible deviation band of system response under feedback control. An angular position servo uses a resolver as its feedback sensor. If peak-to-peak oscillations of the servo load under steady-state conditions have to be limited to no more than two degrees, what is the worst tolerable resolution of the resolver? Note that, in practice, the feedback sensor should have a resolution better (smaller) than this worst value.

2.33. An actuator (e.g., electric motor, hydraulic piston-cylinder) is used to drive a terminal device (e.g., gripper, hand, wrist with active remote center compliance) of a robotic manipulator. The terminal device functions as a force generator. A schematic diagram

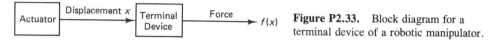

Figure P2.33. Block diagram for a terminal device of a robotic manipulator.

for the system is shown in figure P2.33. Show that the displacement error e_x is related to the force error e_f through

$$e_f = \frac{x}{f}\frac{df}{dx}e_x$$

The actuator is known to be 100 percent accurate for practical purposes, but there is an initial position error δx_o (at $x = x_o$). Obtain a suitable transfer relation $f(x)$ for the terminal device so that the force error e_f remains constant throughout the dynamic range of the device.

2.34. Consider, again, the mechanical tachometer shown in figure 2.10 (example 2.3). Write expressions for sensitivy and bandwidth for the device. Using the example, show that the two performance ratings, sensitivity and bandwidth, generally conflict. Discuss ways to improve the sensitivity of this mechanical tachometer.

REFERENCES

BRIGNELL, J. F., and RHODES, G. M. *Laboratory On-Line Computing.* Wiley, New York, 1975.

BROCH, J. T. *Mechanical Vibration and Shock Measurements.* Brüel and Kjaer, Naerum, Denmark, 1980.

CRANDALL, S. H., KARNOPP, D. C., KURTZ, E. F., JR., and PRIDMORE-BROWN, D. C. *Dynamics of Mechanical and Electromechanical Systems.* McGraw-Hill, New York, 1968.

DALLY, J. W., RILEY, W. F., and MCCONNELL, K. G. *Instrumentation for Engineering Measurements.* Wiley, New York, 1984.

DESILVA, C. W. *Dynamic Testing and Siesmic Qualification Practice.* Lexington Books, Lexington, Mass., 1983.

————. "Motion Sensors in Industrial Robots." *Mechanical Engineering* 107(6): 40–51, June 1985.

DOEBELIN, E. O. *Measurement Systems,* 3d ed. McGraw-Hill, New York, 1983.

HARRIS, C. M., and CREDE, C. E. *Shock and Vibration Handbook,* 2d ed. McGraw-Hill, New York, 1976.

HERCEG, E. E. *Handbook of Measurement and Control.* Schaevitz Engineering, Pennsauken, N.J., 1972.

ROSENBERG, R. C., and KARNOPP, D. C. *Introduction to Physical Systems Dynamics.* McGraw-Hill, New York, 1983.

3

Analog Sensors for Motion Measurement

3.1 INTRODUCTION

Measurement of plant outputs is essential for feedback control. Output measurements are also useful in performance evaluation of a process. Furthermore, in learning systems (e.g., teach-repeat operation of robotic manipulators), measurements are made and stored in the computer for subsequent use in operating the system. Input measurements are needed in feedforward control. It is evident, therefore, that the measurement subsystem is an important part of a control system.

The measurement subsystem in a control system contains sensors and transducers that detect measurands and convert them into acceptable signals—typically, voltages. These voltage signals are then appropriately modified using signal-conditioning hardware such as filters, amplifiers, demodulators, and analog-to-digital converters. Impedance matching might be necessary to connect sensors and transducers to signal-conditioning hardware.

Accuracy of sensors, transducers, and associated signal-conditioning devices is important in control system applications for two main reasons. The measurement system in a feedback control system is situated in the feedback path of the control system. Even though measurements are used to compensate for the poor performance in the open-loop system, any errors in measurements themselves will enter directly into the system and cannot be corrected if they are unknown. Furthermore, it can be shown that sensitivity of a control system to parameter changes in the measurement system is direct. This sensitivity cannot be reduced by increasing the loop gain, unlike in the case of sensitivity to the open-loop components. Accordingly, the design strategy for closed-loop (feedback) control is to make the measurements very accurate and to employ a suitable controller to reduce other types of errors.

Most sensor-transducer devices used in feedback control applications are analog components that generate analog output signals. This is the case even in real-time direct digital control systems. When analog transducers are used in digital control applications, however, some type of analog-to-digital conversion (ADC) is needed to obtain a digital representation of the measured signal. The resulting digital signal is subsequently conditioned and processed using digital means.

In this chapter, we shall study several analog sensor-transducer devices that are commonly used in control system instrumentation. We will not attempt to present an exhaustive discussion of all types of sensors; rather, we will consider a representative selection. Analog motion-measuring devices will be considered in this chapter. Force, torque, and tactile sensors will be discussed in chapter 4, and digital motion transducers will be studied in chapter 5.

In the sensor stage, the signal being measured is felt as the "response of the sensor element." This is converted by the transducer into the transmitted (or measured) quantity. In this respect, the output of a measuring device can be interpreted as the "response of the transducer." In control system applications, this output is typically (and preferably) an electrical signal. We shall limit our discussion to such measuring devices. Note that it is somewhat redundant to consider electrical-to-electrical sensors-transducers as measuring devices, particularly in control system studies, because electrical signals need conditioning only before they are fed into a controller or to a drive system. In this sense, electrical-to-electrical transduction should be considered a "conditioning" task rather than a "measuring" function.

3.2 MOTION TRANSDUCERS

By motion, we mean the four kinematic variables:

- Displacement (including position, distance, proximity, and size or gage)
- Velocity
- Acceleration
- Jerk

Note that each variable is the time derivative of the preceding one. Motion measurements are extremely useful in controlling mechanical responses and interactions in dynamic systems. Numerous examples can be cited of situations in which motion measurements are used for control purposes. The rotating speed of a work piece and the feed rate of a tool are measured in controlling machining operations. Displacements and speeds (both angular and translatory) at joints (revolute and prismatic) of robotic manipulators or kinematic linkages are used in controlling manipulator trajectory. In high-speed ground transit vehicles, acceleration and jerk measurements can be used for active suspension control to obtain improved ride quality. Angular speed is a crucial measurement that is used in the control of rotating machinery, such as turbines, pumps, compressors, motors, and generators in power-generating plants. Proximity sensors (to measure displacement) and accelerometers (to measure acceleration) are the two most common types of measuring devices used in machine protection systems for condition monitoring, fault detection, diagnosis, and on-line (often real-time) control of large and complex machinery. The accelerometer is often the only measuring device used in controlling dynamic test rigs. Displacement measurements are used for valve control in process applications. Plate thickness (or gage) is continuously monitored by the automatic gage control (AGC) system in steel rolling mills.

A one-to-one relationship may not always exist between a measuring device and a measured variable. For example, although strain gages are devices that measure strains (and, hence, stresses and forces), they can be adapted to measure displacements by using a suitable front-end auxiliary sensor element, such as a cantilever (or spring). Furthermore, the same measuring device may be used to measure different variables through appropriate data interpretation techniques. For example, piezoelectric accelerometers with built-in microelectronic integrated circuitry are marketed as piezoelectric velocity transducers. Resolver signals that provide angular displacements are differentiated to get angular velocities. Pulse-generating (or digital) transducers, such as optical encoders and digital tachometers, can serve as both displacement transducers and velocity transducers, depending on whether the absolute number of pulses generated is counted or the pulse rate is measured. Note that pulse rate can be measured either by counting the number of pulses during a unit interval of time or by gating a high-frequency clock signal through the pulse width. Furthermore, in principle, any force sensor can be used as an acceleration sensor, velocity sensor, or displacement sensor, depending on whether an inertia element (converting acceleration into force), a damping element (converting velocity into force), or a spring element (converting displacement into force), respectively, is used as the *front-end auxiliary sensor*.

We might question the need for separate transducers to measure the four kinematic variables—displacement, velocity, acceleration, and jerk—because any one variable is related to any other through simple integration or differentiation. It should be possible, in theory, to measure only one of these four variables and use either analog processing (through analog circuit hardware) or digital processing (through a dedicated processor) to obtain any of the remaining motion variables. The feasibility of this approach is highly limited, however, and it depends crucially on several factors, including the following:

1. The nature of the measured signal (e.g., steady, highly transient, periodic, narrow-band, broad-band)
2. The required frequency content of the processed signal (or the frequency range of interest)
3. The signal-to-noise ratio (SNR) of the measurement
4. Available processing capabilities (e.g., analog or digital processing, limitations of the digital processor, and interface, such as the speed of processing, sampling rate, and buffer size)
5. Controller requirements and the nature of the plant (e.g., time constants, delays, hardware limitations)
6. Required accuracy in the end objective (on which processing requirements and hardware costs will depend)

For instance, differentiation of a signal (in the time domain) is often unacceptable for noisy and high-frequency narrow-band signals. In any event, costly signal-conditioning hardware might be needed for preprocessing prior to differentiating a signal. As a rule of thumb, in low-frequency applications (on the order of 1 Hz), displacement measurements generally provide good accuracies. In intermediate-frequency applications (less than 1 kHz), velocity measurement is usually favored. In measuring high-frequency motions with high noise levels, acceleration measurement is preferred. Jerk is particularly useful in ground transit (ride quality), manufacturing (forging, rolling, and similar impact-type operations), and shock isolation (delicate and sensitive equipment) applications.

Our discussion of motion transducers will be limited mainly to the following types of devices:

- Potentiometers (resistively coupled devices)
- Variable-inductance transducers (electromagnetically coupled devices)
- Eddy current transducers
- Variable-capacitance transducers
- Piezoelectric transducers

The underlying concepts of these devices can be extended to many other kinds of transducers. Other types, such as optical sensors, will be introduced. Force and torque sensors, including strain gage devices, will be studied in chapter 4.

3.3 POTENTIOMETERS

The potentiometer, or *pot,* is a displacement transducer. This active transducer consists of a uniform coil of wire or a film of high-resistive material—such as carbon, platinum, or conductive plastic—whose resistance is proportional to its length. A fixed voltage v_{ref} is applied across the coil (or film) using an external, constant DC voltage supply. The transducer output signal v_o is the DC voltage between the movable contact (wiper arm) sliding on the coil and one terminal of the coil, as shown schematically in figure 3.1a. Slider displacement x is proportional to the output voltage:

$$v_o = kx \tag{3.1}$$

This relationship assumes that the output terminals are open-circuit; that is, infinite-impedance load (or resistance in the present DC case) is present at the output terminals, so that the output current is zero. In actual practice, however, the load (the circuitry into which the pot signal is fed—e.g., conditioning, processing, or control

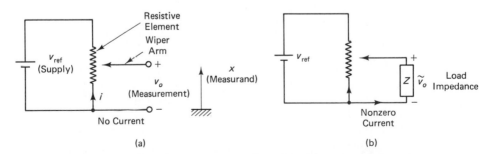

Figure 3.1. (a) Schematic diagram of a potentiometer. (b) Potentiometer loading.

circuitry) has a finite impedance. Consequently, the output current (the current through the load) is nonzero, as shown in figure 3.1b. The output voltage thus drops to $\tilde{v}_o$, even if the reference voltage v_{ref} is assumed to remain constant under load variations (i.e., the voltage source has zero output impedance); this consequence is known as the *loading effect* of the transducer. Under these conditions, the linear relationship given by equation 3.1 would no longer be valid. This causes an error in the displacement reading. Loading can affect the transducer reading in two ways: by changing the reference voltage (i.e., loading the voltage source) and by loading the transducer. To reduce these effects, a voltage source that is not seriously affected by load variations (e.g., a regulated or stabilized power supply that has low output impedance) and data acquisition circuitry (including signal-conditioning circuitry) that has high input impedance should be used.

The resistance of a potentiometer should be chosen with care. On the one hand, an element with high resistance is preferred because this results in reduced power dissipation for a given voltage, which has the added benefit of reduced thermal effects. On the other hand, increased resistance increases the output impedance of the potentiometer and results in loading nonlinearity error (see example 3.1) unless the load resistance is also increased proportionately. Low-resistance pots have resistances less than 10 Ω. High-resistance pots can have resistances on the order of 100 kΩ. Conductive plastics can provide high resistances—typically about 100 Ω/mm—and are increasingly used in potentiometers. Reduced friction (low mechanical loading), reduced wear, reduced weight, and increased resolution are advantages of using conductive plastics in potentiometers.

Potentiometers that measure angular (rotary) displacements are more common and convenient, because in conventional designs of rectilinear (translatory) potentiometers, the length of the resistive element has to be increased in proportion to the measurement range or stroke. Figure 3.2 presents schematic representations of

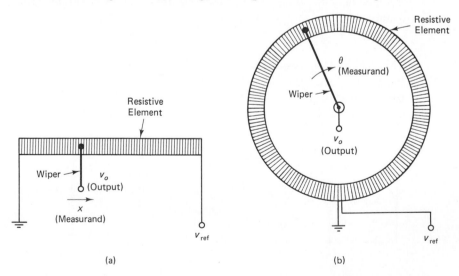

Figure 3.2. Practical potentiometer configurations for measuring (a) rectilinear motions and (b) angular motions.

translatory and rotatory potentiometers. Helix-type rotatory potentiometers are available for measuring absolute angles exceeding 360 degrees. The same function may be accomplished with a standard single-cycle rotary pot simply by including a counter to record full 360-degree rotations.

Note that angular displacement transducers, such as rotatory potentiometers, can be used to measure large rectilinear displacements on the order of 100 inches. A cable extension mechanism may be employed to accomplish this. A light cable wrapped around a spool that moves with the rotary element of the transducer is the cable extension mechanism. The free end of the cable is attached to the moving object, and the potentiometer housing is mounted on a stationary structure. The device is properly calibrated so that as the object moves, the rotation count and fractional rotation measure will directly provide the rectilinear displacement. A spring-loaded recoil device, such as a spring motor, will wind the cable back when the object moves toward the transducer.

Example 3.1

Consider the rotary potentiometer shown in figure 3.3. Discuss the significance of the loading nonlinearity error caused by a purely resistive load connected to the pot.

Solution   For a general position θ of the pot slider arm, suppose that the resistance in the output (pick-off) segment of the coil is R_θ. Note that, assuming a uniform coil,

$$R_\theta = \frac{\theta}{\theta_{max}} R_c \tag{i}$$

where R_c is the total resistance of the potentiometer coil.

The current balance at the sliding contact (node) point gives

$$\frac{v_{ref} - v_o}{R_c - R_\theta} = \frac{v_o}{R_\theta} + \frac{v_o}{R_L} \tag{ii}$$

where R_L is the load resistance.

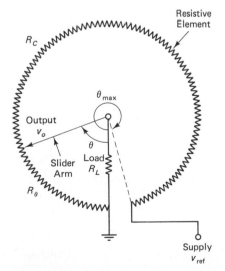

Figure 3.3.   A rotary potentiometer with a resistive load.

Multiply throughout this equation by R_c and use (i); thus,

$$\frac{v_{\text{ref}} - v_o}{1 - \theta/\theta_{\text{max}}} = \frac{v_o}{\theta/\theta_{\text{max}}} + \frac{v_o}{R_L/R_c}$$

By using straightforward algebra, we have

$$\frac{v_o}{v_{\text{ref}}} = \left[\frac{(\theta/\theta_{\text{max}})(R_L/R_c)}{(R_L/R_c) + (\theta/\theta_{\text{max}}) - (\theta/\theta_{\text{max}})^2}\right] \tag{3.2}$$

Equation 3.2 is plotted in figure 3.4. Loading error appears to be high for low values of the R_L/R_c ratio. Good accuracy is possible for $R_L/R_c > 10$, particularly for small values of $\theta/\theta_{\text{max}}$. Hence, to reduce loading error in pots,

1. Increase R_L/R_c (increase load impedance, reduce coil impedance).
2. Use pots to measure small values of $\theta/\theta_{\text{max}}$ (or calibrate only a small segment of the element for linear reading).

The loading nonlinearity error is defined by

$$e = \frac{(v_o/v_{\text{ref}} - \theta/\theta_{\text{max}})}{\theta/\theta_{\text{max}}} 100\% \tag{3.3}$$

The error at $\theta/\theta_{\text{max}} = 0.5$ is tabulated in table 3.1. Note that this error is always negative. Using only a segment of the resistance element as the range of the potentiometer is similar to adding two end resistors to the elements. It is known that this tends to linearize the pot (see problem 3.5). If the load resistance is known to be small, a voltage

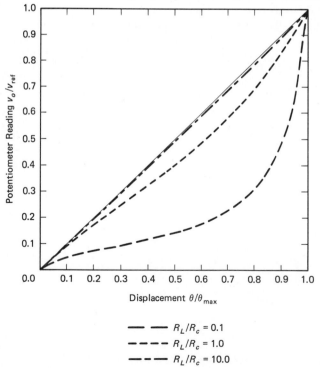

Figure 3.4. Loading nonlinearity in a potentiometer.

- — · — $R_L/R_c = 0.1$
- — — — $R_L/R_c = 1.0$
- — · · — $R_L/R_c = 10.0$

TABLE 3.1 LOADING NONLINEARITY ERROR IN
POTENTIOMETERS, FOR $\theta/\theta_{max} = 0.5$.

Load resistance ratio R_L/R_c	Loading nonlinearity error e
0.1	−71.4%
1.0	−20%
10.0	−2.4%

follower may be used at the potentiometer output to virtually eliminate loading error. This arrangement provides a high load impedance to the pot and a low impedance at the output of the amplifier.

The force required to move the slider arm comes from the motion source, and the resulting energy is dissipated through friction. This energy conversion, unlike pure mechanical-to-electrical conversions, involves relatively high forces, and the energy is wasted rather than being converted into the output signal of the transducer. Furthermore, the electrical energy from the reference source is also dissipated through the resistor coil (or film), resulting in an undesirable temperature rise. These are two obvious disadvantages of this resistively coupled transducer. Another disadvantage is the finite *resolution* in coil-type pots.

Coils, instead of straight wire, are used to increase the resistance per unit travel of the slider arm. But the slider contact jumps from one turn to the next in this case. Accordingly, the resolution of a coil-type potentiometer is determined by the number of turns in the coil. For a coil that has N turns, the resolution r, expressed as a percentage of the output range, is given by

$$r = \frac{100}{N}\%$$

(3.4)

Resolutions better (smaller) than 0.1 percent (i.e., 1,000 turns) are available with coil potentiometers. Infinitesimal (incorrectly termed infinite) resolutions are now possible with high-quality resistive film potentiometers that use conductive plastics, for example. In this case, the resolution is limited by other factors, such as mechanical limitations and signal-to-noise ratio. Nevertheless, resolutions on the order of 0.01 mm are possible with good rectilinear potentiometers.

Example 3.2

The range of a coil-type potentiometer is 10 cm. If the wire diameter is 0.1 mm, determine the resolution of the device.

Solution In this problem

$$N = 10 \times 10/0.1 = 1,000 \text{ turns}$$

$$\text{resolution} = \text{wire diameter} = 0.1 \text{ mm}$$

From equation 3.4,

$$\text{percentage resolution} = \frac{100}{1{,}000}\% = 0.1\%$$

Note that this value is equal to wire diameter/range $\times$ 100%.

Selection of a potentiometer involves many considerations. A primary factor is the required resolution for the particular application. Power consumption, loading, and size are also important factors. The following design example highlights some of these considerations.

Example 3.3

A mobile robot uses a potentiometer attached to the drive wheel to record its travel during the "teach" mode in a teach-repeat operation. The required resolution for robot motion is 1 mm, and the diameter of the drive wheel of the robot is 20 cm. Examine the design considerations for a standard (single-coil) rotary potentiometer to be used in this application.

Solution Assume that the potentiometer is directly connected (without gears) to the drive wheel. The required resolution for the pot is

$$\frac{0.1}{\pi \times 20} \times 100\% = 0.16\%$$

This resolution is feasible with a coil-type rotary pot.

From equation 3.4 the number of turns in the coil = 100/0.16 = 625 turns. Assuming an average pot diameter of 10 cm and denoting the wire diameter by d, we have

$$\text{pot circumference} = \pi \times 10 = 625 \times d$$

or

$$d = 0.5 \text{ mm}$$

Now, taking the resistance of the potentiometer to be 5 Ω and the resistivity of the wire to be 4 $\mu\Omega$ cm, the diameter D of the core of the coil is given by

$$\frac{4 \times 10^{-6} \times \pi D \times 625}{\pi (0.05/2)^2} = 5 \ \Omega$$

Note: Resistivity = (resistance) $\times$ (cross-section area)/(length).
Hence, D = 1.25 cm.

The *sensitivity* of a potentiometer represents the change (Δv_o) in the output signal associated with a given small change ($\Delta\theta$) in the measurand (the displacement). This is usually nondimensionalized, using the actual value of the output signal (v_o) and the actual value of the displacement (θ). For a rotary potentiometer in particular, the sensitivity S is given by

$$S = \frac{\Delta v_o}{\Delta\theta} \tag{3.5}$$

Or, in the limit,

$$S = \frac{\partial v_o}{\partial \theta} \qquad\qquad (3.6)$$

These relations may be nondimensionalized by multiplying by θ/v_o. An expression for S may be obtained by simply substituting equation 3.2 into equation 3.6.

Some limitations and disadvantages of potentiometers as displacement measuring devices are as follows:

1. The force needed to move the slider (against friction and arm inertia) is provided by the displacement source. This mechanical loading distorts the measured signal itself.
2. High-frequency (or highly transient) measurements are not feasible because of such factors as slider bounce, friction and inertia resistance, and induced voltages in the wiper arm and primary coil.
3. Variations in the supply voltage cause error.
4. Electrical loading error can be significant when the load resistance is low.
5. Resolution is limited by the number of turns in the coil and by the coil uniformity. This limits small displacement measurements.
6. Wearout and heating up (with associated oxidation) in the coil (film) and slider contact cause accelerated degradation.

There are several advantages associated with potentiometer devices, however, including the following:

1. They are relatively less costly.
2. Potentiometers provide high-voltage (low-impedance) output signals, requiring no amplification in most applications. Transducer impedance can be varied simply by changing the coil resistance and supply voltage.

Example 3.4

A rectilinear potentiometer was tested with its slider arm moving horizontally. It was found that at a speed of 1 cm/s, a driving force of 7×10^{-4} N was necessary to maintain the speed. At 10 cm/s, a force of 3×10^{-3} N was necessary. The slider weighs 5 gm, and the potentiometer stroke is ± 8 cm. If this potentiometer is used to measure the damped natural frequency of a simple mechanical oscillator of mass 10 kg, stiffness 10 N/m, and damping constant 2 N/m/s, estimate the percentage error due to mechanical loading. Justify your procedure for the estimation of damping.

Solution Suppose that the mass, stiffness, and damping constant of the simple oscillator are denoted by M, K, and B, respectively. The equation of free motion of the simple oscillator is given by

$$M\ddot{y} + B\dot{y} + Ky = 0$$

where y denotes the displacement of the mass from the static equilibrium position. This equation is of the form

$$\ddot{y} + 2\zeta\omega_n\dot{y} + \omega_n^2 y = 0$$

where ω_n is the undamped natural frequency of the oscillator and ζ is the damping ratio. By direct comparison, it is seen that

$$\omega_n = \sqrt{\frac{K}{M}} \quad \text{and} \quad \zeta = \frac{B}{2\sqrt{MK}}$$

The damped natural frequency is

$$\omega_d = \sqrt{1 - \zeta^2}\,\omega_n \qquad \text{for } 0 < \zeta < 1$$

Hence,

$$\omega_d = \sqrt{\left(1 - \frac{B^2}{4MK}\right)\frac{K}{M}}$$

Now, if the wiper arm mass and damping constant of the potentiometer are denoted by m and b, respectively, the measured damped natural frequency (using the potentiometer) is given by

$$\tilde{\omega}_d = \sqrt{\left[1 - \frac{(B + b)^2}{4(M + m)K}\right]\frac{K}{(M + m)}}$$

Assuming linear viscous friction (which is, of course, not very realistic), the damping constant b of the potentiometer may be estimated as

$$b = \text{damping force/steady state velocity of the wiper}$$

For the present example,

$$b_1 = 7 \times 10^{-4}/1 \times 10^{-2} \text{ N/m/s} = 7 \times 10^{-2} \text{ N/m/s} \quad \text{at 1 cm/s}$$

$$b_2 = 3 \times 10^{-3}/10 \times 10^{-2} \text{ N/m/s} = 3 \times 10^{-2} \text{ N/m/s} \quad \text{at 10 cm/s}$$

We should use some form of interpolation to estimate b for the actual measuring conditions. Let us estimate the average velocity of the wiper. The natural frequency of the oscillator is

$$\omega_n = \sqrt{\frac{10}{10}} = 1 \text{ rad/s} = \frac{1}{2\pi} \text{ Hz}$$

The wiper travels a maximum distance of $4 \times 8 \text{ cm} = 32 \text{ cm}$ in one cycle. Hence, the average operating speed of the wiper may be estimated as $32/(2\pi)$ cm/s, which is approximately equal to 5 cm/s. Therefore, the operating damping constant may be estimated as the average of b_1 and b_2:

$$b = 5 \times 10^{-2} \text{ N/m/s}$$

With the foregoing numerical values:

$$\omega_d = \sqrt{\left(1 - \frac{2^2}{4 \times 10 \times 10}\right)\frac{10}{10}} = 0.99499 \text{ rad/s}$$

$$\tilde{\omega}_d = \sqrt{\left(1 - \frac{2.05^2}{4 \times 10.005 \times 10}\right)\frac{10}{10.005}} = 0.99449 \text{ rad/s}$$

$$\text{percentage error} = \left[\frac{\tilde{\omega}_d - \omega_d}{\omega_d}\right] \times 100\% = -0.05\%$$

Although pots are primarily used as displacement transducers, they can be adapted to measure other types of signals, such as pressure and force, using appropriate auxiliary sensor (front-end) elements. For instance, a bourdon tube or bellows may be used to convert pressure into displacement, and a cantilever element may be used to convert force or moment into displacement.

3.4 VARIABLE-INDUCTANCE TRANSDUCERS

Motion transducers that employ the principle of electromagnetic induction are termed variable-inductance transducers. When the flux linkage (defined as magnetic flux density times the number of turns in the conductor) through an electrical conductor changes, a voltage is induced in the conductor. This, in turn, generates a magnetic field that opposes the primary field. Hence, a mechanical force is necessary to sustain the change of flux linkage. If the change in flux linkage is brought about by a relative motion, the mechanical energy is directly converted (induced) into electrical energy. This is the basis of electromagnetic induction, and it is the principle of operation of electrical generators and variable-inductance transducers. Note that in these devices, the change of flux linkage is caused by a mechanical motion, and mechanical-to-electrical energy transfer takes place under near-ideal conditions. The induced voltage or change in inductance may be used as a measure of the motion. Variable-inductance transducers are generally electromechanical devices coupled by a magnetic field.

There are many different types of variable-inductance transducers. Three primary types can be identified:

1. Mutual-induction transducers
2. Self-induction transducers
3. Permanent-magnet transducers

Variable-inductance transducers that use a nonmagnetized ferromagnetic medium to alter the reluctance (magnetic resistance) of the flux path are known as *variable-reluctance transducers*. Some of the mutual-induction transducers and most of the self-induction transducers are of this type. Permanent-magnet transducers are not considered variable-reluctance transducers.

Mutual-Induction Transducers and Differential Transformers

The basic arrangement of a mutual-induction transducer constitutes two coils, the *primary winding* and the *secondary winding*. One of the coils (primary winding) carries an AC excitation that induces a steady AC voltage in the other coil (secondary winding). The level (amplitude, rms value, etc.) of the induced voltage depends on the flux linkage between the coils. In mutual-induction transducers, a change in the flux linkage is effected by one of two common techniques. One technique is to move an object made of ferromagnetic material within the flux path. This changes the reluctance of the flux path, with an associated change of the flux linkage in the secondary coil. This is the operating principle of the linear variable differential transformer (LVDT), the rotary variable differential transformer (RVDT), and the mutual-induction proximity probe. All of these are, in fact, variable-reluctance transducers. The other common way to change the flux linkage is to move one coil with respect to the other. This is the operating principle of the resolver and the synchro-transformer. These are not variable-reluctance transducers, however.

The motion can be measured by using the secondary signal in several ways. For example, the AC signal in the secondary winding may be demodulated by rejecting the carrier frequency (primary-winding excitation frequency) and directly measuring the resulting signal, which represents the motion. This method is particularly suitable for measuring transient motions. Alternatively, the amplitude or the rms (root-mean-square) value of the secondary (induced) voltage may be measured. Another method is to measure the change of inductance in the secondary circuit directly, by using a device such as an inductance bridge circuit.

The linear variable differential transformer (LVDT). The LVDT is a displacement measuring device that overcomes most of the shortcomings of the potentiometer. It is considered a passive transducer because the measured displacement provides energy for "changing" the induced voltage, even though an external power supply is used to energize the primary coil which in turn induces a steady carrier voltage in the secondary coil. The LVDT is a variable-reluctance transducer of the mutual induction type. A selection of commercially available LVDTs is shown in figure 3.5a. In its simplest form, the LVDT consists of a cylindrical, insulating, non-

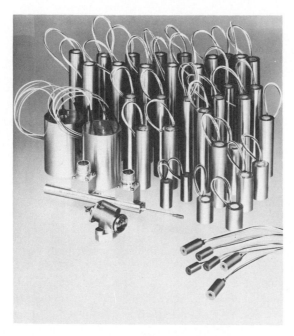

Figure 3.5. (a) A collection of commercially available LVDTs (courtesy of Robinson-Halpern Co.). (b) Schematic diagram of an LVDT. (c) A typical operating curve.

(a)

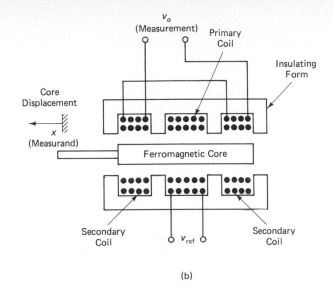

(b)

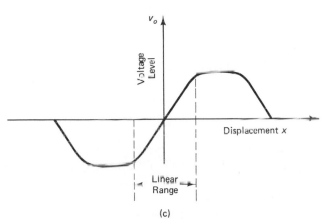

(c)

Figure 3.5. (*continued*)

magnetic form that has a primary coil in the midsegment and a secondary coil symmetrically wound in the two end segments, as depicted schematically in figure 3.5b. The primary coil is energized by an AC supply of voltage v_{ref}. This will generate, by mutual induction, an AC of the same frequency in the secondary winding. A core made of ferromagnetic material is inserted coaxially into the cylindrical form without actually touching it, as shown. As the core moves, the reluctance of the flux path changes. Hence, the degree of flux linkage depends on the axial position of the core. Since the two secondary coils are connected in series opposition (as shown in figure 3.6), so that the potentials induced in the two secondary coil segments oppose each other, it is seen that the net induced voltage is zero when the core is centered between the two secondary winding segments. This is known as the *null position*. When the core is displaced from this position, a nonzero induced voltage will be generated. At steady state, the amplitude v_o of this induced voltage is proportional, in the linear (operating) region, to the core displacement x (see figure 3.5c). Consequently, v_o may be used as a measure of the displacement. Note that because of op-

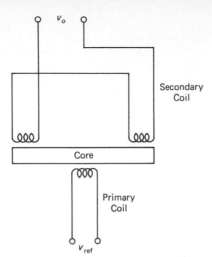

Figure 3.6. Series opposition connection of secondary windings.

posed secondary windings, the LVDT provides the direction as well as the magnitude of displacement. If the output signal is not demodulated, the direction is determined by the phase angle between the primary (reference) voltage and the secondary (output) voltage, including the carrier signal.

For an LVDT to measure transient motions accurately, the frequency of the reference voltage (the carrier frequency) has to be at least ten times larger than the largest significant frequency component in the measured motion. For quasi-dynamic displacements and slow transients on the order of a few hertz, a standard AC supply (at 60 Hz line frequency) is adequate. The performance (particularly sensitivity and accuracy) is known to improve with the excitation frequency, however. Since the amplitude of the output signal is proportional to the amplitude of the primary signal, the reference voltage should be regulated to get accurate results. In particular, the power source should have a low output impedance.

An error known as *null voltage* is present in some differential transformers. This manifests itself as a nonzero reading at the null position (i.e., at zero displacement). This is usually 90° out of phase from the main output signal and, hence, is known as *quadrature error*. Nonuniformities in the windings (unequal impedances in the two segments of the secondary winding) are a major reason for this error. The null voltage may also result from harmonic noise components in the primary signal and nonlinearities in the device. Null voltage is usually negligible (typically about 0.1 percent of the full scale). This error can be eliminated from the measurements by employing appropriate signal-conditioning and calibration practices.

The output signal from a differential transformer is normally not in phase with the reference voltage. Inductance in the primary winding and the leakage inductance in the secondary windings are mainly responsible for this phase shift. Since demodulation involves extraction of the modulating signal by rejecting the carrier frequency component from the secondary signal, it is important to understand the size of this phase shift.

Example 3.5

An equivalent circuit for a differential transformer is shown in figure 3.7. The resistance in the primary winding is denoted by R_p, and the corresponding inductance is denoted by L_p. The total resistance of the secondary winding is R_s. The net leakage inductance, due to magnetic flux leakage, in the two segments is denoted by L_ℓ. The load resistance is R_L and the load inductance is L_L. Derive an expression for the phase shift in the output signal.

Solution   The magnetizing voltage in the primary coil is given by

$$v_{\text{ref}}\left[\frac{j\omega L_p}{R_p + j\omega L_p}\right]$$

in the frequency domain. Suppose that the segment of the secondary winding that has full flux linkage through the ferromagnetic core receives an induced voltage that is proportional to the primary magnetizing voltage. The other segment receives only a portion of this voltage, depending on the displacement x of the core. The net induced voltage is proportional to x and is given by

$$v_{\text{ref}}\left[\frac{j\omega L_p}{R_p + j\omega L_p}\right]kx$$

where k is a proportionality constant. It follows that the output voltage v_o at the load is given by

$$v_o = \left[\frac{j\omega L_p}{R_p + j\omega L_p}\right]\left[\frac{R_L + j\omega L_L}{(R_L + R_s) + j\omega(L_L + L_\ell)}\right]kx \qquad (3.7)$$

This corresponds to a phase *lead* at the output, given by

$$\phi = 90^\circ - \tan^{-1}\frac{\omega L_p}{R_p} + \tan^{-1}\frac{\omega L_L}{R_L} - \tan^{-1}\frac{\omega(L_L + L_\ell)}{R_L + R_s} \qquad (3.8)$$

Note that the level of dependence of the phase shift on the load and secondary circuit can be reduced by increasing the load impedance.

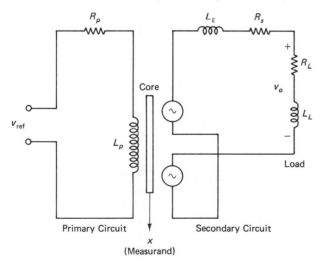

Primary Circuit

x
(Measurand)

Secondary Circuit

Figure 3.7.   Equivalent circuit for a differential transformer.

Signal conditioning associated with differential transformers includes filtering and amplification. Filtering is needed to improve the signal-to-noise ratio of the output signal. Amplification is necessary to increase the signal strength for data acquisition and processing. Since the reference frequency (carrier frequency) is embedded in the output signal, it is also necessary to interpret the output signal properly, particularly for transient motions. Two methods are commonly used to interpret the amplitude-modulated output signal from a differential transformer: rectification and demodulation.

Block diagram representations of these two procedures are given in figure 3.8. In the first method—rectification—the AC output from the differential transformer is rectified to obtain a DC signal. This signal is amplified and then low-pass filtered to eliminate any high-frequency noise components. The amplitude of the resulting signal provides the transducer reading. In this method, phase shift in the LVDT out-

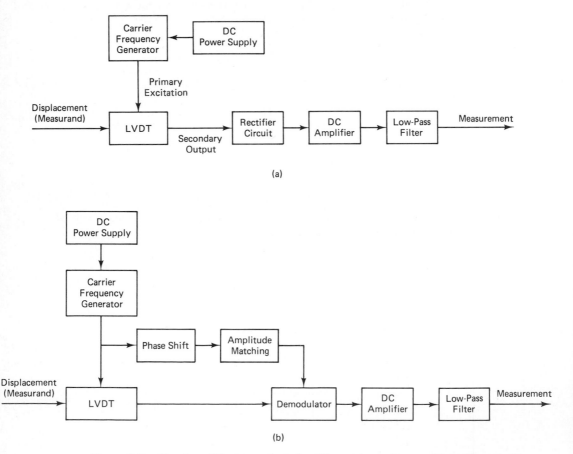

Figure 3.8. Signal-conditioning methods for differential transformers: (a) rectification; (b) demodulation.

put has to be checked separately to determine the direction of motion. In the second method—demodulation—the carrier frequency component is rejected from the output signal by comparing it with a phase-shifted and amplitude-adjusted version of the primary (reference) signal. Note that phase shifting is necessary because the output signal is not in phase with the reference signal. The modulating signal extracted in this manner is subsequently amplified and filtered. As a result of advances in miniature integrated circuit technology, differential transformers with built-in microelectronics for signal conditioning are commonly available today. DC differential transformers have built-in oscillator circuits to generate the carrier signal powered by a DC supply. The supply voltage is usually on the order of 25 V, and the output voltage is about 5 V.

Advantages of the LVDT include the following:

1. It is essentially a noncontacting device with no frictional resistance. Near-ideal electromechanical energy conversion and light-weight core result in very small resistive forces. Hysteresis (both magnetic hysteresis and mechanical backlash) is negligible.
2. It has low output impedance, typically on the order of 100 Ω. (Signal amplification is usually not needed.)
3. Directional measurements (positive/negative) are obtained.
4. It is available in small size (e.g., 1 cm long with maximum travel of 2 mm).
5. It has a simple and robust construction (inexpensive and durable).
6. Fine resolutions are possible (theoretically, infinitesimal resolution; practically, much better than a coil potentiometer).

The rotatory variable differential transformer (RVDT). The RVDT operates using the same principle as the LVDT, except that in an RVDT, a rotating ferromagnetic core is used. The RVDT is used for measuring angular displacements. A schematic diagram of the device is shown in figure 3.9a, and a typical operating curve is shown in figure 3.9b. The rotating core is shaped such that a reasonably wide linear operating region is obtained. Advantages of the RVDT are essentially the same as those cited for the LVDT.

Since the RVDT measures angular motions directly, without requiring nonlinear transformations (which is the case in resolvers, as will be discussed later in this chapter), its use is convenient in angular position servos. The linear range is typically ±40°, with a nonlinearity error less than 1 percent.

In variable-inductance devices, the induced voltage is generated through the rate of change of the magnetic flux linkage. Therefore, displacement readings are distorted by velocity; similarly, velocity readings are affected by acceleration. For the same displacement value, the transducer reading will depend on the velocity at that displacement. This error is known to increase with the ratio (cyclic velocity of the core)/(carrier frequency). Hence, these *rate errors* can be reduced by increasing carrier frequency.

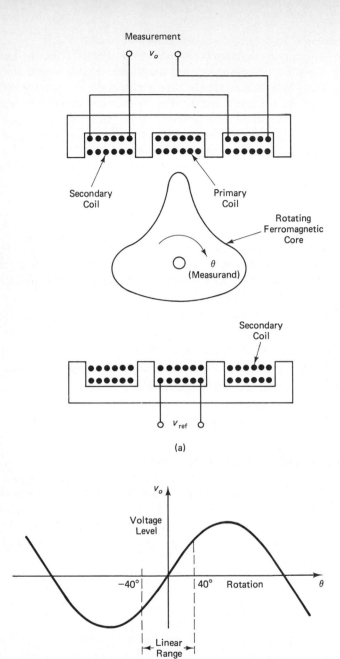

Figure 3.9. (a) Schematic diagram of the RVDT. (b) Operating curve.

The mutual-induction proximity sensor. This displacement transducer also operates on the mutual-induction principle. A simplified schematic diagram of such a device is shown in Figure 3.10a. The insulating "E core" carries the primary

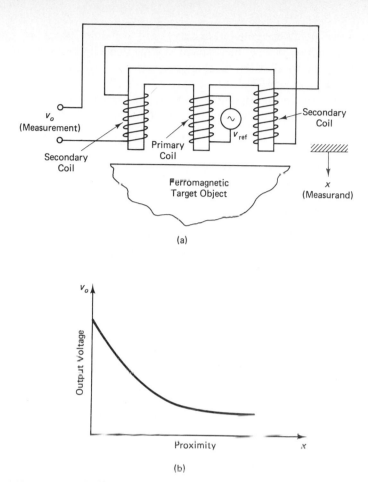

Figure 3.10. (a) Schematic diagram of the mutual-induction proximity sensor. (b) Operating curve.

winding in its middle limb. The two end limbs carry secondary windings that are connected in series. Unlike the LVDT and the RVDT, the two voltages induced in the secondary winding segments are additive in this case. The region of the moving surface (target object) that faces the coils has to be made of ferromagnetic material so that as it moves, the magnetic reluctance and the flux linkage will change. This, in turn, changes the induced voltage in the secondary windings, and this change is a measure of the displacement. Note that, unlike the LVDT, which has an "axial" displacement configuration, the proximity probe has a "transverse" displacement configuration. Hence, it is particularly suitable for measuring transverse displacements or proximities of moving objects (e.g., transverse motion of a beam or whirling shaft). We can see from the operating curve shown in figure 3.10b that the displacement–voltage relation of a proximity probe is nonlinear. Hence, these proximity sensors should be used only for measuring very small displacements, unless accurate nonlinear calibration curves are available. Since the proximity sensor is a noncontacting device, mechanical loading is negligible. Because a ferromagnetic object is used to alter the reluctance of the flux path, the mutual-induction proximity sensor is a variable-reluctance device.

Proximity sensors are used in a wide variety of applications pertaining to noncontacting displacement sensing and dimensional gaging. Some typical applications are

1. Measurement and control of the gap between a robotic welding torch head and the work surface
2. Gaging the thickness of metal plates in manufacturing operations (e.g., rolling and forming)
3. Detecting surface irregularities in machined parts
4. Angular speed measurement at steady state, by counting the number of rotations per unit time
5. Level detection (e.g., in the filling, bottling, and chemical process industries)

Some mutual-induction displacement transducers depend on relative motion between the primary coil and the secondary coil to produce a change in flux linkage. Two such devices are the resolver and the synchro-transformer. These are not variable-reluctance transducers because they do not employ a ferromagnetic moving element.

The resolver. This mutual-induction transducer is widely used for measuring angular displacements. A simplified schematic diagram of the resolver is shown in figure 3.11. The *rotor* contains the primary coil. It consists of a single two-pole winding element energized by the AC supply voltage v_{ref}. The rotor is directly attached to the object whose rotation is being measured. The *stator* consists of two sets

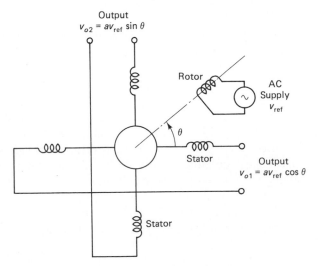

Figure 3.11. Operating schematic diagram of the resolver.

of windings placed 90° apart. If the angular position of the rotor with respect to one pair of stator windings is denoted by θ, the induced voltage in this pair of windings is given by

$$v_{o1} = av_{\text{ref}} \cos \theta \qquad (3.9)$$

The induced voltage in the other pair of windings is given by

$$v_{o2} = av_{\text{ref}} \sin \theta \qquad (3.10)$$

Note that these are amplitude-modulated signals; the carrier signal v_{ref} is modulated by the motion θ. The constant parameter a depends primarily on geometric and material characteristics of the device. Either of the two output signals v_{o1} and v_{o2} may be used to determine the angular position in the first quadrant ($0 \le \theta \le 90°$). Both signals are needed, however, to determine the displacement (direction as well as magnitude) in all four quadrants ($0 \le \theta \le 360°$) without causing any ambiguity. For instance, the same sine value is obtained for both $90° + \theta$ and $90° - \theta$ (i.e., a positive rotation and a negative rotation from the 90° position), but the corresponding cosine values have opposite signs, thus providing the proper direction. As for differential transformers, transient displacement signals can be extracted by demodulating the modulated outputs. This is accomplished by filtering out the carrier signal, thereby extracting the modulating signal.

Example 3.6

The two output signals v_{o1} and v_{o2} of the resolver are termed *quadrature signals*. Show how these quadrature signals could be demodulated to obtain the speed of rotation directly.

Solution If the angular speed is denoted by ω, the quadrature signals, at steady speed, may be expressed as

$$v_{o1} = av_{\text{ref}} \cos \omega t$$

$$v_{o2} = av_{\text{ref}} \sin \omega t$$

By differentiating these signals, we get

$$\dot{v}_{o1} = -av_{\text{ref}} \omega \sin \omega t$$

$$\dot{v}_{o2} = av_{\text{ref}} \omega \cos \omega t$$

Hence,

$$\omega = \dot{v}_{o2}/v_{o1}$$

or

$$\omega = -\dot{v}_{o1}/v_{o2}$$

Each relation provides direction as well as magnitude of the speed.

An alternative form of resolver uses two AC voltages 90° out of phase, generated from a digital signal generator board, to power the two stator windings. The rotor is the secondary winding in this case. The phase shift of the induced voltage determines the angular position of the rotor. An advantage of this arrangement is that it does not require slip rings and brushes to energize the windings. But it will need some mechanism to pick off the output signal from the rotor.

The output signals of a resolver are nonlinear (trigonometric) functions of the angle of rotation. (Historically, resolvers were used to compute trigonometric functions or to resolve a vector into orthogonal components.) In robot control applications, this is sometimes viewed as a blessing. For computed torque control of robotic manipulators, for example, trigonometric functions of the joint angles are needed in order to compute the required joint torques. Consequently, when resolvers are used to measure joint angles in manipulators, there is an associated reduction in processing time because the trigonometric functions are available as direct measurements.

The primary advantages of the resolver include

1. Fine resolution and high accuracy
2. Low output impedance (high signal levels)
3. Small size (e.g., 10 mm diameter)
4. Simple and robust construction

Its main limitations are

1. Nonlinear output signals (an advantage in some applications where trigonometric functions of rotations are needed)
2. Bandwidth limited by supply frequency
3. Slip rings and brushes needed (which adds mechanical loading and also creates wearout, oxidation, and thermal and noise problems)

The synchro-transformer. The synchro is somewhat similar in operation to the resolver. The main differences are that the synchro employs two identical rotor-stator pairs, and each stator has three sets of windings that are placed 120° apart around the rotor shaft. A schematic diagram for this arrangement is shown in figure 3.12. Both rotors have single-phase windings, and—contrary to popular belief—the synchro is essentially a single-phase device. One of the rotors is energized with an

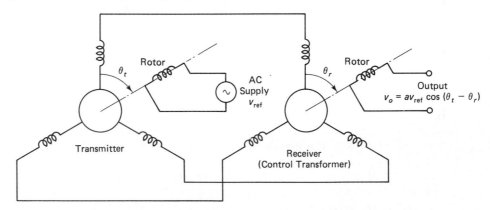

Figure 3.12. Schematic diagram of a synchro-transformer.

AC supply voltage v_{ref}. This induces voltages in the three winding segments of the corresponding stator. These voltages have different amplitudes that depend on the angular position of the rotor but are in phase. This drive rotor-stator pair is known as the *transmitter*. The other rotor-stator pair is known as the *receiver* or the *control transformer*. Windings of the transmitter stator are connected correspondingly to the windings of the receiving stator, as shown in figure 3.12. This induces a voltage v_o in the rotor of the receiver. Suppose that the angle between the drive rotor and one set of windings in its stator is denoted by θ_t. The resultant magnetic field on the receiver stator will make the same angle with the corresponding winding of that stator. If the receiver rotor is aligned with this direction (i.e., $\theta_r = \theta_t$), then the induced voltage v_o will be maximum. If the receiver rotor is placed at 90° to this resultant magnetic field, then $v_o = 0$. Therefore, an appropriate expression for the synchro output is

$$v_o = av_{ref} \cos (\theta_t - \theta_r) \qquad (3.11)$$

Synchros are operated near $\theta_r = \theta_t + 90°$, where the output voltage is zero. Hence, we define a new angle θ such that

$$\theta_r = \theta_t + 90° - \theta \qquad (3.12)$$

As a result, equation 3.11 becomes

$$v_o = av_{ref} \sin \theta \qquad (3.13)$$

Synchro-transformers can be used to measure relative displacements between two rotating objects. For measuring absolute displacements, one of the rotors is attached to the rotating member (e.g., the shaft) while the other rotor is fixed to a stationary member (e.g., the bearing). As is clear from previous discussion, a zero reading corresponds to the case where the two rotors are 90° apart.

Synchros have been used extensively in position servos, particularly for the position control of rotating objects. Typically, the input command is applied to the transmitter rotor. The receiver rotor is attached to the object that is being controlled. The initial physical orientations of the two rotors should ensure that for a given command, the desired position of the object corresponds to zero output voltage v_o—that is, when the two rotors are 90° apart. In this manner, v_o can be used as the position error signal, which is fed into the control circuitry that generates a drive signal so as to compensate for the error (e.g., using proportional plus derivative control). For small angles θ, the output voltage may be assumed proportional to the angle. For large angles, inverse sine should be taken. Note that ambiguities arise when the angle θ exceeds 90°. Hence, synchro readings should be limited to ±90°. In this range, the synchro provides directional measurements. For example, direction may be obtained by determining the phase shift between the output signal and the reference signal. Furthermore, demodulation is required to extract transient measurements from the output signal. This is accomplished, as usual, by suppressing the carrier from the modulated signal.

The advantages and disadvantages of the synchro are essentially the same as those of the resolver. In particular, quadrature error (null voltage) may be present because of impedance nonuniformities in the winding segments. Furthermore, velocity error (i.e., velocity-dependent displacement readings) is also a possibility. This may be reduced by increasing the carrier frequency, as in the case of the differential transformer and the resolver.

Self-Induction Transducers

These transducers are based on the principle of self-induction. Unlike mutual-induction transducers, only a single coil is employed. This coil is activated by an AC supply voltage v_{ref}. The current produces a magnetic flux, which is linked with the coil. The level of flux linkage (or self-inductance) can be varied by moving a ferromagnetic object within the magnetic field. This changes the reluctance of the flux path and the inductance in the coil. This change is a measure of the displacement of the ferromagnetic object. The change in inductance is measured using an inductance-measuring circuit (e.g., an inductance bridge). Note that self-induction transducers are usually variable-reluctance devices.

A typical self-induction transducer is a self-induction proximity sensor. A schematic diagram of this device is shown in figure 3.13. This device can be used as a displacement sensor for transverse displacements. For instance, the distance between the sensor tip and the ferromagnetic surface of a moving object, such as a beam or shaft, can be measured. Other applications include those mentioned for mutual-induction proximity sensors. High-speed displacement measurements can result in velocity error (rate error) when variable-inductance displacement sensors (including self-induction transducers) are used. This effect may be reduced, as in other AC-powered variable-inductance sensors, by increasing the carrier frequency.

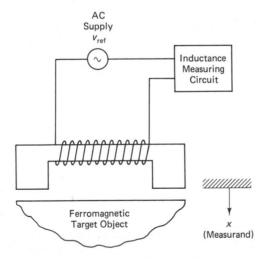

Figure 3.13. Schematic diagram of a self-induction proximity sensor.

3.5 PERMANENT-MAGNET TRANSDUCERS

In discussing this third type of variable-inductance transducer, we will first consider the permanent-magnet DC velocity sensors (DC tachometers). A distinctive feature of permanent-magnet transducers is that they have a permanent magnet to generate a uniform and steady magnetic field. A relative motion between the magnetic field and an electrical conductor induces a voltage that is proportional to the speed at which the conductor crosses the magnetic field. In some designs, a unidirectional magnetic field generated by a DC supply (i.e., an electromagnet) is used in place of a permanent magnet. Nevertheless, this is generally termed a permanent-magnet transducer.

The principle of electromagnetic induction between a permanent magnet and a conducting coil is used in speed measurement by permanent-magnet transducers. Depending on the configuration, either rectilinear speeds or angular speeds can be measured. Schematic diagrams of the two configurations are shown in figure 3.14. Note that these are passive transducers, because the energy for the output signal v_o is derived from the motion (measured signal) itself. The entire device is usually enclosed in a steel casing to isolate it from ambient magnetic fields.

In the rectilinear velocity transducer (figure 3.14a), the conductor coil is wrapped on a core and placed centrally between two magnetic poles, which produce a cross-magnetic field. The core is attached to the moving object whose velocity must be measured. The velocity v is proportional to the induced voltage v_o. A moving-magnet and fixed-coil arrangement can also be used, thus eliminating the

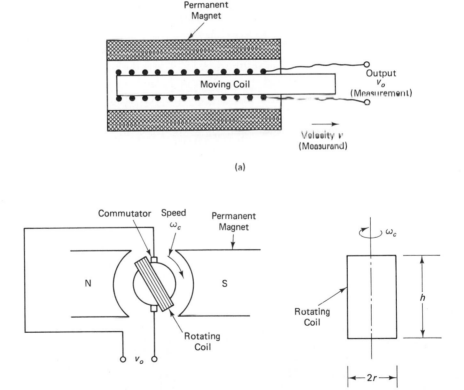

Figure 3.14. Permanent-magnet transducers: (a) rectilinear velocity transducer; (b) DC tachometer-generator.

need for any sliding contacts (slip rings and brushes) for the output leads, thereby reducing mechanical loading error, wearout, and related problems. The tachogenerator (or tachometer) is a very common permanent-magnet device. The principle of operation of a DC tachogenerator is shown in Figure 3.14b. The rotor is directly connected to the rotating object. The output signal that is induced in the rotating coil is picked up as DC voltage v_o using a suitable *commutator* device—typically consisting of a pair of low-resistance carbon brushes—that is stationary but makes contact with the rotating coil through split slip rings so as to maintain the positive direction of induced voltage throughout each revolution. (See chapter 7 for a discussion of this commutation arrangement.) The induced voltage is given by

$$v_o = (2nhr\beta)\omega_c \tag{3.14}$$

for a coil of height h and width $2r$ that has n turns, moving at an angular speed ω_c in a uniform magnetic field of flux density β. This proportionality between v_o and ω_c is used to measure the angular speed ω_c.

When tachometers are used to measure transient velocities, some error will result from the rate (acceleration) effect. This error generally increases with the maximum significant frequency that must be retained in the transient velocity signal. Output distortion can also result because of reactive (inductive and capacitive) loading of the tachometer. Both types of error can be reduced by increasing the load impedance.

As illustration, consider the equivalent circuit for a tachometer with an impedance load, as shown in figure 3.15. The induced voltage $k\omega_c$ is represented by a voltage source. Note that the constant k depends on the coil geometry, the number of turns, and the magnetic flux density (see equation 3.14). Coil resistance is denoted by R, and leakage inductance is denoted by L_ℓ. The load impedance is Z_L. From straightforward circuit analysis in the frequency domain, the output voltage at the load is given by

$$v_o = \left[\frac{Z_L}{R + j\omega L_\ell + Z_L}\right] k\omega_c \tag{3.15}$$

It can be seen that because of the leakage inductance, the output signal attenuates more at higher frequencies ω of the velocity transient. In addition, loading error is present. If Z_L is much larger than the coil impedance, however, the ideal proportionality, as given by

$$v_o = k\omega_c \tag{3.16}$$

is achieved.

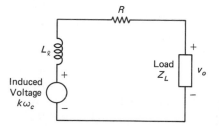

Figure 3.15. Equivalent circuit for a tachometer with an impedance load.

Some tachometers operate in a different manner. *Digital tachometers* generate voltage pulses at a frequency proportional to the angular speed. These will be discussed separately, as digital transducers, in chapter 5.

AC Permanent-Magnet Tachometers

These devices have a permanent magnet rotor and two separate sets of stator windings. One set of windings is energized using an AC reference voltage. Induced voltage in the other set of windings is the tachometer output. When the rotor is stationary or moving in a quasi-static manner, the output voltage is a constant-amplitude signal much like the reference voltage. As the rotor moves at a finite speed, an additional induced voltage that is proportional to the rotor speed is generated in the secondary winding. The net output is an amplitude-modulated signal whose amplitude is proportional to the rotor speed. For transient velocities, it will be necessary to demodulate this signal in order to extract the transient velocity signal (i.e., the modulating signal) from the modulated output. The direction of velocity is determined from the phase angle of the modulated signal with respect to the carrier signal.

For low-frequency applications (5 Hz or less), a standard AC supply (60 Hz) may be used to power AC tachometers. For moderate-frequency applications, a 400 Hz supply is widely used. Typical sensitivity of an AC permanent-magnet tachometer is on the order of 75 mV/rad/s.

AC Induction Tachometers

These tachometers are similar in construction to the two-phase induction motors described in chapter 7. The stator arrangement is identical to that of the AC permanent-magnet tachometer. The rotor, however, has windings that are shorted and not energized by an external source. One of the stator windings is energized with an AC supply. This induces a voltage in the rotor windings, and it has two components. One component is due to the direct transformer action of the supply AC. The other component is induced by the speed of rotation of the rotor and is proportional to the speed. The nonenergized stator winding provides the output of the tachometer. Voltage induced in the output stator winding is due to both the primary stator winding and the rotor winding. As a result, the tachometer output has a carrier AC component and a modulating component that is proportional to the speed of rotation, as in the case of the AC permanent-magnet tachometer.

The main advantage of AC tachometers over their DC counterparts is the absence of slip-ring brush devices. In particular, the signal from a DC tachometer usually has a voltage ripple, known as commutator ripple, which is generated as the split ends of the slip ring pass over the brushes. The frequency of the commutator ripple depends on the speed of operation; consequently, filtering it out using a notch filter is difficult. Also, there are problems with frictional loading and contact bounce in DC tachometers that are absent in AC tachometers. It is known, however, that the output from an AC tachometer is somewhat nonlinear (saturation effect) at high speeds. Furthermore, for measuring transient speeds, signal demodulation would be necessary. Another disadvantage of AC tachometers is that the output signal level depends on the supply voltage; hence, a stabilized voltage source that has very small output impedance is necessary for accurate measurements.

3.6 EDDY CURRENT TRANSDUCERS

If a conducting (i.e., low-resistivity) medium is subjected to a fluctuating magnetic field, eddy currents are generated in the medium. The strength of eddy currents increases with the strength of the magnetic field and the frequency of the magnetic flux. This principle is used in eddy current proximity sensors. Eddy current sensors may be used as either dimensional gaging devices or displacement sensors.

A schematic diagram of an eddy current proximity sensor is shown in figure 3.16a. Unlike variable-inductance proximity sensors, the target object of the eddy

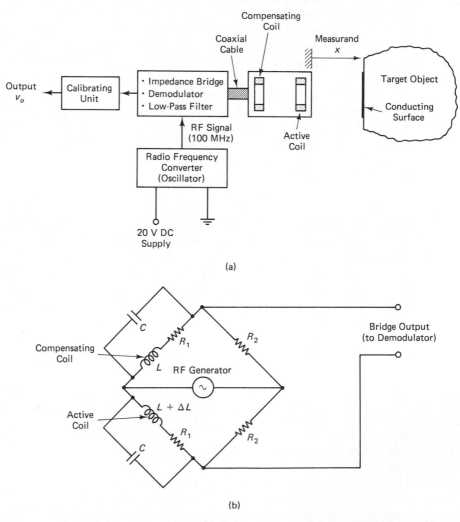

Figure 3.16. Eddy current proximity sensor: (a) schematic diagram; (b) impedance bridge.

current sensor does not have to be made of a ferromagnetic material. A conducting target object is needed, but a thin film conducting material—such as household aluminum foil glued onto a nonconducting target object—would be adequate. The probe head has two identical coils, which form two arms of an impedance bridge. The coil closer to the probe face is the *active coil*. The other coil is the *compensating coil*. It compensates for ambient changes, particularly thermal effects. The other two arms of the bridge consist of purely resistive elements (see figure 3.16b). The bridge is excited by a radio-frequency voltage supply. The frequency may range from 1 MHz to 100 MHz. This signal is generated from a radio-frequency converter (an oscillator) that is typically powered by a 20 V DC supply. In the absence of the target object, the output of the impedance bridge is zero, which corresponds to the balanced condition. When the target object is moved close to the sensor, eddy currents are generated in the conducting medium because of the radio-frequency magnetic flux from the active coil. The magnetic field of the eddy currents opposes the primary field that generates these currents. Hence, the inductance of the active coil increases, creating an imbalance in the bridge. The resulting output from the bridge is an amplitude-modulated signal containing the radio-frequency carrier. This signal is demodulated by removing the carrier. The resulting signal (modulating signal) measures transient displacement of the target object. Low-pass filtering is used to remove high-frequency leftover noise in the output signal once the carrier is removed. For large displacements, the output is not linearly related to the displacement. Furthermore, the sensitivity of the eddy current probe depends nonlinearly on the nature of the conducting medium, particularly the resistivity. For example, for low resistivities, sensitivity increases with resistivity; for high resistivities, sensitivity decreases with resistivity. A calibrating unit is usually available with commercial eddy current sensors to accommodate various target objects and nonlinearities. The gage factor is usually expressed in volts/millimeter. Note that eddy current probes can also be used to measure resistivity and surface hardness (which affects resistivity) in metals.

The facial area of the conducting medium on the target object has to be slightly larger than the frontal area of the eddy current probe head. If the target object has a curved surface, its radius of curvature has to be at least four times the diameter of the probe. These are not serious restrictions, because the typical diameter of the probe head is about 2 mm. Eddy current sensors are medium-impedance devices; 1,000 Ω output impedance is typical. Sensitivity is on the order of 5 V/mm. Since the carrier frequency is very high, eddy current devices are suitable for highly transient displacement measurements—for example, bandwidths up to 100 kHz. Another advantage of the eddy current sensor is that it is a noncontacting device; there is no mechanical loading on the moving (target) object.

3.7 VARIABLE-CAPACITANCE TRANSDUCERS

Capacitive transducers are commonly used to measure small transverse displacements, large rotations, and fluid levels. They may also be employed to measure angular velocities. In addition to analog capacitive sensors, digital (pulse-generating) capacitive tachometers are also available.

Capacitance C of a two-plate capacitor is given by

$$C = \frac{kA}{x} \tag{3.17}$$

where A is the common (overlapping) area of the two plates, x is the gap width between the two plates, and k is a constant that depends on dielectric properties of the medium between the two plates. A change in any one of these three parameters may be used in the sensing process. Schematic diagrams for measuring devices that use this feature are shown in figure 3.17. In figure 3.17a, angular displacement of one of the plates causes a change in A. In figure 3.17b, a transverse displacement of one of the plates changes x. Finally, in figure 3.17c, a change in k is produced as the fluid level between the capacitor plates changes. In all cases, the associated change in capacitance is measured directly or indirectly and is used to estimate the measurand. A popular method is to use a capacitance bridge circuit to measure the change in capacitance, in a manner similar to how an inductance bridge is used to measure changes in inductance (see figure 3.16b). Other methods include measuring a change in such quantities as charge (using a charge amplifier), voltage (using a high input impedance device in parallel), and current (using a very low impedance device in series) that result from the change in capacitance in a suitable circuit. An alternative method is to make the capacitor a part of an inductance–capacitance (L–C) oscillator circuit; the natural frequency of the oscillator ($1/\sqrt{LC}$) measures the capacitance. (Incidentally, this method may also be used to measure inductance.)

Capacitive Displacement Sensors

For the arrangement shown in figure 3.17a, since the common area A is proportional to the angle of rotation θ, equation 3.17 may be written as

$$C = K\theta \tag{3.18}$$

where K is a sensor constant. This is a linear relationship between C and θ. The capacitance may be measured by any convenient method. The sensor is linearly calibrated to give the angle of rotation.

The sensitivity of this angular displacement sensor is

$$S = \frac{\partial C}{\partial \theta} = K \tag{3.19}$$

which is constant throughout the measurement. This is expected because the sensor relationship is linear. Note that in the nondimensional form, the sensitivity of the sensor is unity, implying "direct" sensitivity.

For the arrangement shown in figure 3.17b, the sensor relationship is

$$C = \frac{K}{x} \tag{3.20}$$

The constant K has a different meaning here. The corresponding sensitivity is given by

$$S = \frac{\partial C}{\partial x} = -\frac{K}{x^2} \tag{3.21}$$

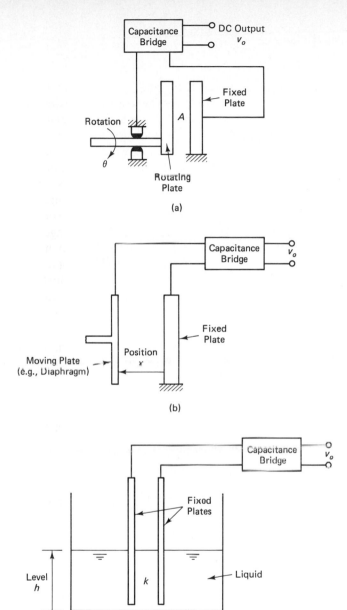

Figure 3.17. Schematic diagrams of capacitive sensors: (a) capacitive rotation sensor; (b) capacitive displacement sensor; (c) capacitive liquid level sensor.

Again, the sensitivity is unity (negative) in the nondimensional form, which indicates direct sensitivity of the sensor. Note that equation 3.20 is a nonlinear relationship. A simple way to linearize this transverse displacement sensor is to use an inverting amplifier, as shown in figure 3.18. Note that C_{ref} is a fixed, reference capacitance. Since the gain of the operational amplifier is very high, the voltage at point A is zero for most practical purposes. Furthermore, since the input impedance

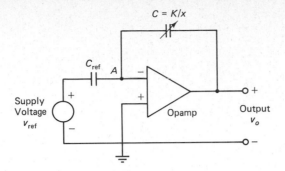

Figure 3.18. Inverting amplifier circuit used to linearize the capacitive transverse displacement sensor.

of the opamp is also very high, the current through the input leads is negligible. Accordingly, the charge balance equation for node point A is

$$v_{\text{ref}} C_{\text{ref}} + v_o C = 0$$

Now, in view of equation 3.20, we get the following linear relationship for the output voltage v_o in terms of the displacement x:

$$v_o = -\frac{v_{\text{ref}} C_{\text{ref}}}{K} x \qquad (3.22)$$

Hence, measurement of v_o gives the displacement through linear calibration. The sensitivity of the device can be increased by increasing v_{ref} and C_{ref}. The reference voltage could be DC as well as AC. With an AC reference voltage, the output voltage is a modulated signal that has to be demodulated to measure transient displacements.

Example 3.7

Consider the circuit shown in figure 3.19. Examine how this arrangement could be used to measure displacements.

Solution The current through the capacitor is the same as that through the resistor, assuming that a very high impedance device is used to measure the output voltage v_o. Thus,

$$i = \frac{d}{dt}(Cv_o) = \frac{v_{\text{ref}} - v_o}{R} \qquad (3.23)$$

For example, if a transverse displacement capacitor is used, from equation 3.20 we have

$$x = RKv_o \Big/ \int_0^t (v_{\text{ref}} - v_o)\, dt \qquad (3.24)$$

This is a nonlinear differential relationship. To measure x, we need to measure the output voltage and perform an integration by either analog or digital means. Furthermore,

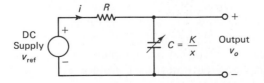

Figure 3.19. Capacitive displacement sensor example.

since $v_o = v_{ref}$ and $v_o = 0$ at steady state, it follows that this procedure cannot be used to make steady-state (or quasi-static) measurements. This situation can be corrected by using an AC source as the supply. If the supply frequency is ω, the frequency domain transfer function between the supply and the output is given by

$$\frac{v_o}{v_{ref}} = \frac{1}{[1 + RKj\omega/x]} \tag{3.25}$$

Now the displacement x may be determined by measuring either the signal amplification (i.e., amplitude ratio or magnitude) M at the output or the phase "lag" ϕ of the output signal. The corresponding relations are

$$x = \frac{RK\omega}{\sqrt{1/M^2 - 1}} \tag{3.26}$$

and

$$x = RK\omega/\tan \phi \tag{3.27}$$

Note that the differential equation of the circuit is not linear unless x is constant. The foregoing transfer-function relations are no longer valid if the displacement is transient. Nevertheless, reasonably accurate results are obtained if the displacement is slowly varying.

Capacitive Angular Velocity Sensor

The schematic diagram for an angular velocity sensor that uses a rotating-plate capacitor is shown in figure 3.20. Since the current sensor has negligible resistance, the voltage across the capacitor is almost equal to the supply voltage v_{ref}, which is constant. It follows that the current in the circuit is given by

$$i = \frac{d}{dt}(Cv_{ref}) = v_{ref}\frac{dC}{dt}$$

which, in view of equation 3.18, may be expressed as

$$\frac{d\theta}{dt} = \frac{i}{Kv_{ref}} \tag{3.28}$$

This is a linear relationship for the angular velocity in terms of the measured current i. Care must be exercised to guarantee that the current-measuring device does not interfere with the basic circuit.

An advantage of capacitance transducers is that because they are noncontacting devices, mechanical loading effects are negligible. There is some loading due to inertial and frictional resistance in the moving plate. This can be eliminated by using the moving object itself to function as the moving plate. Variations in the dielectric properties due to humidity, temperature, pressure, and impurities introduce errors.

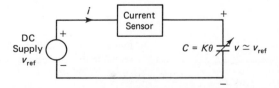

Figure 3.20. Rotating-plate capacitive angular velocity sensor.

A capacitance bridge circuit can compensate for these effects. Extraneous capacitances, such as cable capacitance, can produce erroneous readings in capacitive sensors. This problem can be reduced by using a charge amplifier to condition the sensor signal. One drawback of capacitance displacement sensors is low sensitivity. For a transverse displacement transducer, the sensitivity is typically less than one picofarad (pF) per millimeter (1 pF = 10^{-12} F). This problem is not serious, because high supply voltages and amplifier circuitry can be used to increase the sensor sensitivity.

3.8 PIEZOELECTRIC TRANSDUCERS

Some substances, such as barium titanate and single-crystal quartz, can generate an electrical charge and associated potential difference when they are subjected to mechanical stress or strain. This piezoelectric effect is used in piezoelectric transducers. Direct application of the piezoelectric effect is found in pressure and strain measuring devices, and many indirect applications also exist. They include piezoelectric accelerometers and velocity sensors and piezoelectric torque sensors and force sensors. It is also interesting to note that piezoelectric materials deform when subjected to a potential difference (or charge). Some delicate test equipment (e.g., in vibration testing) use piezoelectric actuating elements (reverse piezoelectric action) to create fine motions. Also, piezoelectric valves (e.g., flapper valves), directly actuated using voltage signals, are used in pneumatic and hydraulic control applications and in ink-jet printers of microcomputers. Miniature stepper motors based on the reverse piezoelectric action are available.

Consider a piezoelectric crystal in the form of a disc with two electrodes plated on the two opposite faces. Since the crystal is a dielectric medium, this device is essentially a capacitor, which may be modeled by equation 3.17. Accordingly, a piezoelectric sensor may be represented as a charge source with a series capacitive impedance (figure 3.21) in equivalent circuits. The impedance from the capacitor is given by

$$Z = \frac{1}{j\omega C} \qquad (3.29)$$

As is clear from equation 3.29, the output impedance of piezoelectric sensors is very high, particularly at low frequencies. For example, a quartz crystal may present an

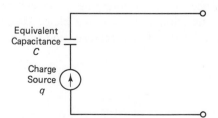

Figure 3.21. Equivalent circuit representation of a piezoelectric sensor.

impedance of several megohms at 100 Hz, increasing hyperbolically with decreasing frequencies. This is one reason why piezoelectric sensors have a limitation on the useful lower frequency. The other reason is the charge leakage.

Sensitivity

The sensitivity of a piezoelectric crystal may be represented either by its *charge sensitivity* or by its *voltage sensitivity*. Charge sensitivity is defined as

$$S_q = \frac{\partial q}{\partial F} \tag{3.30}$$

where q denotes the generated charge and F denotes the applied force. For a crystal with surface area A, equation 3.30 may be expressed as

$$S_q = \frac{1}{A}\frac{\partial q}{\partial p} \tag{3.31}$$

where p is the stress (normal or shear) or pressure applied to the crystal surface. Voltage sensitivity S_v is given by the change in voltage due to a unit increment in pressure (or stress) per unit thickness of the crystal. Thus, in the limit, we have

$$S_v - \frac{1}{d}\frac{\partial v}{\partial p} \tag{3.32}$$

where d denotes the crystal thickness.

Now, since

$$\delta q = C\,\delta v \tag{3.33}$$

the following relationship between charge sensitivity and voltage sensitivity is obtained:

$$S_q - kS_v \tag{3.34}$$

Note that k is the dielectric constant of the crystal capacitor, as defined by equation 3.17.

Example 3.8

A barium titanate crystal has a charge sensitivity of 1.5 picocoulombs per newton (pC/N). (*Note:* 1 pC = 1 $\times$ 10^{-12} coulombs; coulombs = farads $\times$ volts.) The dielectric constant for the crystal is 1.25×10^{-8} farads per meter (F/m). What is the voltage sensitivity of the crystal?

Solution   The voltage sensitivity of the crystal is given by

$$S_v = \frac{1.5 \text{ pC/N}}{1.25 \times 10^{-8} \text{ F/m}} = \frac{1.5 \times 10^{-12} \text{ C/N}}{1.25 \times 10^{-8} \text{ F/m}}$$

or

$$S_v = 1.2 \times 10^{-4} \text{ V.m/N}$$

Piezoelectric Accelerometer

Since we are concerned only with motion transducers in this chapter, we will discuss a piezoelectric motion transducer—the piezoelectric accelerometer—in more detail. A piezoelectric velocity transducer is simply a piezoelectric accelerometer with a built-in integrating amplifier in the form of a miniature integrated circuit.

Accelerometers are acceleration-measuring devices. It is known from Newton's second law that a force (f) is necessary to accelerate a mass (or inertia element), and its magnitude is given by the product of mass (M) and acceleration (a). This product (Ma) is commonly termed *inertia force*. The rationale for this terminology is that if a force of magnitude Ma were applied to the accelerating mass in the direction opposing the acceleration, then the system could be analyzed using static-equilibrium considerations. This is known as *d'Alembert's principle* (figure 3.22). The force that causes acceleration is itself a measure of the acceleration (mass is kept constant). Accordingly, mass can serve as a front-end element to convert acceleration into a force. This is the principle of operation of common accelerometers. There are many different types of accelerometers, ranging from strain gage devices to those that use electromagnetic induction. For example, force that causes acceleration may be converted into a proportional displacement using a spring element, and this displacement may be measured using a convenient displacement sensor. Examples of this type are differential-transformer accelerometers, potentiometer accelerometers, and variable-capacitance accelerometers. Alternatively, the strain at a suitable location of a member that was deflected due to inertia force may be determined using a strain gage. This method is used in strain gage accelerometers. Vibrating-wire accelerometers use the accelerating force to tension a wire. The force is measured by detecting the natural frequency of vibration of the wire (which is proportional to the square root of tension). In servo force-balance (or null-balance) accelerometers, the inertia element is restrained from accelerating by detecting its motion and feeding back a force (or torque) to exactly cancel out the accelerating force (torque). This feedback force is known, for instance, by knowing the motor current, and it is a measure of the acceleration.

The advantages of piezoelectric accelerometers (also known as *crystal accelerometers*) over the other devices are their light weight and high-frequency response (up to about 1 MHz). However, piezoelectric transducers are inherently high output impedance devices that generate small voltages (on the order of 1 mV). For this reason, special impedance-transforming amplifiers (e.g., charge amplifiers) have to be employed to condition the output signal and to reduce loading error (see chapter 2).

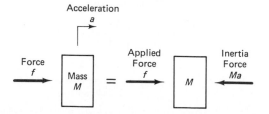

Figure 3.22. Illustration of d'Alembert's principle.

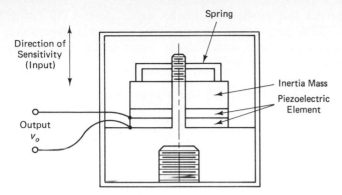

Figure 3.23. A compression-type piezoelectric accelerometer.

A schematic diagram for a compression-type piezoelectric accelerometer is shown in figure 3.23. The crystal and the inertia mass are restrained by a spring of very high stiffness. Consequently, the fundamental natural frequency or resonant frequency of the device becomes high (typically 20 kHz). This gives a reasonably wide useful range (typically up to 5 kHz). The lower limit of the useful range (typically 1 Hz) is set by factors such as the limitations of the signal-conditioning systems, the mounting methods, the charge leakage in the piezoelectric element, the time constant of the charge-generating dynamics, and the signal-to-noise ratio. A typical frequency response curve for a piezoelectric accelerometer is shown in figure 3.24.

In compression-type crystal accelerometers, the inertia force is sensed as a compressive normal stress in the piezoelectric element. There are also piezoelectric accelerometers that sense inertia force as a shear strain or tensile strain. For an accelerometer, acceleration is the signal that is being measured (the measurand). Hence, accelerometer sensitivity is commonly expressed in terms of electrical charge per unit acceleration or voltage per unit acceleration (compare this with equations 3.31 and 3.32). Acceleration is measured in units of acceleration due to gravity (g), and charge is measured in picocoulombs (pC), which are units of 10^{-12} coulombs (C). Typical accelerometer sensitivities are 10 pC/g and 5 mV/g. Sensitivity depends

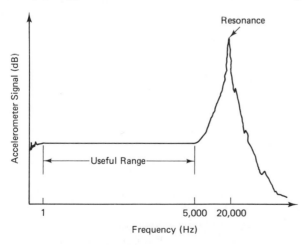

Figure 3.24. A typical frequency response curve for a piezoelectric accelerometer.

on the piezoelectric properties and on the mass of the inertia element. If a large mass is used, the reaction inertia force on the crystal will be large for a given acceleration, thus generating a relatively large output signal. Large accelerometer mass results in several disadvantages, however. In particular,

1. The accelerometer mass distorts the measured motion variable (mechanical loading effect).
2. A heavy accelerometer has a lower resonant frequency and, hence, a lower useful frequency range (figure 3.24).

For a given accelerometer size, improved sensitivity can be obtained by using the shear-strain configuration. In this configuration, several shear layers can be used (e.g., in a delta arrangement) within the accelerometer housing, thereby increasing the shear area and, hence, the sensitivity in proportion to the shear area. Another factor that should be considered in selecting an accelerometer is its cross-sensitivity or transverse sensitivity (see chapter 2). Cross-sensitivity primarily results from manufacturing irregularities of the piezoelectric element, such as material unevenness and incorrect orientation of the sensing element. Cross-sensitivity should be less than the maximum error (percentage) that is allowed for the device (typically 1 percent).

The technique employed to mount the accelerometer to an object can significantly affect the useful frequency range of the accelerometer. Some common mounting techniques are

1. Screw-in base
2. Glue, cement, or wax
3. Magnetic base
4. Spring-base mount
5. Hand-held probe

Drilling holes in the object can be avoided by using the second through fifth methods, but the useful range can decrease significantly when spring-base mounts or hand-held probes are used (typical upper limit of 500 Hz). The first two methods usually maintain the full useful range, whereas the magnetic attachment method reduces the upper frequency limit to some extent (typically 1.5 kHz).

Piezoelectric signals cannot be read using low-impedance devices. The two primary reasons for this are

1. High output impedance in the sensor results in small output signal levels and large loading errors.
2. The charge can quickly leak out through the load.

The charge amplifier is the commonly used signal-conditioning device for piezoelectric sensors. (Its operation was discussed in chapter 2.) Because of impedance transformation, the impedance at the charge amplifier output becomes much smaller than the output impedance of the piezoelectric sensor. This virtually eliminates loading error. Also, by using a charge amplifier circuit with a large time constant, charge leakage speed can be decreased.

Example 3.9

Consider a piezoelectric sensor and charge amplifier combination, as represented by the circuit in figure 3.25. Examine how the charge leakage rate is slowed down by using this arrangement. Sensor capacitance, feedback capacitance of the charge amplifier, and feedback resistance of the charge amplifier are denoted by C, C_f, and R_f, respectively. The capacitance of the cable that connects the sensor to the charge amplifier is denoted by C_c.

Solution   For an opamp of gain K, the voltage at its negative input is $-v_o/K$, where v_o is the voltage at the amplifier output. Note that the positive input of the opamp is grounded (zero potential). Current balance at point A gives

$$\dot{q} + C_c \frac{\dot{v}_o}{K} + C_f\left(\dot{v}_o + \frac{\dot{v}_o}{K}\right) + \frac{v_o + v_o/K}{R_f} = 0$$

Since gain K is very large (typically 10^5 to 10^9) compared to unity, this differential equation may be approximated to

$$R_f C_f \frac{dv_o}{dt} + v_o = -R_f \frac{dq}{dt} \qquad (3.35)$$

The corresponding transfer function is

$$\frac{v_o(s)}{q(s)} = -\frac{R_f s}{[R_f C_f s + 1]} \qquad (3.36)$$

where s is the Laplace variable. Now, in the frequency domain ($s = j\omega$), we have

$$\frac{v_o(j\omega)}{q(j\omega)} = -\frac{R_f j\omega}{[R_f C_f j\omega + 1]} \qquad (3.37)$$

Note that the output is zero at zero frequency ($\omega = 0$). Hence, a piezoelectric sensor cannot be used for measuring constant (DC) signals. At very high frequencies, on the other hand, the transfer function approaches the constant value $-1/C_f$, which is the calibration constant for the device.

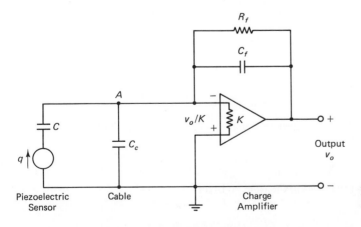

Figure 3.25.   A piezoelectric sensor and charge amplifier combination.

From equation (3.35), which represents a first-order system, it is clear that the time constant τ_c is

$$\tau_c = R_f C_f \tag{3.38}$$

Suppose that the charge amplifier is properly calibrated (by the factor $-1/C_f$) so that the frequency transfer function (equation 3.37) can be written as

$$G(j\omega) = \frac{j\tau_c\omega}{[j\tau_c\omega + 1]} \tag{3.39}$$

Magnitude M of this transfer function is given by

$$M = \frac{\tau_c\omega}{\sqrt{\tau_c^2\omega^2 + 1}} \tag{3.40}$$

As $\omega \to \infty$, note that $M \to 1$. Hence, at infinite frequency, there is no error. Measurement accuracy depends on the closeness of M to 1. Suppose that we want the accuracy to be better than a specified value M_o. Accordingly, we must have

$$\frac{\tau_c\omega}{\sqrt{\tau_c^2\omega^2 + 1}} > M_o$$

or

$$\tau_c\omega > \frac{M_o}{\sqrt{1 - M_o^2}} \tag{3.41}$$

If the required lower frequency limit is ω_{min}, the time constant requirement is

$$\tau_c > \frac{M_o}{\omega_{min}\sqrt{1 - M_o^2}} \tag{3.42}$$

or

$$R_f C_f > \frac{M_o}{\omega_{min}\sqrt{1 - M_o^2}} \tag{3.43}$$

It follows that a specified lower limit on frequency of operation, for a specified level of accuracy, may be achieved by increasing the time constant (i.e., by increasing R_f, C_f, or both). For instance, accuracy better than 99 percent is obtained if

$$\frac{\tau_c\omega}{\sqrt{\tau_c^2\omega^2 + 1}} > 0.99$$

or

$$\tau_c\omega > 7.0$$

The minimum frequency of a transient signal that can tolerate this level of accuracy is

$$\omega_{min} = \frac{7.0}{\tau_c}$$

Now ω_{min} can be set by adjusting the time constant.

In theory, it is possible to measure velocity by first converting velocity into a force using a viscous damping element and measuring the resulting force using a piezoelectric sensor. This principle may be used to develop a piezoelectric velocity transducer. Commercial piezoelectric velocity transducers use a piezoelectric accelerometer and an integrating amplifier, however. The difficulty of practical implementation of an ideal velocity–force transducer is the main reason for using the accelerometer approach to measure velocities using piezoelectric devices. A piezoelectric velocity transducer is shown in figure 3.26a, and a schematic diagram of the arrangement is shown in figure 3.26b. The size of the overall unit is about a cube of side 2 cm. With double integration, a piezoelectric displacement transducer

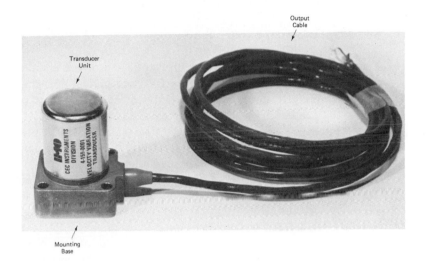

(a)

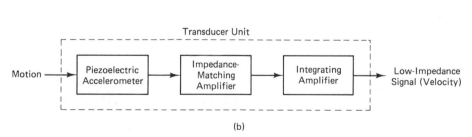

(b)

Figure 3.26. A piezoelectric velocity transducer: (a) photograph (courtesy of IMO Delaval, Inc., CEC Instruments Division); (b) schematic diagram.

is obtained. Alternatively, an ideal spring element that converts displacement into a force may be employed to stress the piezoelectric element, resulting in a displacement transducer. Such devices are usually not practical for low-frequency (few hertz) applications because of the poor low-frequency characteristics of piezoelectric elements.

Typical specifications for some of the motion sensors discussed in this chapter are given in table 3.2. (Strain gages will be discussed in the next chapter.)

3.9 OTHER TYPES OF SENSORS

This section introduces several types of sensors, not discussed previously, that are useful in motion sensing (or position sensing).

Fiber Optic Sensors and Lasers

The characteristic component in a fiber optic sensor is a bundle of glass fibers (typically a few hundred) that can carry light. Each optical fiber may have a diameter on the order of 0.01 mm. There are two basic types of fiber optic sensors. In one type—the "indirect" type—the optical fiber acts only as the medium in which sensed light is transmitted. In this type, the sensing element itself does not consist of optical fibers. In the second type—the "direct" type—the optical fiber bundle acts as the sensing element. When the conditions of the sensed medium change, the light-propagation properties of the optical fibers change, providing a measurement of the change in conditions. Examples of the first type of sensor include fiber optic position sensors and tactile sensors. The second type of sensor is found, for example, in fiber optic gyroscopes and fiber optic hydrophones.

A schematic representation of a fiber optic position sensor (or proximity sensor or displacement sensor) is shown in figure 3.27. The optical fiber bundle is divided into two groups: transmitting fibers and receiving fibers. Light from the light source is transmitted along the first bundle of fibers to the target object whose position is being measured. Light reflected onto the receiving fibers by the surface of the target object is carried to a photodetector. The intensity of the light received by the photodetector will depend on position x of the target object. In particular, if $x = 0$,

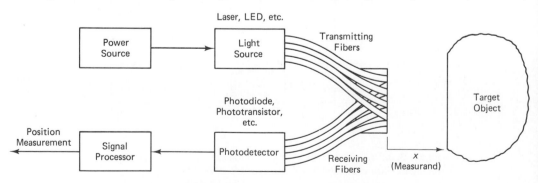

Figure 3.27. A fiber optic position sensor.

TABLE 3.2 TYPICAL SPECIFICATIONS FOR ANALOG MOTION TRANSDUCERS

Transducer	Measurand	Measurand frequency Max	Measurand frequency Min	Output impedance	Typical resolution	Accuracy	Sensitivity
Potentiometer	Displacement	5 Hz	DC	Low	0.1 mm	0.1%	200 mV/mm
LVDT	Displacement	2,500 Hz	DC	Low	0.001 mm or less	0.3%	50 mV/mm
Resolver	Angular displacement	500 Hz (limited by ref. frequency)	DC	Low	2 min	0.2%	10 mV/deg
Tachometer	Velocity	700 Hz	DC	Moderate	0.2 mm/s	0.5%	5 mV/mm/s 100 mV/rad/s
Eddy current proximity sensor	Displacement	100 kHz	DC	Moderate	0.001 mm, 0.05% full scale	0.5%	5 V/mm
Piezoelectric accelerometer	Acceleration (and velocity, etc.)	25 kHz	1 Hz	High	1 mm/s^2	1%	0.5 mV/m/s^2
Semiconductor strain gage	Strain (displacement, acceleration, etc.)	1 kHz (limited by fatigue)	DC	200 Ω	1 – 10 $\mu\epsilon$ (1 $\mu\epsilon$ = 10^{-6} unity strain)	1%	1 V/ϵ, 2,000 $\mu\epsilon$ max

the transmitting bundle will be completely blocked off and the light intensity at the receiver will be zero. As x is increased, the received light intensity will increase, because more and more light will be reflected onto the receiving bundle tip. This will reach a peak at some value of x. When x is increased beyond that value, more and more light will be reflected outside the receiving bundle; hence, the intensity of the received light will decrease. The light source could be a laser (structured light) or some other type, such as a light-emitting diode (LED). The light sensor (photodetector) could be some element such as a photodiode or a photo field effect transistor (photo FET). Very fine resolutions better than 1×10^{-6} can be obtained using a fiber optic position sensor.

The advantages of fiber optics include insensitivity to electrical and magnetic noise (due to optical coupling) and safe operation in explosive and hazardous environments. Furthermore, mechanical loading and wear problems do not exist because fiber optic position sensors are noncontacting devices with no moving parts. The disadvantages include direct sensitivity to variations in the intensity of the light source and dependence on ambient conditions (dust, moisture, smoke, etc.).

The fiber optic gyroscope.   This is an angular speed sensor that uses fiber optics; contrary to the implication of its name, however, it is not actually a gyroscope. Two loops of optical fibers wrapped around a cylinder are used in this sensor. One loop carries a monochromatic light (or laser) beam in the clockwise direction; the other loop carries a beam from the same light (laser) source in the counterclockwise direction. Since the laser beam traveling in the direction of rotation of the cylinder has a higher frequency than that of the other beam, the difference in frequencies of the two laser beams received at a common location will measure the angular speed of the cylinder. Note that the length of the optical fiber in each loop can be about 100 m. Angular displacements can be measured with the same sensor simply by counting the number of cycles and clocking fractions of cycles. Acceleration can be determined by digitally determining the rate of change of speed.

The laser Doppler interferometer.   The laser (*light amplification by stimulated emission of Radiation*) produces electromagnetic radiation in the ultraviolet, visible, or infrared bands of the spectrum. A laser can provide a single-frequency (*monochromatic*) light source. Furthermore, the electromagnetic radiation in a laser is coherent in the sense that all waves generated have constant phase angles. The laser uses oscillations of atoms or molecules of various elements. The helium-neon (HeNe) laser and the semiconductor laser are commonly used in industrial applications.

As noted earlier, the laser is useful in fiber optics, but it can also be used directly in sensing and gaging applications. The laser Doppler interferometer is one such sensor. It is useful in the accurate measurement of small displacements—for example, in strain measurements. To understand the operation of this device, we should explain two phenomena: the Doppler effect and light wave interference. Consider a wave source (e.g., a light source or sound source) that is moving with respect to a receiver (observer). If the source moves toward the receiver, the frequency of the received wave appears to have increased; if the source moves away from the receiver, the frequency of the received wave appears to have decreased. The change in frequency is proportional to the velocity of the source relative to the receiver. This phenomenon is known as the *Doppler effect*. Now consider a monochromatic (single-frequency) light wave of frequency f (say, 5×10^{14} Hz) emitted by a laser source. If this ray is reflected by a target object and received by a light detector, the frequency of the received wave would be

$$f_2 = f + \Delta f \tag{3.44}$$

The frequency increase Δf will be proportional to the velocity v of the target object, which is assumed positive when moving toward the light source. Hence,

$$\Delta f = cv \tag{3.45}$$

Now by comparing the frequency f_2 of the reflected wave with the frequency

$$f_1 = f \tag{3.46}$$

of the original wave, we can determine Δf and, hence, the velocity v of the target object.

The change in frequency Δf due to the Doppler effect can be determined by observing the fringe pattern due to light wave interference. To understand this, consider the two waves

$$v_1 = a \sin 2\pi f_1 \tag{3.47}$$

and

$$v_2 = a \sin 2\pi f_2 \tag{3.48}$$

If we add these two waves, the resulting wave would be

$$v = v_1 + v_2 = a(\sin 2\pi f_1 + \sin 2\pi f_2)$$

which can be expressed as

$$v = 2a \sin \pi(f_2 + f_1) \cos \pi(f_2 - f_1) \tag{3.49}$$

It follows that the combined signal beats at the beat frequency $\Delta f / 2$. When f_2 is very close to f_1 (i.e., when Δf is small compared to f), these beats will appear as dark and light lines (fringes). This is known as *wave interference*. Note that Δf can be determined by two methods:

1. By measuring the spacing of the fringes
2. By counting the beats in a given time interval or by timing successive beats using a high-frequency clock signal

The velocity of the target object is determined in this manner. Displacement can be obtained simply by digital integration (or by accumulating the count). A schematic diagram for the laser Doppler interferometer is shown in figure 3.28. Industrial interferometers usually employ a helium-neon laser that has waves of two frequencies close together. In that case, the arrangement shown in figure 3.28 has to be modified to take into account the two frequency components.

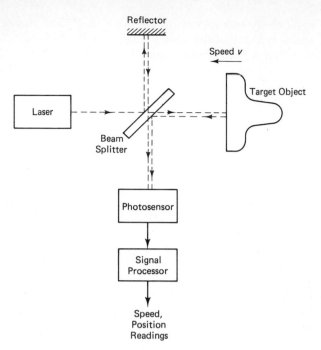

Figure 3.28. A laser Doppler interferometer for measuring velocity and displacement.

Ultrasonic Sensors

Ultrasound waves are pressure waves, just like sound waves, but their frequencies are higher than the audible frequencies. Ultrasonic sensors are used in many applications, including medical imaging, ranging systems for cameras with autofocusing capability, level sensing, and speed sensing. Ultrasound can be generated according to several principles. For example, high-frequency (gigahertz) oscillations in piezoelectric crystals subjected to electrical potentials are used to generate very high-frequency ultrasound. Another method is to use the magnetostrictive property of ferromagnetic material. Ferromagnetic materials deform when subjected to magnetic fields. Resonant oscillations generated by this principle can produce ultrasonic waves. Another method of generating ultrasound is to apply a high-frequency voltage to a metal-film capacitor. A microphone can serve as an ultrasound detector (receiver).

In distance (proximity, displacement) measurement using ultrasound, a burst of ultrasound is projected at the target object, and the time taken for the echo to be received is clocked. A signal processor computes the position of the target object, possibly compensating for environmental conditions. This configuration is shown in figure 3.29. Alternatively, the velocity of the target object can be measured, using the Doppler effect, by measuring (clocking) the change in frequency between the transmitted wave and the received wave. Position measurements with fine resolution (e.g., a fraction of a millimeter) can be provided using the ultrasonic method. Since the speed of ultrasonic wave propagation depends on the temperature of the medium

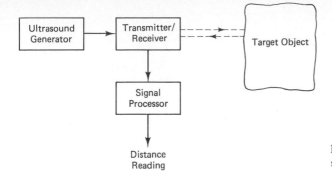

Figure 3.29.   An ultrasonic position sensor.

(typically air), errors will enter into ultrasonic readings unless the sensor is compensated for temperature variations.

Gyroscopic Sensors

Consider a rigid body spinning about an axis at angular speed ω. If the moment of inertia of the body about that axis is J, the angular momentum H about the same axis is given by

$$H = j\omega \tag{3.50}$$

Newton's second law (torque = rate of change of angular momentum) tells us that to rotate (precess) the spinning axis slightly, a torque has to be applied, because precession causes a change in the spinning angular momentum vector (the magnitude remains constant but the direction changes), as shown in figure 3.30a. This is the principle of operation of a gyroscope. Gyroscopic sensors are commonly used in control systems for stabilizing vehicle systems.

Consider the gyroscope shown in figure 3.30b. The disk is spun about frictionless bearings using a torque motor. Since the gimbal (the framework on which the disk is supported) is free to turn about frictionless bearings on the vertical axis, it will remain fixed with respect to an inertial frame, even if the bearing housing (the main structure in which the gyroscope is located) rotates. Hence, the relative angle between the gimbal and the bearing housing (angle θ in the figure) can be measured, and this gives the angle of rotation of the main structure. In this manner, angular displacements in systems such as aircraft, space vehicles, ships, and land vehicles can be measured and stabilized with respect to an inertial frame. Note that bearing friction introduces an error that has to be compensated for, perhaps by recalibration before a reading is taken.

The *rate gyro*—the same arrangement shown in figure 3.30b, with a slight modification—can be used to measure angular speeds. In this case, the gimbal is not free; it is restrained by a torsional spring. A viscous damper is provided to suppress any oscillations. By analyzing this gyro as a mechanical tachometer (see example 2.3 and problem 3.31), we will note that the relative angle of rotation θ gives the angular speed of the structure about the gimbal axis.

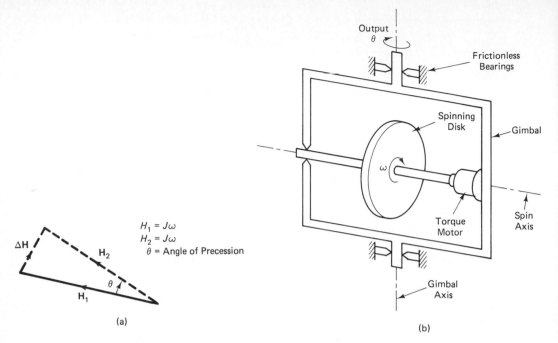

$H_1 = J\omega$
$H_2 = J\omega$
θ = Angle of Precession

(a)

(b)

Figure 3.30. (a) Illustration of the gyroscopic torque needed to change the direction of an angular momentum vector. (b) A simple single-axis gyroscope for sensing angular diplacements.

3.10 A DESIGN CRITERION FOR CONTROL SYSTEMS

In the beginning of this chapter, we stated that a general design criterion can be proposed for feedback control systems in terms of the accuracy of the measurement system. Since accuracy is affected by parameter changes in the control system components and by the influence of external disturbances, we are able to develop a preliminary design criterion simply by analyzing the sensitivity of a general feedback control system to parameter changes and to external disturbances.

Consider the block diagram of a typical feedback control system, shown in figure 3.31. In the usual notation:

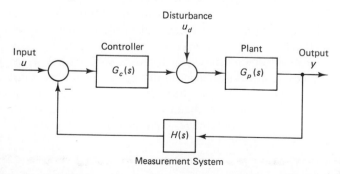

Figure 3.31. Block diagram representation of a feedback control system.

Analog Sensors for Motion Measurement   Chap. 3

$G_p(s)$ = transfer function of the plant (or the system to be controlled)

$G_c(s)$ = transfer function of the controller (including compensators)

$H(s)$ = transfer function of the output feedback system (including the measurement system)

u = system input command

u_d = external disturbance input

y = system output

Using straightforward algebraic manipulation, it can be shown that the following input/output relationship holds:

$$y = \left[\frac{G_c G_p}{1 + G_c G_p H} \right] u + \left[\frac{G_p}{1 + G_c G_p H} \right] u_d \qquad (3.51)$$

The closed-loop transfer function $\tilde{G}$ is given by y/u, with $u_d = 0$. Thus,

$$\tilde{G} = \frac{G_c G_p}{[1 + G_c G_p H]} \qquad (3.52)$$

The sensitivity of the system to a change in some parameter K may be expressed as the ratio of the change in output to the change in the parameter, or $\Delta y / \Delta K$. In the nondimensional form, this sensitivity is given by

$$S_k = \frac{K}{y} \frac{\Delta y}{\Delta K} \qquad (3.53)$$

Since $y = \tilde{G} u$, with $u_d = 0$, it follows that for a given input u,

$$\frac{\Delta y}{y} = \frac{\Delta \tilde{G}}{\tilde{G}}$$

Consequently, equation 3.53 may be expressed as

$$S_k = \frac{K}{\tilde{G}} \frac{\Delta \tilde{G}}{\Delta K} \qquad (3.54)$$

or, in the limit,

$$S_k = \frac{K}{\tilde{G}} \frac{\partial \tilde{G}}{\partial K} \qquad (3.55)$$

Now, from equation 3.52, we are able to determine expressions for the control system sensitivity to changes in various components in the control system. Specifically, by straightforward partial differentiation of equation 3.52, we get

$$S_{Gp} = \frac{1}{[1 + G_c G_p H]} \qquad (3.56)$$

$$S_{Gc} = \frac{1}{[1 + G_c G_p H]} \qquad (3.57)$$

$$S_H = -\frac{G_c G_p H}{[1 + G_c G_p H]} \tag{3.58}$$

It is clear from these three relations that as the gain of the loop (i.e., $G_c G_p H$, with $s = 0$) is increased, the sensitivity of the control system to changes in the plant and controller decreases, but the sensitivity to changes in the feedback (measurement) system becomes (negative) unity. Furthermore, it is clear from equation 3.51 that the effect of the disturbance input can be reduced by increasing the gain of $G_c H$. By combining these observations, the following design criterion can be stipulated for feedback control systems:

1. Make the measurement system (H) very accurate and stable.
2. Increase the loop gain to reduce the sensitivity of the control system to changes in the plant and controller.
3. Increase the gain of $G_c H$ to reduce the influence of external disturbances.

In practical situations, the plant G_p is usually fixed and cannot be modified. Furthermore, once an accurate measurement system is chosen, H is essentially fixed. Hence, most of the design freedom is available with respect to G_c only. It is virtually impossible to achieve all the design requirements simply by increasing the gain of G_c. The dynamics (i.e., the entire transfer function) of G_c also have to be properly designed in order to obtain the desired performance of a control system.

PROBLEMS

3.1. Giving examples, discuss situations in which measurement of more than one type of kinematic variables using the same measuring device is
(a) An advantage
(b) A disadvantage

3.2 Giving examples for suitable auxiliary front-end elements, discuss the use of a force sensor to measure
(a) Displacement
(b) Velocity
(c) Acceleration

3.3 Write the expression for loading nonlinearity error (percentage) in a rotary potentiometer in terms of the angular displacement, maximum displacement (stroke), potentiometer element resistance, and load resistance. Plot the percentage error as a function of the fractional displacement for the three cases $R_L/R_c = 0.1$, 1.0, and 10.0.

3.4. Determine the angular displacement of a rotary potentiometer at which the loading nonlinearity error is the largest.

3.5. A potentiometer circuit with element resistance R_c and equal end resistors R_e is shown in figure P3.5. Derive the necessary input/output relations. Show that the end resistors can produce a linearizing effect in the potentiometer. At half the maximum reading of the potentiometer shown in figure P3.5, calculate the percentage loading error for the three values of the resistance ratio $R_c/R_e = 0.1$, 1.0, and 10.0, assuming that the load resistance R_L is equal to the element resistance. Compare the results with the corresponding value for $R_e = 0$. Finally, choose a suitable value for R_c/R_e and plot the curve of percentage loading error versus fractional displacement x/x_{max}. From the graph, estimate the maximum loading error.

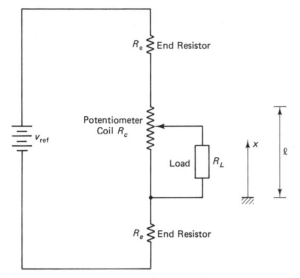

Figure P3.5. Potentiometer circuit with end resistors.

3.6. Derive an expression for the sensitivity of a rotatory potentiometer as a function of displacement. Plot the corresponding curve in the nondimensional form for the three load values given by $R_L/R_c = 0.1$, 1.0, and 10.0. Where does the maximum sensitivity occur? Verify your observation using the analytical expression.

3.7. The data acquisition system connected at the output of a differential transformer (say, an LVDT) has a very high resistive load. Obtain an expression for the phase lead of the output signal (at the load) of the differential transformer, with reference to the supply to the primary winding of the transformer, in terms of the impedance of the primary winding only.

3.8. At the null position, the impedances of the two secondary winding segments of an LVDT were found to be equal in magnitude but slightly unequal in phase. Show that the quadrature error (null voltage) is about 90° out of phase with reference to the predominant component of the output signal under open-circuit conditions. *Hint:* This may be proved either analytically or graphically by considering the difference between two rotating directed lines (phasors) that are separated by a very small angle.

3.9. Discuss factors that limit the lower frequency and upper frequency limits of the output from the following measuring devices:
(a) Potentiometer
(b) LVDT
(c) Resolver
(d) Eddy current proximity sensor
(e) DC tachometer
(f) Piezoelectric transducer

3.10. Joint angles and angular speeds are the two basic measurements used in the low-level control of robotic manipulators. One type of robot arm uses resolvers to measure angles and differentiate these signals (digitally) to obtain angular speeds. A gear system is used to step up the measurement (typical gear ratio, 1:8). Since the gear wheels are ferromagnetic, an alternative measuring device would be a self-induction or mutual-induction proximity sensor located at a gear wheel. This arrangement, known as a pulse tachometer, generates a pulse (or near-sine) signal, which can be used to determine both angular displacement and angular speed. Discuss the advantages and disadvantages of these two arrangements (resolver and pulse tachometer) in this particular application.

3.11. An active suspension system is proposed for a high-speed ground transit vehicle in order to achieve improved ride quality. The system senses jerk (rate of change of acceleration) due to road disturbances and adjusts system parameters accordingly.

    (a) Draw a suitable schematic diagram for the proposed control system and describe appropriate measuring devices.

    (b) Suggest a way to specify the "desired" ride quality for a given type of vehicle. (Would you specify one value of jerk, a jerk range, or a curve with respect to time or frequency?)

    (c) Discuss the drawbacks and limitations of the proposed control system with respect to such factors as reliability, cost, feasibility, and accuracy.

3.12. A vibrating system has an effective mass M, an effective stiffness K, and an effective damping constant B in its primary mode of vibration at point A with respect to coordinate y. Write expressions for the undamped natural frequency, the damped natural frequency, and the damping ratio for this first mode of vibration of the system. A displacement transducer is used to measure the fundamental undamped natural frequency and the damping ratio of the system by subjecting the system to an initial excitation and recording the displacement trace at a suitable location (point A along y) in the system. This trace will provide the period of damped oscillations and the logarithmic decrement of the exponential decay from which the required parameters can be computed using well-known relations. It was found, however, that the mass m of the moving part of the displacement sensor and the associated equivalent viscous damping constant b are not negligible. Using the model shown in figure P3.12, derive expressions for the measured undamped natural frequency and damping ratio. Suppose that $M = 10$ kg, $K = 10$ N/m, and $B = 2$ N/m/s. Consider an LVDT whose core weighs 5 gm and has negligible damping and a potentiometer whose slider arm weighs 5 gm and has an equivalent viscous damping constant of 0.05 N/m/s. Estimate the percentage error of the results for the undamped natural frequency and damping ratio measured using each of these two displacement sensors.

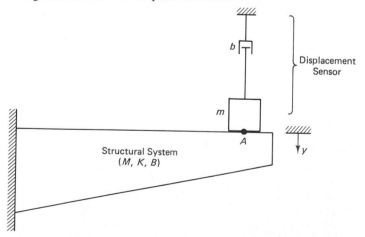

Figure P3.12. The use of a displacement sensor to measure the natural frequency and damping ratio of a structure.

3.13. Why is motion sensing important in trajectory-following control of robotic manipulators? Identify five types of motion sensors that could be used in robotic manipulators.

3.14. Show that the quadrature signals v_{o1} and v_{o2} from a resolver can be demodulated according to

$$\omega = \sqrt{\frac{\dot{v}_{o1}^2 + \dot{v}_{o2}^2}{v_{o1}^2 + v_{o2}^2}}$$

This expression can be used to determine the angular speed. What is the main disadvantage of this method?

3.15. Compare and contrast the principles of operation of the DC tachometer and the AC tachometer (both permanent-magnet and induction types). What are the advantages and disadvantages of these two types of tachometers?

3.16. A design ojective in most control system applications is to achieve small time constants. An exception is the time constant requirements for a piezoelectric sensor. Explain why a large time constant, on the order of 10 s, is desirable for piezoelectric sensors. An equivalent circuit for a piezoelectric accelerometer that uses a quartz crystal as the sensing element is shown in figure P3.16. The charge generated is denoted by q, and the voltage output at the end of the accelerometer cable is v_o. The sensor capacitance is modeled by C, and the overall capacitance experienced at the sensor output, whose primary contribution is due to cable capacitance, is denoted by C_c. The resistance of the electric insulation in the accelerometer is denoted by R. Write a differential equation relating v_o to q. What is the corresponding transfer function? Using this result, show that the accuracy of the accelerometer improves when the sensor time constant is large and when the frequency of the measured acceleration is high. For a quartz crystal sensor with $R = 1 \times 10^{11}$ Ω and $C_c = 1,000$ pF, compute the time constant.

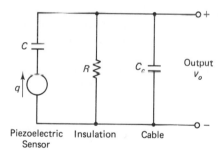

Piezoelectric   Insulation   Cable
Sensor

Figure P3.16. Equivalent circuit for a quartz crystal (piezoelectric) accelerometer.

3.17. Level sensors are used in a wide variety of applications, including soft drink bottling, food packaging, and monitoring of storage vessels, mixing tanks, and pipelines. Consider the following types of level sensors, and briefly explain the principle of operation of each type in level sensing. Also, what are the limitations of each type?
  (a) Capacitive sensors
  (b) Inductive sensors
  (c) Ultrasonic sensors
  (d) Vibration sensors

3.18. Applications of accelerometers are found in the following areas:
  (a) Transit vehicles (automobiles, aircraft, ships, etc.)
  (b) Power cable monitoring

(c) Robotic manipulator control

(d) Building structures

(e) Shock and vibration testing

(f) Position and velocity sensing

Describe one direct use of acceleration measurement in each application area.

3.19. A standard accelerometer that weighs 100 gm is mounted on a test object that has an equivalent mass of 3 kg. Estimate the accuracy in the first natural frequency of the object measured using this arrangement, considering mechanical loading due to accelerometer mass alone. If a miniature accelerometer that weighs 0.5 gm is used instead, what is the resulting accuracy? A strain gage accelerometer uses a semiconductor strain gage mounted at the root of a cantilever element, with the seismic mass mounted at the free end of the cantilever. Suppose that the cantilever element has a square cross-section with dimensions 1.5×1.5 mm^2. The equivalent length of the cantilever element is 25 mm, and the equivalent seismic mass is 0.2 gm. If the cantilever is made of an aluminum alloy with Young's modulus $E = 69 \times 10^9$ N/m^2, estimate the useful frequency range of the accelerometer in hertz. *Hint:* When force F is applied to the free end of a cantilever, the deflection y at that location may be approximated by the formula

$$y = \frac{F\ell^3}{3EI}$$

where ℓ = cantilever length

I = second moment area of the cantilever cross-section about the bending axis = $1/12\ bh^3$

b = cross-section width

h = cross-section height

3.20. Standard rectilinear displacement sensors such as the LVDT and the potentiometer are used to measure displacements up to 10 in.; within this limit, accuracies as high as ± 0.2 percent can be obtained. For measuring large displacements on the order of 100 in., cable extension displacement sensors that have an angular displacement sensor as the basic sensing unit could be used. One type of rectilinear displacement sensor has a rotatory potentiometer and a light cable that wraps around a spool that rotates with the wiper arm of the pot. In using this sensor, the free end of the cable is connected to the moving member whose displacement is to be measured. The sensor housing is mounted on a stationary platform, such as the support structure of the system being monitored. A spring motor winds the cable back as the cable retracts. Using suitable sketches, describe the operation of this displacement sensor. Discuss the shortcomings of this device.

3.21. It is known that some of the factors that should be considered in selecting an LVDT for a particular application are linearity, sensitivity, response time, size and weight of core, size of the housing, primary excitation frequency, output impedance, phase change between primary and secondary voltages, null voltage, stroke, and environmental effects (temperature compensation, magnetic shielding, etc.). Explain why and how each of these factors is an important consideration.

3.22. The signal-conditioning system for an LVDT has the following components: power supply, oscillator, synchronous demodulator, filter, and voltage amplifier. Using a block diagram, show how these components are connected to the LVDT. Describe the purpose of each component. A high-performance LVDT has a linearity rating of 0.01

percent in its output range of 0.1–1.0 V AC. The response time of the LVDT is known to be 10 ms. What should be the frequency of the primary excitation?

3.23. Describe three different types of proximity sensors. In some applications, it is required to sense only two-state positions (e.g., presence or absence, go or no-go). Proximity sensors can be used in such applications, and in that context they are termed proximity switches. For example, consider a parts-handling application in automated manufacturing in which a robot end effector grips a part and picks it up to move it from a conveyor to a machine tool. We can identify four separate steps in the gripping process. Explain how proximity switches can be used for sensing in each of these four tasks:
  (a) Make sure that the part is at the expected location on the conveyor.
  (b) Make sure that the gripper is open.
  (c) Make sure that the end effector has moved to the correct location so that the part is in between the gripper fingers.
  (d) Make sure that the part did not slip when the gripper was closed.

3.24. Discuss the relationships among displacement sensing, distance sensing, position sensing, and proximity sensing. Explain why the following characteristics are important in using some types of motion sensors:
  (a) Material of the moving (or target) object
  (b) Shape of the moving object
  (c) Size (including mass) of the moving object
  (d) Distance (large or small) of the target object
  (e) Nature of motion (transient or not, what speed, etc.) of the moving object
  (f) Environmental conditions (humidity, temperature, magnetic fields, dirt, lighting conditions, shock and vibration, etc.)

3.25. Consider the following types of position sensors: inductive, capacitive, eddy current, fiber optic, and ultrasonic. For the following conditions, indicate which of these types are not suitable and explain why.
  (a) Environment with variable humidity
  (b) Target object made of aluminum
  (c) Target object made of steel
  (d) Target object made of plastic
  (e) Target object several feet away from the sensor location
  (f) Environment with significant temperature fluctuations
  (g) Smoke-filled environment

3.26. In some industrial processes, it is necessary to sense the condition of a system at one location and, depending on that condition, activate an operation at a location far from that location. For example, in a manufacturing environment, when the count of finished parts exceeds some value, as sensed in the storage area, a milling machine could be shut down or started. A proximity switch could be used for sensing, but since activation of the remote process usually requires a current that is larger than the rated load of a proximity switch, one would have to use a relay circuit that is operated by the proximity switch. One such arrangement is shown in figure P3.26. Note that the relay circuit can be used to operate a device such as a valve, a motor, a pump, or a heavy-duty switch. Discuss an application of the arrangement shown in figure P3.26 in the food-packaging industry. A mutual-induction proximity sensor with the following ratings is used in this application:

• Sensor diameter = 1 cm
• Sensing distance = 1 mm

- Supply to primary winding = 110 AC at 60 Hz
- Load current rating (in secondary) = 200 mA

Discuss the limitations of this proximity sensor.

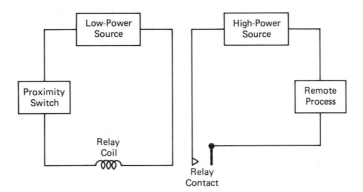

Figure P3.26. Proximity switch–operated relay circuit.

3.27. Applications of piezoelectric sensors are numerous: pushbutton devices and switches, pressure and force sensing, robotic tactile sensing, accelerometers, glide testing of computer disk drive heads, excitation sensing in dynamic testing, respiration sensing in medical diagnostics, and graphics input devices for computers. Discuss the advantages and disadvantages of piezoelectric sensors. What is cross-sensitivity of a sensor? Indicate how the anisotropy of piezoelectric crystals (i.e., charge sensitivity quite large along one particular axis) is useful in reducing cross-sensitivity problems in a piezoelectric sensor.

3.28. Compression molding is used in making parts of complex shapes and varying sizes. Typically, the mold consists of two platens, the bottom platen fixtured to the press table and the top platen operated by a hydraulic press. Metal or plastic sheets—for example, for the automotive industry—can be compression-molded in this manner. The main requirement in controlling the press is to position the top platen accurately with respect to the bottom platen (say, with a 0.001 in. tolerance), and it has to be done quickly (say, in a few seconds). How many degrees of freedom have to be sensed (how many position sensors are needed) in controlling the mold? Suggest typical displacement measurements that would be made in this application and the types of sensors that could be employed. Indicate sources of error that cannot be perfectly compensated for in this application.

3.29. Seam tracking in robotic arc welding needs accurate position control under dynamic conditions. The welding seam has to be followed by the welding torch accurately. Typically, the position error should not exceed 0.2 mm. A proximity sensor could be used for sensing the gap between the welding torch and the welded part. It is necessary to install the sensor on the robot end effector so that it tracks the seam at some distance (typically 1 in.) ahead of the welding torch. Explain why this is important. If the speed of welding is not constant and the distance between the torch and the proximity sensor is fixed, what kind of compensation would be necessary in controlling the end effector position? Sensor sensitivity of several volts per millimeter is required in this position control application. What type of proximity sensor would you recommend?

3.30. Discuss advantages and disadvantages of fiber optic sensors. Consider the fiber optic position sensor shown in figure 3.27. Sketch the curve of light intensity received versus x. In which region of this curve would you prefer to operate the sensor, and what are the corresponding limitations?

3.31. Using a suitable sketch, analyze a single-axis rate gyro. Obtain a relationship between the gimbal angle θ and the angular velocity Ω of the mounting structure (say, a missile) about the gimbal axis. Use the following parameters:

J = moment of inertia of the gyroscopic disk about the spinning axis

ω = angular speed of spin

k = torsional stiffness of the gimbal restraint

Assume that Ω is constant and the conditions are steady. How would you improve the sensitivity of this device? Discuss any problems associated with the suggested methods of sensitivity improvement and ways to reduce them.

3.32. As a result of advances in microelectronics, piezoelectric sensors (such as accelerometers and impedance heads) are now available with built-in charge amplifiers in a single integral package. When such units are employed, additional signal conditioning is usually not necessary. An external power supply unit is needed, however, to provide power for the amplifier circuitry. Discuss the advantages and disadvantages of a piezoelectric sensor with built-in microelectronics for signal conditioning. A piezoelectric accelerometer is connected to a charge amplifier. An equivalent circuit for this arrangement is shown in figure 3.25.
   (a) Obtain a differential equation for the output v_o of the charge amplifier, with acceleration a as the input, in terms of the following parameters: S_a = charge sensitivity of the accelerometer (charge/acceleration); R_f = feedback resistance of the charge amplifier; τ_c = time constant of the system (charge amplifier).
   (b) If an acceleration pulse of magnitude a_o and duration T is applied to the accelerometer, sketch the time response of the amplifier output v_o. Show how this response varies with τ_c. Using this result, show that larger the τ_c, the more accurate the measurement.

3.33. Give typical values for the output impedance and the time constant of the following measuring devices:
   (a) Potentiometer
   (b) Differential transformer
   (c) Resolver
   (d) Piezoelectric accelerometer
   A resistance temperature detector (RTD) has an output impedance of 500 Ω. If the loading error has to be maintained near 5 percent, estimate a suitable value for the load impedance.

3.34. The manufacturer of an ultrasonic gage states that the device has applications in measuring cold roll steel thickness, determining parts positions in robotic assembly, lumber sorting, measurement of particle board and plywood thickness, ceramic tile dimensional inspection, sensing the fill level of food in a jar, pipe diameter gaging, rubber tire positioning during fabrication, gaging of fabricated automotive components, edge detection, location of flows in products, and parts identification. Discuss whether the following types of sensors are also equally suitable for some or all of the foregoing applications. In each case where you think that a particular sensor is not suitable for a given application, give reasons to support your claim.

(a) Fiber optic position sensors

(b) Self-induction proximity sensors

(c) Eddy current proximity sensors

(d) Capacitive gages

(e) Potentiometers

(f) Differential transformers

3.35. An angular motion sensor that operates somewhat like a conventional resolver has been developed at Wright State University. The rotor of this resolver is a permanent magnet. A $\frac{3}{4}$ in. diameter Alnico-2 disk magnet, diametrically magnetized as a two-pole rotor, has been used. Instead of the two sets of stationary windings placed at 90° in a conventional resolver, two Hall-effect sensors (see chapter 5) placed at 90° around the permanent-magnet rotor are used for detecting quadrature signals. Describe the operation of this modified resolver and explain how this device could be used to measure angular motions continuously. Compare this device with a conventional resolver, giving advantages and disadvantages.

3.36. The optical potentiometer shown schematically in figure P3.36 is a displacement sensor. A layer of photoresistive material is sandwiched between a layer of regular resistive material and a layer of conductive material. The layer of resistive material has a total resistance of R_c, and it is uniform (i.e., it has a constant resistance per unit length). The photoresistive layer is practically an electrical insulator when no light is projected on it. The displacement of the moving object (whose displacement is being measured) causes a moving light beam to be projected on a rectangular area of the photoresistive layer. This light-projected area attains a resistance of R_p, and this resistance joins the resistive layer that is above the photoresistive layer and the conductive layer that is below the photoresistive layer. The supply voltage to the potentiometer is v_{ref}, and the length of the resistive layer is L. If the light spot is projected at a distance x from one end of the resistive element, as shown in figure P3.36, obtain an expression for the output voltage v_o in terms of x, L, R_c, R_p, and v_{ref}, assuming that the output is open-circuit. If R_p is much larger than R_c, show that v_o is proportional to x.

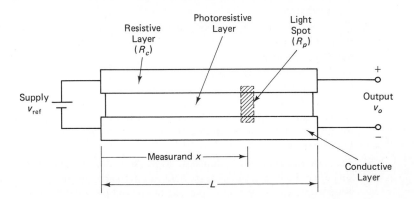

Figure P3.36. An optical potentiometer for displacement sensing.

REFERENCES

BECKWITH, T. G., BUCK, N. L., and MARANGONI, R. D. *Mechanical Measurements,* 3d ed. Addison-Wesley, Reading, Mass., 1982.

DALLY, J. W., RILEY, W. F., and McCONNELL, K. G. *Instrumentation for Engineering Measurements*. Wiley, New York, 1984.

deSILVA, C. W. *Dynamic Testing and Seismic Qualification Practice*. Lexington Books, Lexington, Mass., 1983.

————."Motion Sensors in Industrial Robots." *Mechanical Engineering* 107(6): 40–51, June 1985.

DOEBELIN, E. O. *Measurement Systems,* 3d ed. McGraw-Hill, New York, 1983.

HERCEG, E. E. *Handbook of Measurement and Control*. Schaevitz Engineering, Pennsauken, N.J., 1972.

LENK, J. D. *Handbook of Microcomputer-Based Instrumentation and Controls*. Prentice-Hall, Englewood Cliffs, N.J., 1984.

Sensors 3(1–12). North American Technology, Peterborough, N.H., 1986.

4

Torque, Force, and Tactile Sensors

4.1 INTRODUCTION

The response of a mechanical system depends on forces and torques applied to the system. Many applications exist in which process performance is specified in terms of forces and torques. Examples include machine-tool operations, such as grinding, cutting, forging, extrusion, and rolling; manipulator tasks, such as parts handling, assembly, engraving, and general fine manipulation; and actuation tasks, such as locomotion. The forces and torques present in dynamic systems are generally functions of time. Performance monitoring and evaluation, failure detection and diagnosis, testing, and control of mechanical systems can depend heavily on accurate measurement of associated forces and torques. One example in which force (and torque) sensing can be very useful is a drilling robot. The drill bit is held at the end effector by the gripper of the robot, and the work piece is rigidly fixed to a support structure by clamps. Although a displacement sensor (such as a potentiometer or a differential transformer) can be used to measure drill motion in the axial direction, this alone does not determine the drill performance. Depending on the material properties of the work piece (e.g., hardness) and the nature of the drill bit (e.g., degree of wear), a small misalignment or slight deviation in feed (axial movement) or speed (rotational speed of the drill) can create large normal (axial) and lateral forces and resistance torques. This can create problems such as excessive vibrations, uneven drilling, excessive tool wear, and poor product quality and eventually may lead to a major mechanical failure. Sensing of axial force or motor torque, for example, and using the information to adjust process variables (speed, feed rate, etc.), or even to provide warning signals and eventually stop the process, can significantly improve the system performance. Another example in which force sensing is useful is in *nonlinear feedback control* of mechanical systems such as robotic manipulators. This application will be discussed later in this chapter.

Since both force and torque are *effort variables*, the term *force* may be used to represent both these variables. This generalization is adopted in this chapter except when discrimination might be necessary—for example, when discussing specific applications.

4.2 FORCE CONTROL

One important application of force (and torque) sensing is in the area of control. Since forces are variables in a mechanical system, their measurement can lead to effective control. There are applications in which force control is invaluable. This is particularly evident in situations where a small error in motion can lead to the generation of large forces, which is the case, for example, in parts assembly operations. In assembly, a slight misalignment (or position error) can cause jamming and generation of damaging forces. As another example, consider precision machining of a hard work piece. A slight error in motion could generate large cutting forces, which might lead to unacceptable product quality or even to rapid degradation of the machine tool. In such situations, measurement and control of forces seem to be an effective way to improve the system performance. In this section, we shall address the force control problem from a generalized and unified point of view. The concepts introduced in this section will be illustrated further by examples throughout the chapter.

Force-Motion Causality

The response of a mechanical control system is not always limited to motion variables. When the objective of a control system is to produce a desired motion, the response variables (outputs) are the associated motion variables. A good example is the response of a spray-painting robot whose end effector is expected to follow a specific trajectory of motion. On the other hand, when the objective is to exert a desired set of forces (or torques)—which is the case in some machining, forging, gripping, engraving, and assembly tasks—the outputs are the associated force variables.

A lumped-parameter mechanical system can be treated as a set of basic mechanical elements (particularly springs, masses, dampers, levers, gyros, force sources, and velocity sources) that are interconnected through ports (or *bonds*) through which power (or energy) flows. Each port actually consists of two terminals. There is an effort variable (force) f and a flow variable (motion, such as velocity) v associated with each port. In particular, consider a port that connects two subsystems A and B, as shown in figure 4.1a. The two subsystems interact, and power flows through the port. Hence, for example, if we assume that A pushes B with a force f, then f is considered input to B. Now, B responds with motion v—the output of B—which, in turn, causes A to respond with force f. Consequently, the input to A is v and the output of A is f. This cause-effect relationship (*causality*) is clearly shown by the block diagram representation in figure 4.1b. A similar argument would lead to the opposite causality if we had started with the assumption that B is

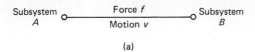

(a)

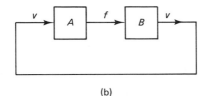

(b)

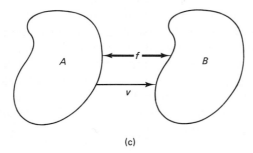

(c)

Figure 4.1. Force-motion causality: (a) bond or port representation; (b) block diagram representation; (c) free-body diagram representation.

pushing A with force f. It follows that one has to make some causal decisions for a dynamic system, based on system requirements, and the rest of the causalities will be decided automatically. It can be shown that conflicts in causality indicate that the energy storage elements (e.g., springs, masses) in the particular system model are not independent. Note that causality is not governed by the positive direction assigned to the variables f and v. This is clear from the free-body diagram representation in figure 4.1c. In particular, if the force f on A (the action) is taken as positive in one direction, then according to Newton's third law, the force f on B (the reaction) is positive in the opposite direction. Similarly, if the velocity v of B relative to A is positive in one direction, then the velocity v of A relative to B is positive in the opposite direction. These directions are unrelated to the input/output (causality) directions. The foregoing discussion shows that forces can be considered inputs (excitations) as well as outputs (responses), depending on the functional requirements of a particular system.

Force Control Problems

Some forces in a control system are actuating (input) forces, and some others are response (output) forces. For example, torques (or forces) driving the joints of a robotic manipulator are considered inputs to the robot. (From the control point of view, however, joint input is the voltage applied to the motor drive amplifier, which produces a field current that generates the motor torque that is resisted by the load torque at the joint.) Gripping or tactile (touch) forces at the end effector of a manipulator, tool tip forces in a milling machine, and forces at the die in a forging machine can be considered output forces that are "exerted" on objects in the outside environment of the system. Output forces should be completely determined by the inputs (motions as well as forces) to the system, however. Unknown input forces and output forces can be measured using appropriate force sensors, and force control may be implemented using these measurements. Three basic types of force control will be discussed here.

Force feedback control.   Consider the system shown in figure 4.2, which is connected to its environment through two ports. Force variable f_A and motion variable v_A (both in the same direction) are associated with port A, and force variable f_B and motion variable v_B (in the same direction) are associated with port B. Suppose that f_A is an input to the system and f_B is an output. It follows that v_B is also an input to the system. In other words, the motion at B is constrained. For example, the point might be completely restrained in the direction of f_B, resulting in the constraint $v_B = 0$. Ideally, if we knew the dynamic behavior of the system—and assuming that there are no extraneous inputs (disturbances and noise)—we would be able to determine analytically the input f_A that will generate a desired f_B for a specified v_B. Then we could control the output force f_B simply by supplying the predetermined f_A and by subjecting B to the specified motion v_B. The corresponding open-loop configuration is shown in figure 4.3a. Since it is practically impossible to achieve accurate system performance with this open-loop control arrangement (inappropriately known as feedforward control), except in a few simple situations, the feedback control loop shown in figure 4.3b has to be added. In this case, the response force f_B is measured and fed back into the controller that will modify the control input signals according to a suitable control law, so as to correct any deviations in f_B from the desired value.

Feedforward force control.   Once again, consider the system shown in figure 4.2, which has two ports, A and B. Here again, suppose that f_A is an actuating (input) force that can be generated according to a given specification (a "known" input). But suppose that f_B is an unknown input force. It could be a disturbance force, such as that resulting from a collision, or a useful force, such as a gripping force whose value is not known. This configuration is shown in figure 4.4a. Since f_B is unknown, it might not be possible to control the system response (v_A and v_B) accurately. One solution is to measure the unknown force f_B, using a suitable force sensor, and feed it forward into the controller. The controller can use this additional

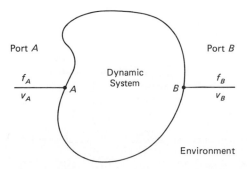

Figure 4.2.   A two-port system.

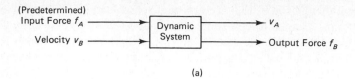

(a)

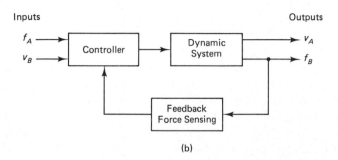

(b)

Figure 4.3. (a) An open-loop (feedforward control) system. (b) A force feedback control system.

information to compensate for the influence of f_B on the system and produce the desired response. This is an example of feedforward control. Sometimes, if an input force to a system (e.g., a joint force or torque of a robotic manipulator) is computed using an analytical model and is supplied to the actuator, the associated control is inappropriately termed feedforward control. It should be termed *computed force/ torque control* or *computed input control,* to be exact.

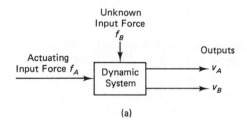

(a)

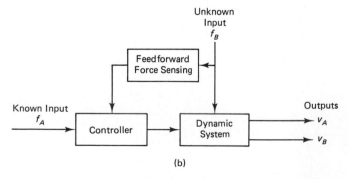

(b)

Figure 4.4. (a) A system with an unknown input force. (b) Feedforward force control.

Torque, Force, and Tactile Sensors   Chap. 4

Impedance control. Consider a mechanical operation in which we push against a spring that has constant stiffness. Here, the value of the force completely determines displacement; similarly, the value of displacement completely determines the force. It follows that, in this example, we are unable to control force and displacement independently at the same time. Also, it is not possible, in this example, to apply a command force that has an arbitrary relationship to displacement. In other words, *stiffness control* is not possible. Now suppose that we push against a complex dynamic system, not a simple spring element. In this case, we should be able to command a pushing force in response to the displacement of the dynamic system so that the ratio of force to displacement varies in a desired manner. This is a stiffness control (or *compliance control*) action. Dynamic stiffness is defined as the ratio (output force)/(input displacement), expressed in the frequency domain (see table 2.2). Mechanical impedance is defined as the ratio (output force)/(input velocity) in the frequency domain. Note that stiffness and impedance both relate force and motion variables in a mechanical system. The objective of impedance control is to make the impedance function equal to some specified function (without separately controlling or independently constraining the associated force variable and velocity variable). Force control and motion control can be considered extreme cases of impedance control (and stiffness control). Specifically, since the objective of force control is to keep the force variable from deviating from a desired level, in the presence of independent variations of the associated motion variable (an input), force control can be considered zero-impedance control if velocity is chosen as the motion variable (or zero-stiffness control if displacement is chosen as the motion variable). Similarly, displacement control can be considered infinite-stiffness control and velocity control can be considered infinite-impedance control.

Impedance control has to be accomplished through active means, in general, by generating forces as specified functions of associated displacements. Impedance control is particularly useful in mechanical manipulation against physical constraints, which is the case in assembly and machining tasks. In particular, very high impedance is naturally present in the direction of a motion constraint, and very low impedance is naturally present in the direction of a free motion. Problems created by using motion control in applications where small motion errors would create large forces can be avoided if stiffness or impedance control is used. Furthermore, the stability of the overall system can be guaranteed and the robustness of the system improved by properly bounding the values of impedance parameters. Since impedance relates an input velocity to an output force, it is a transfer function. The concepts of impedance control can be applied to situations in which the input is not a velocity and the output is not a force. Still, the term *impedance control* is used, even though the corresponding transfer function is, strictly speaking, not an impedance.

Example 4.1

A schematic representation of a single joint of a direct-drive robotic manipulator is shown in figure 4.5a. Note that direct-drive joints have no speed reducers, such as gears; the stator of the drive motor is rigidly attached to one link, and the rotor is rigidly attached to the next link. Motor torque is T_m, and the joint torque that is transmitted to the driven link (link 2) is T_J. Draw a block diagram for the joint, and show

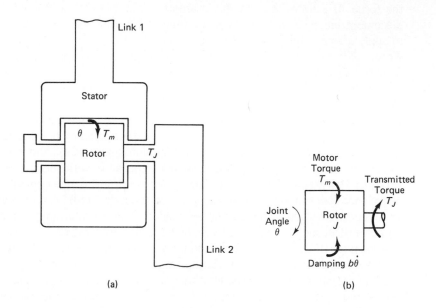

(a)

(b)

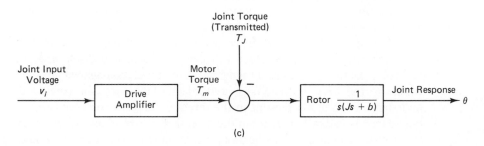

(c)

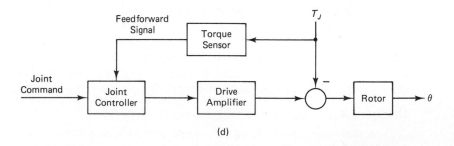

(d)

Figure 4.5. (a) A single joint of a direct-drive arm. (b) Free-body diagram of the motor rotor. (c) Block diagram of the joint. (d) Feedforward control using the joint torque.

that T_J is represented as an input to the joint control system. If T_J is measured directly, using a semiconductor strain gage torque sensor, what type of control would you recommend for improving the manipulator performance? Extend the discussion to the case in which the joint actuator is a hydraulic piston-cylinder mechanism.

Solution   Let us assume linear viscous friction, with damping constant b. If θ denotes the relative rotation of Link 2 with respect to Link 1, the equation of motion for the motor rotor (inertia J) may be written using the free-body diagram shown in figure 4.5b. Thus,

$$J\ddot{\theta} = T_m - b\dot{\theta} - T_J$$

Here, for simplicity, we have assumed that Link 1 is at rest (or moving with constant velocity).

The motor torque T_m is generated from the field current provided by the drive amplifier in response to a command voltage v_i to the joint. Figure 4.5c shows a block diagram for the joint. Note that T_J enters as an input. Hence, an appropriate control method would be to measure the joint torque T_J directly and feed it forward so as to correct any deviations in the joint response θ. This feedforward control structure is shown by the block diagram in figure 4.5d.

The same concepts are true in the case of a hydraulic actuator. Note that the control input is the voltage signal to the hydraulic valve actuator. The pressure in the hydraulic fluid is analogous to the motor torque. Joint torque/force is provided by the force exerted by the piston on the driven link. This force should be measured for feedforward control.

Example 4.2

The control of processes such as machine tools and robotic manipulators may be addressed from the point of view of impedance control. For example, consider a milling machine that performs a straight cut on a work piece, as shown in figure 4.6a. The tool position is stationary, and the machine table that holds the work piece moves along a horizontal axis at speed v, the feed rate. The cutting force in the direction of feed is f. Suppose that the machine table is driven using the speed error, according to the law

$$F = Z_d(V_{\text{ref}} - V) \tag{i}$$

where Z_d denotes the drive impedance of the table and V_{ref} is the reference (command) feed rate. (The uppercase letters are used to represent frequency domain variables of the system.) Cutting impedance Z_w of the work piece satisfies the relation

$$F = Z_w V \tag{ii}$$

Note that Z_w depends on system properties, and we usually don't have direct control over it. The overall system is represented by the block diagram in figure 4.6b. An impedance control problem would be to adjust (or "adapt") the drive impedance Z_d so as to maintain the feed rate near V_{ref} and the cutting force near F_{ref}. We wish to determine an adaptive control law for Z_d.

Solution   The control objective is satisfied by minimizing the objective function

$$J = \frac{1}{2}\left[\frac{F - F_{\text{ref}}}{f_o}\right]^2 + \frac{1}{2}\left[\frac{V - V_{\text{ref}}}{v_o}\right]^2 \tag{iii}$$

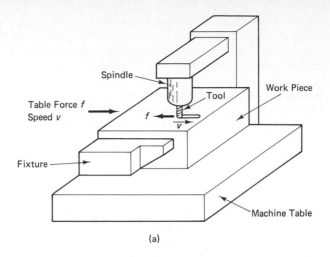

(a)

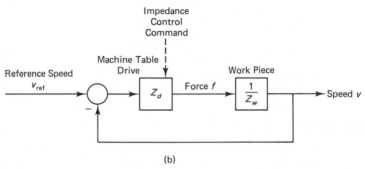

(b)

Figure 4.6. (a) A straight-cut milling operation. (b) Impedance block diagram representation.

where f_o denotes the *force tolerance* and v_o denotes the *speed tolerance*. For example, if we desire stringent control of the feed rate, we need to choose a small value for v_o that corresponds to a heavy weighting on the feed rate term in J.

The optimal solution is given by

$$\frac{\partial J}{\partial Z_d} = 0 = \frac{(F - F_{\text{ref}})}{f_o^2} \frac{\partial F}{\partial Z_d} + \frac{(V - V_{\text{ref}})}{v_o^2} \frac{\partial V}{\partial Z_d} \tag{iv}$$

Now, from equations (i) and (ii), we obtain

$$V = \left[\frac{Z_d}{Z_d + Z_w} \right] V_{\text{ref}} \tag{v}$$

$$F = \left[\frac{Z_d Z_w}{Z_d + Z_w} \right] V_{\text{ref}} \tag{vi}$$

On differentiating equations (v) and (vi), we get

$$\frac{\partial V}{\partial Z_d} = \frac{Z_w}{(Z_d + Z_w)^2} V_{\text{ref}} \tag{vii}$$

and

$$\frac{\partial F}{\partial Z_d} = \frac{Z_w^2}{(Z_d + Z_w)^2} V_{\text{ref}} \qquad \text{(viii)}$$

Next, we substitute equations (vii) and (viii) in (iv) and divide by the common term. Thus,

$$\frac{(F - F_{\text{ref}})}{f_o^2} Z_w + \frac{(V - V_{\text{ref}})}{v_o^2} = 0 \qquad \text{(ix)}$$

Equation (ix) is expanded after substituting equations (v) and (vi) in order to get the required expression for Z_d:

$$Z_d = \left[\frac{Z_o^2 + Z_w Z_{\text{ref}}}{Z_w - Z_{\text{ref}}} \right] \qquad \text{(x)}$$

where

$$Z_o = \frac{f_o}{v_o}$$

and

$$Z_{\text{ref}} = \frac{F_{\text{ref}}}{V_{\text{ref}}}$$

Equation (x) is the impedance control law for the table drive. Specifically, since Z_w—which depends on work piece characteristics, tool bit characteristics, and the rotating speed of the tool bit—is known through a suitable model or might be determined experimentally (identified) by monitoring v and f, and since Z_o and Z_{ref} are specified, we are able to determine the necessary drive impedance Z_d using equation (x). Parameters of the table drive controller— particularly gain—can be adjusted to match this optimal impedance. Unfortunately, exact matching is virtually impossible, because Z_d is generally a function of frequency. If the component bandwidths are high, we may assume that the impedance functions are independent of frequency, and this somewhat simplifies the impedance control task.

Note from equation (ii) that for the ideal case of $V = V_{\text{ref}}$ and $F = F_{\text{ref}}$, we have $Z_w = Z_{\text{ref}}$. Then, from equation (x), it follows that an infinite drive impedance is needed for exact control. This is impossible to achieve in practice, however. Of course, an upper limit for the drive impedance should be set in any practical impedance control scheme.

4.3 STRAIN GAGES

Many types of force/torque sensors are based on strain gage measurements. Although strain gages measure strain, the measurements can be directly related to stress and force. Hence, it is appropriate to discuss strain gages in this chapter. Note, however, that strain gages may be used in a somewhat indirect manner (using auxiliary front-end elements) to measure other types of variables, including displacement and acceleration. Two common types of resistance strain gages will be discussed in this section and the next section. Specific types of force/torque sensors will be studied in the subsequent sections.

Equations for Strain Gage Measurements

The change of electrical resistance in material when mechanically deformed is the property used in resistance-type strain gages. The resistance R of a conductor that has length ℓ and area of cross-section A is given by

$$R = \rho \frac{\ell}{A} \tag{4.1}$$

where ρ denotes the *resistivity* of the material. Taking the logarithm of equation 4.1, we have

$$\log R = \log \rho + \log (\ell/A)$$

Now, taking the differential, we obtain

$$\frac{dR}{R} = \frac{d\rho}{\rho} + \frac{d(\ell/A)}{\ell/A} \tag{4.2}$$

The first term on the right-hand side of equation 4.2 depends on the change in resistivity, and the second term represents deformation. It follows that the change in resistance comes from the change in shape as well as from the change in resistivity of the material. For linear deformations, the two terms on the right-hand side of equation 4.2 are linear functions of strain ϵ; the proportionality constant of the second term, in particular, depends on Poisson's ratio of the material. Hence, the following relationship can be written for a strain gage element:

$$\frac{dR}{R} = S_s \epsilon \tag{4.3}$$

The constant S_s is known as the *sensitivity* or *gage factor* of the strain gage element. The numerical value of this constant ranges from 2 to 6 for most *metallic strain gage* elements and from 40 to 200 for *semiconductor strain gages*. These two types of strain gages will be discussed later.

The change in resistance of a strain gage element, which determines the associated strain (equation 4.3), is measured using a suitable electrical circuit. Many variables—including displacement, acceleration, pressure, liquid level, stress, force, and torque—can be determined using strain measurements. Some variables (e.g., stress, force, and torque) can be determined by measuring the strain of the dynamic object itself at suitable locations. In other situations, an auxiliary front-end device may be required to convert the measurand into a proportional strain. For instance, pressure or displacement may be measured by converting them to a measurable strain using a diaphragm or bending element. Acceleration may be measured by first converting it into an inertia force of a suitable mass (seismic) element, then subjecting a cantilever (strain member) to that inertia force and, finally, measuring the strain at a high-sensitivity location of the cantilever element (see figure 4.7). *Ther-*

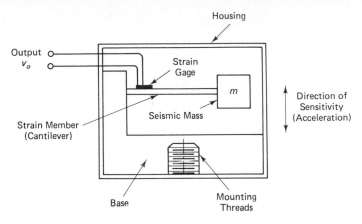

Figure 4.7. A strain gage accelerometer.

mistors are temperature sensors made of semiconductor material whose resistance changes with temperature. *Resistance temperature detectors* (RTDs) operate by the same principle, except that they are made of metals, not of semiconductor material. These temperature sensors, and the piezoelectric sensors discussed in chapter 3, should not be confused with strain gages. Resistance strain gages are based on resistance change due to strain, or the *piezoresistive* property of materials.

Early strain gages were fine metal filaments. Modern strain gages are manufactured primarily as metallic foil (for example, using the copper-nickel alloy known as constantan) or semiconductor elements (e.g., silicon with trace impurity boron). They are manufactured by first forming a thin film (foil) of metal or a single crystal of semiconductor material and then cutting it into a suitable grid pattern, either mechanically or by using photoetching (chemical) techniques. This process is much more economical and is more precise than making strain gages with metal filaments. The strain gage element is formed on a backing film of electrically insulated material (e.g., plastic). This element is cemented onto the member whose strain is to be measured. Alternatively, a thin film of insulating ceramic substrate is melted onto the measurement surface, on which the strain gage is mounted directly. The direction of sensitivity is the major direction of elongation of the strain gage element (figure 4.8a). To measure strains in more than one direction, multiple strain gages (e.g., various rosette configurations) are available as single units. These units have more than one direction of sensitivity. Principal strains in a given plane (the surface of the object on which the strain gage is mounted) can be determined by using these multiple strain gage units. Typical foil-type gages are shown in figure 4.8b, and a semiconductor strain gage is shown in figure 4.8c.

A direct way to obtain strain gage measurement is to apply a constant DC voltage across a series-connected strain gage element and a suitable resistor and to measure the output voltage v_o across the strain gage under open-circuit conditions (using a voltmeter with high input impedance). A schematic diagram of this arrangement is shown in figure 4.9a. It is known as a *potentiometer circuit* or *ballast circuit*. This arrangement has several weaknesses. Any ambient temperature variation will directly introduce some error because of associated change in the strain gage re-

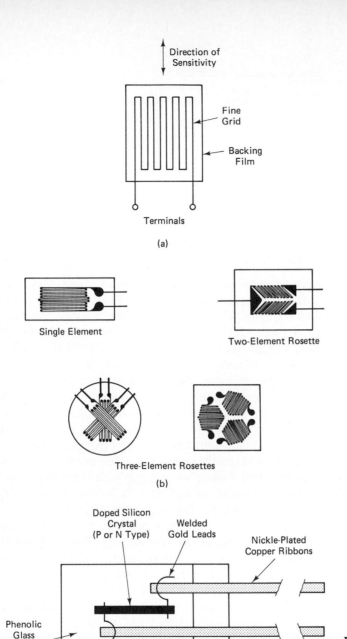

Direction of
Sensitivity

Fine
Grid

Backing
Film

Terminals

(a)

Single Element

Two-Element Rosette

Three-Element Rosettes

(b)

Doped Silicon
Crystal
(P or N Type)

Welded
Gold Leads

Nickle-Plated
Copper Ribbons

Phenolic
Glass
Backing
Plate

(c)

Figure 4.8. (a) Strain gage nomenclature. (b) Typical foil-type strain gages. (c) A semiconductor strain gage.

sistance and the resistance of the connecting circuitry. Also, measurement accuracy will be affected by possible variations in the supply voltage v_{ref}. Furthermore, the electrical loading error will be significant unless the load impedance is very high. Perhaps the most serious disadvantage of this circuit is that the change in signal due to strain is usually a very small percentage of the total signal level in the circuit output.

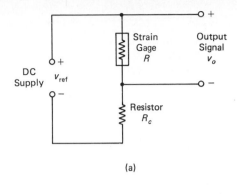

(a)

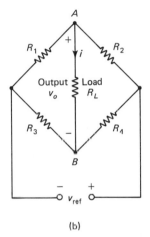

(b)

Figure 4.9. (a) A potentiometer circuit (ballast circuit) for strain gage measurements. (b) A Wheatstone bridge circuit for strain gage measurements.

With very high load impedance, the output signal from the ballast circuit is given by

$$v_o = \frac{R}{(R + R_c)} v_{\text{ref}} \tag{4.4}$$

Note that this is a large signal on the order of the supply voltage v_{ref}. However, strain is measured by the change in output voltage caused by the change in resistance of the strain gage. The applicable relationship is obtained by taking the differential of equation 4.4; thus,

$$\delta v_o = \frac{R_c}{(R + R_c)^2} v_{\text{ref}} \, \delta R \tag{4.5}$$

Since the voltage δv_o is only a very small fraction of the total output voltage v_o, large errors would be present in noisy readings. This problem can be reduced to some extent by decreasing v_o, which may be accomplished by increasing the resistance R_c. This, however, reduces sensitivity of the circuit, as is clear from equation 4.5. Any changes in the strain gage resistance due to ambient changes will directly enter the strain gage reading unless R and R_c have identical coefficients with respect to ambient changes. This may be shown by taking the differential of equation 4.4 with respect to both R and R_c.

Example 4.3

Show that, up to first order, ambient effects are compensated in the ballast circuit if the series resistor R_c and the strain gage resistance R have identical temperature coefficients of resistance. What happens if higher-order changes are included in the analysis?

Solution   Since only first-order effects are considered, take the differential of equation 4.5 with respect to R and R_c:

$$\delta v_o = \frac{R_c}{(R + R_c)^2} v_{\text{ref}} \, \delta R - \frac{R}{(R + R_c)^2} v_{\text{ref}} \, \delta R_c$$

$$= \frac{v_{\text{ref}}}{(R + R_c)^2} [R_c \, \delta R - R \, \delta R_c]$$

The assumption here is that the higher-order terms (δR^2, δR_c^2, etc.) are too small to have a significant effect on the output signal v_o. If R and R_c have identical temperature coefficients, they change by amounts proportional to their individual magnitudes because of temperature changes. Hence, we have $\delta R = kR$ and $\delta R_c = kR_c$. This gives $\delta v_o = 0$, indicating that ambient changes do not affect v_o in the first-order approximation. Now consider the higher-order effects. If ambient changes alone are considered, we have the exact relationship

$$\delta v_o = \left[\frac{R + \delta R}{R + \delta R + R_c + \delta R_c}\right] v_{\text{ref}} - \left[\frac{R}{R + R_c}\right] v_{\text{ref}}$$

Now, with $\delta R = kR$ and $\delta R_c = kR_c$, we again get $\delta v_o = 0$. Now suppose that the strain gage element changes resistance by $\delta R'$ as a result of strain in the measured object. Then the overall change in the output signal, including higher-order effects, is

$$\delta \tilde{v}_o = \left[\frac{R + \delta R + \delta R'}{R + \delta R + \delta R' + R_c + \delta R_c}\right] v_{\text{ref}} - \left[\frac{R}{R + R_c}\right] v_{\text{ref}}$$

Note that even with $R = R_c$ and $\delta R = \delta R_c$, the terms containing δR and δR_c do not cancel out in this case. They do cancel out if only the first-order terms are retained, however. Hence, exact compensation is not realized when higher-order effects are included.

A more favorable circuit for use in strain gage measurements is the *Wheatstone bridge,* shown in figure 4.9b. One or more of the four resistors R_1, R_2, R_3, and R_4 in the circuit may represent strain gages. To obtain the output relationship for the Wheatstone bridge circuit, assume that the load impedance R_L is very high. Hence, the load current i is negligibly small. Then the potentials at nodes A and B are

$$v_A = \frac{R_1}{(R_1 + R_2)} v_{\text{ref}} \quad \text{and} \quad v_B = \frac{R_3}{(R_3 + R_4)} v_{\text{ref}}$$

and the output voltage $v_o = v_A - v_B$ is given by

$$v_o = \left[\frac{R_1}{(R_1 + R_2)} - \frac{R_3}{(R_3 + R_4)}\right] v_{\text{ref}} \tag{4.6}$$

Now, by using straightforward algebra, we get

$$v_o = \frac{(R_1 R_4 - R_2 R_3)}{(R_1 + R_2)(R_3 + R_4)} v_{ref} \qquad (4.7)$$

When this output voltage is zero, the bridge is said to be "balanced." It follows from equation 4.7 that for a balanced bridge,

$$\frac{R_1}{R_2} = \frac{R_3}{R_4} \qquad (4.8)$$

Note that equation 4.8 is valid for any value of R_l, not just for large R_L, because when the bridge is balanced, current i will be zero, even for small R_L.

Bridge Sensitivity

Strain gage measurements are calibrated with respect to a balanced bridge. When the strain gages in the bridge deform, the balance is upset. If one of the arms of the bridge has a variable resistor, it can be changed to restore balance. The amount of this change measures the amount by which the resistance of the strain gages changed, thereby measuring the applied strain. This is known as the *null-balance method* of strain measurement. This method is inherently slow because of the time required to balance the bridge each time a reading is taken. Hence, the null-balance method is generally not suitable for dynamic measurements. This approach to strain measurement can be speeded up by using servo balancing, whereby the output error signal is fed back into an actuator that automatically adjusts the variable resistance so as to restore the balance.

A more common method, which is particularly suitable for making dynamic readings from a strain gage bridge, is to measure the output voltage resulting from the imbalance caused by the deformation of active strain gages in the bridge. To determine the *calibration constant* of a strain gage bridge, the sensitivity of the bridge output to changes in the four resistors in the bridge should be known. For small changes in resistance, this may be determined using the differential relation (or, equivalently, the first-order approximation for the Taylor series expansion):

$$\delta v_o = \sum_{i=1}^{4} \frac{\partial v_o}{\partial R_i} \delta R_i \qquad (4.9)$$

The partial derivatives are obtained directly from equation 4.6. Specifically,

$$\frac{\partial v_o}{\partial R_1} = \frac{R_2}{(R_1 + R_2)^2} v_{ref} \qquad (4.10)$$

$$\frac{\partial v_o}{\partial R_2} = -\frac{R_1}{(R_1 + R_2)^2} v_{ref} \qquad (4.11)$$

$$\frac{\partial v_o}{\partial R_3} = -\frac{R_4}{(R_3 + R_4)^2} v_{ref} \qquad (4.12)$$

$$\frac{\partial v_o}{\partial R_4} = \frac{R_3}{(R_3 + R_4)^2} v_{ref} \qquad (4.13)$$

The required relationship is obtained by substituting equations 4.10 through 4.13 into 4.9; thus, we have

$$\frac{\delta v_o}{v_{ref}} = \frac{(R_2\, \delta R_1 - R_1\, \delta R_2)}{(R_1 + R_2)^2} - \frac{(R_4\, \delta R_3 - R_3\, \delta R_4)}{(R_3 + R_4)^2} \qquad (4.14)$$

This result is subject to equation 4.8, because changes are measured from the balanced condition. Note from equation 4.14 that if all four resistors are identical (in value and material), resistance changes due to ambient effects cancel out among the first-order terms (δR_1, δR_2, δR_3, δR_4), producing no net effect on the output voltage from the bridge. Closer examination of equation 4.14 will show that only the adjacent pairs of resistors (e.g., R_1 with R_2 and R_3 with R_4) have to be identical in order to achieve this environmental compensation. Even this requirement can be relaxed. As a matter of fact, compensation is achieved if R_1 and R_2 have the same temperature coefficient and if R_3 and R_4 have the same temperature coefficient.

Example 4.4

Suppose that R_1 represents the only active strain gage and R_2 represents an identical "dummy" gage in figure 4.9b. The other two elements of the bridge are *bridge-completion resistors,* which do not have to be identical to the strain gages. For a balanced bridge, we must have $R_3 = R_4$, but not necessarily equal to the resistance of the strain gage. Let us determine the output of the bridge.

Solution In this example, only R_1 changes. Hence, from equation 4.14, we have

$$\frac{\delta v_o}{v_{ref}} = \frac{\delta R}{4R} \qquad (4.15)$$

where R denotes the strain gage resistance.

The bridge constant. Equation 4.15 assumes that only one resistance (strain gage) in the Wheatstone bridge (Figure 4.9b) is active. Numerous other activating combinations are possible, however—for example, tension in R_1 and compression in R_2, as in the case of two strain gages mounted symmetrically at 45° about the axis of a shaft in torsion. In this manner, the overall sensitivity of a strain gage bridge can be increased. It is clear from equation 4.14 that if all four resistors in the bridge are active, the best sensitivity is obtained if, for example, R_1 and R_4 are in tension and R_2 and R_3 are in compression, so that all four differential terms have the same sign. If more than one strain gage is active, the bridge output may be expressed as

$$\frac{\delta v_o}{v_{ref}} = k\frac{\delta R}{4R} \qquad (4.16)$$

where

$$k = \frac{\text{bridge output in the general case}}{\text{bridge output if only one strain gage is active}}$$

This constant is known as the *bridge constant*. The larger the bridge constant, the better the sensitivity of the bridge.

Example 4.5

A strain gage load cell (force sensor) consists of four identical strain gages, forming a Wheatstone bridge, that are mounted on a rod that has square cross-section. One opposite pair of strain gages is mounted axially and the other pair is mounted in the transverse direction, as shown in figure 4.10a. To maximize the bridge sensitivity, the strain gages are connected to the bridge as shown in figure 4.10b. Determine the bridge constant k in terms of *Poisson's ratio* ν of the rod material.

Solution   Suppose that $\delta R_1 = \delta R$. Then, for the given configuration, we have

$$\delta R_2 = -\nu\,\delta R$$

$$\delta R_3 = -\nu\,\delta R$$

$$\delta R_4 = \delta R$$

Note that from the definition of Poisson's ratio

$$\text{transverse strain} = (-\nu) \times \text{longitudinal strain}$$

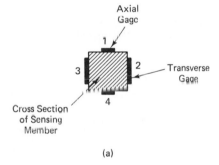

(a)

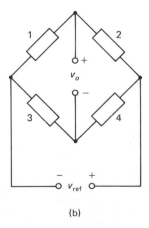

(b)

Figure 4.10.   An example of four active strain gages: (a) mounting configuration on the load cell; (b) bridge circuit.

Now, it follows from equation 4.14 that

$$\frac{\delta v_o}{v_{\text{ref}}} = 2(1 + \nu)\frac{\delta R}{4R} \qquad (4.17)$$

according to which the bridge constant is given by

$$k = 2(1 + \nu)$$

The calibration constant. The calibration constant C of a strain gage bridge relates the strain that is measured to output of the bridge. Specifically,

$$\frac{\delta v_o}{v_{\text{ref}}} = C\epsilon \qquad (4.18)$$

Now, in view of equations 4.3 and 4.16, the calibration constant may be expressed as

$$C = \frac{k}{4}S_s \qquad (4.19)$$

where k is the *bridge constant* and S_s is the *sensitivity* or *gage factor* of the strain gage. Ideally, the calibration constant should remain constant over the measurement range of the bridge (i.e., independent of strain ϵ and time t) and should be stable with respect to ambient conditions. In particular, there should not be any creep, nonlinearities such as hysteresis, or thermal effects.

Example 4.6

A schematic diagram of a strain gage accelerometer is shown in figure 4.11a. A point mass of weight W is used as the acceleration sensing element, and a light cantilever with rectangular cross-section, mounted inside the accelerometer casing, converts the inertia force of the mass into a strain. The maximum bending strain at the root of the cantilever is measured using four identical active semiconductor strain gages. Two of the strain gages (A and B) are mounted axially on the top surface of the cantilever, and the remaining two (C and D) are mounted on the bottom surface, as shown in figure 4.11b. In order to maximize the sensitivity of the accelerometer, indicate the manner in which the four strain gages—A, B, C, and D—should be connected to a Wheatstone bridge circuit. What is the bridge constant of the resulting circuit?

Obtain an expression relating applied acceleration a (in units of g, which denotes acceleration due to gravity) to bridge output δv_o (measured using a bridge balanced at zero acceleration) in terms of the following parameters:

W = weight of the seismic mass at the free end of the cantilever element
E = Young's modulus of the cantilever
ℓ = length of the cantilever
b = cross-section width of the cantilever
h = cross-section height of the cantilever
S_s = sensitivity (gage factor) of each strain gage
v_{ref} = supply voltage to the bridge

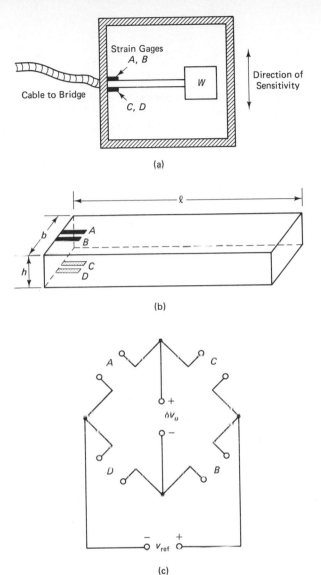

Figure 4.11. A strain gage accelerometer example: (a) schematic diagram; (b) strain gage mounting configuration; (c) bridge connection.

If $W = 0.02$ lb, $E = 10 \times 10^6$ lbf/in.^{2}, $\ell = 1$ in., $b = 0.1$ in., $h = 0.05$ in., $S_s = 200$, and $v_{\text{ref}} = 20$ V, determine the sensitivity of the accelerometer in mV/g.

If the yield strength of the cantilever element is 10×10^3 lbf/in.^{2}, what is the maximum acceleration that could be measured using the accelerometer?

Is the cross-sensitivity (i.e., the sensitivity in the two directions orthogonal to the direction of sensitivity shown in figure 4.11a) small with your arrangement of the strain gage bridge? Explain.

Hint: For a cantilever subjected to force F at the free end, the maximum stress at the root is given by

$$\sigma = \frac{6F\ell}{bh^2}$$

with the present notation.

Solution   Clearly, the bridge sensitivity is maximized by connecting the strain gages A, B, C, and D to the bridge as shown in figure 4.11c. This follows from equation 4.14, noting that the contributions from the four strain gages are positive when δR_1 and δR_4 are positive and δR_2 and δR_3 are negative. The bridge constant for the resulting arrangement is $k = 4$. Hence, from equation 4.16,

$$\frac{\delta v_o}{v_{\text{ref}}} = \frac{\delta R}{R}$$

or, from equations 4.18 and 4.19,

$$\frac{\delta v_o}{v_{\text{ref}}} = S_s \epsilon$$

Also,

$$\epsilon = \frac{\sigma}{E} = \frac{6F\ell}{Ebh^2}$$

where F denotes the inertia force;

$$F = \frac{W}{g}\ddot{x} = Wa$$

Note that $\ddot{x}$ is the acceleration in the direction of sensitivity and $\ddot{x}/g = a$ is the acceleration in units of g.

Thus,

$$\epsilon = \frac{6W\ell}{Ebh^2}a$$

or

$$\delta v_o = \frac{6W\ell}{Ebh^2}S_s v_{\text{ref}}a$$

Now, with the given values,

$$\frac{\delta v_o}{a} = \frac{6 \times 0.02 \times 1 \times 200 \times 20}{10 \times 10^6 \times 0.1 \times (0.05)^2}\ \text{V}/g$$

$$= 0.192\ \text{V}/g = 192\ \text{mV}/g$$

$$\frac{\epsilon}{a} = \frac{1}{S_s v_{\text{ref}}}\frac{\delta v_o}{a} = \frac{0.192}{200 \times 20}\ \text{strain}/g$$

$$= 48 \times 10^{-6}\ \epsilon/g = 48\ \mu\epsilon/g$$

$$\text{yield strain} = \frac{\text{yield strength}}{E} = \frac{10 \times 10^3}{10 \times 10^6} = 1 \times 10^{-3}\ \text{strain}$$

Hence,

$$\text{Number of } g\text{'s to yielding} = \frac{1 \times 10^{-3}}{48 \times 10^{-6}} g = 20.8g$$

Cross-sensitivity comes from accelerations in the two directions (y and z) orthogonal to the direction of sensitivity (x). In the lateral (y) direction, the inertia force causes lateral bending. This will produce equal tensile (or compressive) strains in B and D and equal compressive (or tensile) strains in A and C. According to the bridge circuit, we see that these contributions cancel each other. In the axial (z) direction, the inertia force causes equal tensile (or compressive) stresses in all four strain gages. These also will cancel out, as is clear from the following relationship for the bridge:

$$\frac{\delta v_o}{v_{\text{ref}}} = \frac{(R_C \, \delta R_A - R_A \, \delta R_C)}{(R_A + R_C)^2} - \frac{(R_B \, \delta R_D - R_D \, \delta R_B)}{(R_D + R_B)^2}$$

with

$$R_A = R_B = R_C = R_D = R$$

which gives

$$\frac{\delta v_o}{v_{\text{ref}}} = \frac{(\delta R_A - \delta R_C - \delta R_D + \delta R_B)}{4R}$$

It follows that this arrangement is good with respect to cross-sensitivity problems.

Data Acquisition

As noted earlier, the two common methods of measuring strains using a Wheatstone bridge circuit are (1) the null-balance method and (2) the imbalance output method. One possible scheme for using the first method is shown in figure 4.12. In this particular arrangement, two bridge circuits are used. The *active bridge* contains the ac-

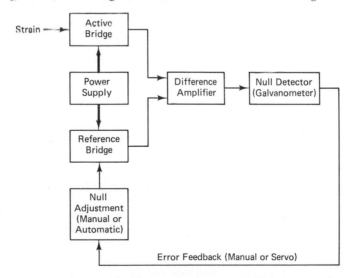

Figure 4.12. The null-balance method of strain measurement using a bridge circuit.

tive strain gages, dummy gages, and bridge-completion resistors. The *reference bridge* has four resistors, one of which is micro-adjustable, either manually or automatically. The outputs from the two bridges are fed into a difference amplifier, which provides an amplified difference of the two signals. This error signal is indicated on a null detector, such as a galvanometer. Initially, both bridges are balanced. When the measurement system is in use, the active gages are subjected to the strain that is being measured. This upsets the balance, giving a net output, which is indicated on the null detector. In manual operation of the null-balance mechanism, the resistance knob in the reference bridge is adjusted carefully until the galvanometer indicates a null reading. The knob can be calibrated to indicate the measured strain directly. In servo operation, which is much faster than the manual method, the error signal is fed into an actuator that automatically adjusts the variable resistor in the reference bridge until the null balance is achieved. Actuator movement measures the strain.

The manual null-balance method is appropriate for making static readings. For measuring dynamic strains, either the servo null-balance method or the imbalance output method should be employed. A schematic diagram for the imbalance output method is shown in figure 4.13. In this method, the output from the active bridge is directly measured as a voltage signal and calibrated to provide the measured strain. Figure 4.13 shows the use of an AC bridge. In this case, the bridge is powered by an AC voltage. The supply frequency should be about ten times the maximum frequency of interest in the dynamic strain signal (bandwidth). A supply frequency on the order of 1 kHz is typical. This signal is generated by an oscillator and is fed into the bridge. The transient component of the output from the bridge is very small (typically less than 1 mV and possibly a few microvolts). This signal has to be amplified, demodulated (especially if the signals are transient), and filtered to provide the strain reading. The calibration constant of the bridge should be known in order to convert the output voltage to strain.

Strain gage bridges powered by DC voltages are very common. They have the advantages of simplicity with regard to necessary circuitry and portability. The advantages of AC bridges include improved stability (reduced drift) and accuracy and reduced power consumption.

Accuracy Considerations

Foil gages are available with resistances as low as 50 Ω and as high as several kilohms. The power consumption of the bridge decreases with increased resistance. This has the added advantage of decreased heat generation. Bridges with a high range of measurement (e.g., a maximum strain of 0.01 m/m) are available. The ac-

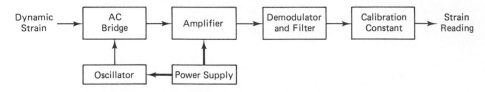

Figure 4.13. Measurement of dynamic strains using an AC bridge.

curacy depends on the linearity of the bridge, environmental (particularly temperature) effects, and mounting techniques. For example, zero shift, due to strains produced when the cement that is used to mount the strain gage dries, will result in calibration error. Creep will introduce errors during static and low-frequency measurements. Flexibility and hysteresis of the bonding cement will bring about errors during high-frequency strain measurements. Resolutions on the order of 1 μm/m (i.e., one *microstrain*) are common.

As noted earlier, the cross-sensitivity of a strain gage is the sensitivity to strains that are orthogonal to the measured strain. This cross-sensitivity should be small (say, less than 1 percent of the direct sensitivity). Manufacturers usually provide cross-sensitivity factors for their strain gages. This factor, when multiplied by the cross strain present in a given application, gives the error in the strain reading due to cross-sensitivity.

Often, measurements of strains in moving members are needed for control purposes. Examples include real-time monitoring and failure detection in machine tools, measurement of power, measurement of force and torque for feedforward and feedback control in dynamic systems, and tactile sensing using instrumented hands in industrial robots. If the motion is small or the device has a limited stroke, strain gages mounted on the moving member can be connected to the signal-conditioning circuitry and the power source using coiled flexible cables. For large motions, particularly in rotating shafts, some form of commutating arrangement has to be used. Slip rings and brushes are commonly used for this purpose. When AC bridges are used, a mutual-induction device (rotary transformer) may be used, with one coil located on the moving member and the other coil stationary. To accommodate and compensate for errors (e.g., losses and glitches in the output signal) caused by commutation, it is desirable to place all four arms of the bridge, rather than just the active arms, on the moving member.

4.4 SEMICONDUCTOR STRAIN GAGES

In some low-strain applications (e.g., dynamic torque measurement), the sensitivity of foil gages is not adequate to produce an acceptable strain gage signal. Semiconductor (SC) strain gages are particularly useful in such situations. The strain element of an SC strain gage is made of a single crystal of *piezoresistive* material such as silicon, doped with a trace impurity such as boron. A typical construction is shown in figure 4.14. The sensitivity (gage factor) of an SC strain gage is about two orders of magnitude higher than that of a metallic foil gage (typically, 40 to 200). The resistivity is also higher, providing reduced power consumption and heat generation. Another advantage of SC strain gages is that they deform elastically to fracture. In particular, mechanical hysteresis is negligible. Furthermore, they are smaller and lighter, providing less cross-sensitivity, reduced distribution error (i.e., improved spatial resolution), and negligible error due to mechanical loading. The maximum strain measurable using a semiconductor strain gage is typically 0.003 m/m (i.e., 3000 $\mu\epsilon$). Strain gage resistance can be several hundred ohms (typically, 120 Ω or 350 Ω).

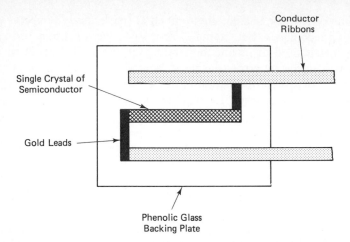

Single Crystal of
Semiconductor

Conductor
Ribbons

Gold Leads

Phenolic Glass
Backing Plate

Figure 4.14. Component details of a
semiconductor strain gage.

There are several disadvantages associated with semiconductor strain gages,
however, which can be interpreted as advantages of foil gages. Undesirable characteristics of SC gages include the following:

1. The strain–resistance relationship is more nonlinear.
2. They are brittle and difficult to mount on curved surfaces.
3. The maximum strain that can be measured is an order of magnitude smaller
   (typically, less than 0.01 m/m).
4. They are more costly.
5. They have a much higher temperature sensitivity.

The first disadvantage is illustrated in figure 4.15. There are two types of
semiconductor strain gages: the P-type and the N-type. In P-type strain gages, the
direction of sensitivity is along the (1, 1, 1) crystal axis, and the element produces a
"positive" (P) change in resistance in response to a positive strain. In N-type strain
gages, the direction of sensitivity is along the (1, 0 , 0) crystal axis, and the element
responds with a "negative" (N) change in resistance to a positive strain. In both
types, the response is nonlinear and can be approximated by the quadratic relationship

$$\frac{\delta R}{R} = S_1 \epsilon + S_2 \epsilon^2 \qquad (4.20)$$

The parameter S_1 represents the *linear sensitivity,* which is positive for P-type gages
and negative for N-type gages. Its magnitude is usually somewhat larger for P-type
gages, corresponding to better sensitivity. The parameter S_2 represents the degree of
nonlinearity, which is usually positive for both types of gages. Its magnitude, however, is typically a little smaller for P-type gages. It follows that P-type gages are
less nonlinear and have higher strain sensitivities. The nonlinear relationship given
by equation 4.20 or the nonlinear characteristic curve (figure 4.15) should be used
when measuring moderate to large strains with semiconductor strain gages. Otherwise, the nonlinearity error would be excessive.

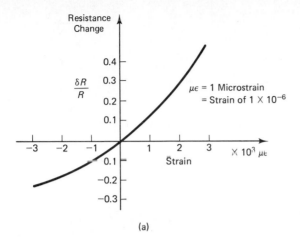

(a)

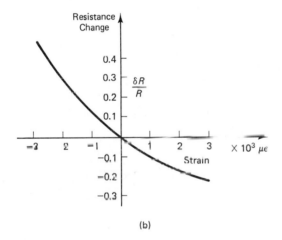

Figure 4.15. Nonlinear behavior of a semiconductor (silicon/boron) strain gage: (a) a P-type gage; (b) an N-type gage.

(b)

Example 4.7

For a semiconductor strain gage characterized by the quadratic strain–resistance relationship, equation 4.20, obtain an expression for the equivalent gage factor (sensitivity) S_s using least squares error linear approximation and assuming that strains in the range $\pm \epsilon_{max}$ have to be measured. Derive an expression for the percentage nonlinearity.

Taking $S_1 = 117$, $S_2 = 3,600$, and $\epsilon_{max} = 1 \times 10^{-2}$, calculate S_s and the percentage nonlinearity.

Solution The linear approximation of equation 4.20 may expressed as

$$\left[\frac{\delta R}{R}\right]_L = S_s \epsilon$$

The error is given by

$$e = \frac{\delta R}{R} - \left[\frac{\delta R}{R}\right]_L = S_1 \epsilon + S_2 \epsilon^2 - S_s \epsilon$$

or

$$e = (S_1 - S_s)\epsilon + S_2\epsilon^2 \tag{i}$$

The quadratic integral error is

$$J = \int_{-\epsilon_{max}}^{\epsilon_{max}} e^2 d\epsilon = \int_{-\epsilon_{max}}^{\epsilon_{max}} [(S_1 - S_s)\epsilon + S_2\epsilon^2]^2 \, d\epsilon \tag{ii}$$

We have to determine S_s that will result in minimum J. Hence, we use

$$\frac{\partial J}{\partial S_s} = 0$$

Thus, from equation (ii),

$$\int_{-\epsilon_{max}}^{\epsilon_{max}} (-2\epsilon)[(S_1 - S_s)\epsilon + S_2\epsilon^2] \, d\epsilon = 0$$

On performing the integration, we get

$$S_s = S_1 \tag{4.21}$$

The quadratic curve and the linear approximation are shown in figure 4.16. Note that the maximum error is at $\epsilon = \pm\epsilon_{max}$. The maximum error value is obtained from equation (i), with $S_s = S_1$ and $\epsilon = \pm\epsilon_{max}$, as

$$e_{max} = S_2\epsilon_{max}^2$$

The true change in resistance (nondimensional) from $-\epsilon_{max}$ to $+\epsilon_{max}$ is obtained using equation 4.20; thus,

$$\frac{\Delta R}{R} = (S_1\epsilon_{max} + S_2\epsilon_{max}^2) - (-S_1\epsilon_{max} + S_2\epsilon_{max}^2)$$

$$= 2S_1\epsilon_{max}$$

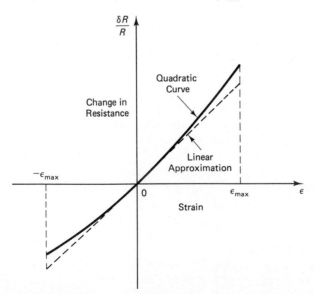

Figure 4.16. Least squares error linear approximation for the characteristic curve of a semiconductor strain gage.

Torque, Force, and Tactile Sensors   Chap. 4

Hence, the percentage nonlinearity is given by

$$N_p = \frac{\text{max error}}{\text{range}} \times 100\%$$

$$= \frac{S_2 \epsilon_{max}^2}{2 S_1 \epsilon_{max}} \times 100\%$$

or

$$N_p = 50 S_2 \epsilon_{max} / S_1 \% \qquad (4.22)$$

Now, with the given numerical values, we have

$$S_s = 117$$

and

$$N_p = 50 \times 3,600 \times 1 \times 10^{-2}/117\%$$

$$= 15.4\%$$

Note that we obtained this high value for nonlinearity because the given strain limits were high. Usually, the linear approximation is adequate for strains up to $\pm 1 \times 10^{-3}$.

The higher temperature sensitivity listed as a disadvantage of semiconductor strain gages may be considered an advantage in some situations. For instance, it is this property of high temperature sensitivity that is used in piezoresistive temperature sensors. Furthermore, since the temperature sensitivity of a semiconductor strain gage can be determined very accurately, more accurate methods can be employed for temperature compensation in strain gage circuitry, and temperature calibration can also be done more accurately. In particular, a passive SC strain gage may be used as an accurate temperature sensor for compensation purposes.

Automatic (Self) Temperature Compensation

When foil gages are used, associated resistance changes with temperature are typically small. Then the linear (first-order) approximation for the contribution from each arm of the bridge to the output signal, as given by equation 4.14, would be satisfactory. These contributions cancel out if we pick strain gage elements and bridge completion resistors properly—for example, R_1 identical to R_2 and R_3 identical to R_4. If this is the case, the only remaining effect of temperature change on the bridge output signal will be due to changes in parameter values k and S_s (see equations 4.18 and 4.19). For foil gages, such changes are also typically negligible. Hence, for small to moderate temperature changes, additional compensation will not be required when foil gage bridge circuits are employed.

In semiconductor gages, not only the resistance change with temperature (and with strain) is larger; the change of S_s with temperature is also larger compared to the corresponding values for foil gages. Hence, the linear approximation given by equation 4.14 might not be accurate under variable temperature conditions; furthermore, the bridge sensitivity could change significantly with temperature. Under such conditions, temperature compensation will be necessary.

Directly measuring temperature and correcting strain gage readings accordingly, using calibration data, is a straightforward way to account for temperature changes. Another method of temperature compensation is described here. This method assumes that the linear approximation given by equation 4.14 is valid; hence, equation 4.18 is applicable.

The resistance R and strain sensitivity (or gage factor) S_s of semiconductor strain gages are highly dependent on the concentration of the trace impurity, in a nonlinear manner. The typical behavior of the temperature coefficients of these two parameters for a P-type semiconductor strain gage is shown in figure 4.17. The *temperature coefficient of resistance* α and the *temperature coefficient of sensitivity* β are defined by

$$R = R_o(1 + \alpha.\Delta T) \tag{4.23}$$

$$S_s = S_{so}(1 + \beta.\Delta T) \tag{4.24}$$

where ΔT denotes the temperature increase. Note from figure 4.17 that β is a negative quantity and that for some dope concentrations, its magnitude is less than the value of the temperature coefficient of resistance (α). This property can be used in self-compensation with regard to temperature.

Consider a constant-voltage bridge circuit with a compensating resistor R_c connected to the supply lead, as shown in figure 4.18a. It can be shown that if R_c is set to a value predetermined on the basis of the temperature coefficients of the strain gages, self-compensation can result. Consider the case where load impedance is very high and the bridge has four identical SC strain gages that have resistance R. In this

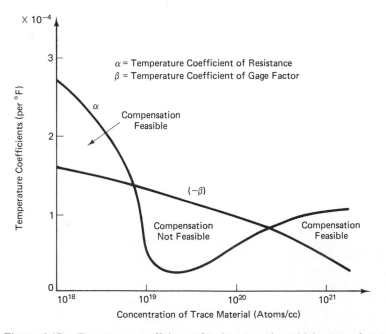

Figure 4.17. Temperature coefficients of resistance and sensitivity (gage factor) for a P-type semicondturor (silicon) strain gage.

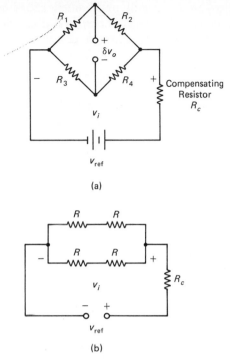

Figure 4.18. A strain gage bridge with a compensating resistor: (a) constant-voltage DC bridge; (b) equivalent circuit with high load impedance.

case, the bridge can be represented by the circuit shown in figure 4.18b. Since series impedances are additive and parallel admittances (inverse of impedance) are additive, the equivalent resistance of the bridge is R. Hence, the voltage supplied to the bridge, allowing for the voltage drop across R_c, is not v_{ref} but v_i, as given by

$$v_i = \frac{R}{(R + R_c)} v_{ref} \tag{4.25}$$

Now, from equation 4.18, we have

$$\frac{\delta v_o}{v_{ref}} = \frac{R}{(R + R_c)} \frac{kS_s}{4} \epsilon \tag{4.26}$$

We will assume that the bridge constant k does not change with temperature. Otherwise, the following procedure will still hold, provided that the calibration constant C is used in place of the strain gage sensitivity S_s (see equation 4.19). For self-compensation, we must have the same output after the temperature changes through ΔT. Hence, from equation 4.26, we have

$$\frac{R_o}{(R_o + R_c)} S_{so} = \frac{R_o(1 + \alpha.\Delta T)}{[R_o(1 + \alpha.\Delta T) + R_c]} S_{so}(1 + \beta.\Delta T)$$

where the subscript o denotes values before the temperature change. Cancellation of the common terms and cross-multiplication gives

$$R_o \beta + R_c(\alpha + \beta) = (R_o + R_c)\alpha\beta \, \Delta T$$

Now, since both $\alpha.\Delta T$ and $\beta.\Delta T$ are usually much smaller than unity, we may neglect the right-hand-side (second-order) term in the preceding equation. This gives the following expression for the compensating resistance:

$$R_c = -\left[\frac{\beta}{\alpha + \beta}\right]R_o \qquad (4.27)$$

Note that this type of compensation is possible because the temperature coefficient of the strain gage sensitivity (β) is negative. The feasible ranges of operation that correspond to positive R_c are indicated in figure 4.17. This method requires that R_c be maintained constant at the chosen value under changing temperature conditions. One way to accomplish this is by selecting a material with negligible temperature coefficient of resistance for R_c. Another way is to locate R_c in a separate, temperature-controlled environment.

4.5 TORQUE SENSORS

Torque and force sensing is useful in many applications, including the following:

1. In robotic tactile and manufacturing applications—such as gripping, surface gaging, and material forming—where exerting an adequate load on an object is the primary purpose of the task
2. In the control of fine motions (e.g., fine manipulation and micromanipulation) and in assembly tasks, where a small motion error can cause large damaging forces or performance degradation
3. In control systems that are not fast enough when motion feedback alone is employed, where force feedback and feedforward force control can be used to improve accuracy and bandwidth
4. In process testing, monitoring, and diagnostic applications, where torque sensing can detect, predict, and identify abnormal operation, malfunction, component failure, or excessive wear (e.g., in monitoring machine tools such as milling machines and drills)
5. In the measurement of power transmitted through a rotating device, where power is given by the product of torque and angular velocity in the same direction
6. In controlling complex nonlinear mechanical systems, where measurement of force and acceleration can be used to estimate unknown nonlinear terms and an appropriate nonlinear feedback can linearize or simplify the system (nonlinear feedback control)

In most applications, torque is sensed by detecting either an effect or the cause of torque. There are also methods for measuring torque directly. Common methods of torque sensing include the following:

1. Measuring strain in a sensing member between the drive element and the driven load, using a strain gage bridge

2. Measuring displacement in a sensing member (as in the first method)—either directly, using a displacement sensor, or indirectly, by measuring a variable, such as magnetic inductance or capacitance, that varies with displacement

3. Measuring reaction in support structure or housing (by measuring a force) and the associated lever arm length

4. In electric motors, measuring the field or armature current that produces motor torque; in hydraulic or pneumatic actuators, measuring actuator pressure

5. Measuring torque directly, using piezoelectric sensors, for example

6. Employing the servo method, balancing the unknown torque with a feedback torque generated by an active device (say, a servomotor) whose torque characteristics are known precisely

7. Measuring the angular acceleration in a known inertia element when the unknown torque is applied

The remainder of this section will be devoted to a discussion of torque measurement using some of these methods. Note that force sensing may be accomplished by essentially the same techniques. For the sake of brevity, however, we limit our treatment primarily to torque sensing. The extension of torque-sensing techniques to force sensing is left as a challenge to the reader.

Strain Gage Torque Sensors

The most straightforward method of torque sensing is to connect a torsion member between the drive unit and the load in series, as shown in figure 4.19, and to measure the torque in the torsion member. If a circular shaft (solid or hollow) is used as the torsion member, the torque–strain relationship becomes relatively simple. A complete development of the relationship is found in standard textbooks on elasticity, solid mechanics, or strength of materials (see Crandall et al., 1972). The basic steps are summarized here.

For a circular shaft in torsion, shear stress varies linearly from the center to the outer circumference, as shown in figure 4.20a. The torque T that causes the shear stress is given by the integral

$$T = \int_A r\tau \, dA \qquad (4.28)$$

where dA is an elemental area at radius r on the shaft cross-section and τ is shear

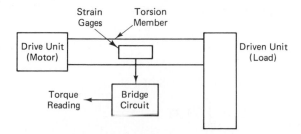

Figure 4.19. Torque sensing using a torsion member.

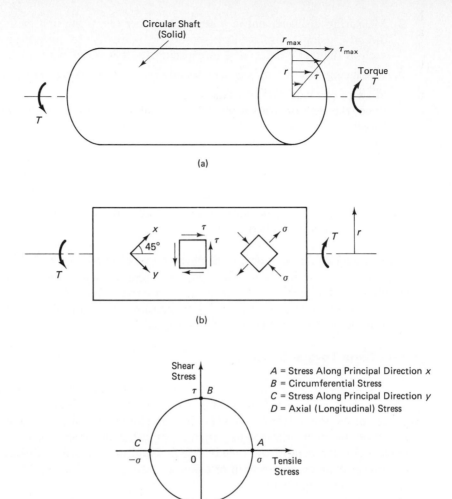

Figure 4.20. (a) Linear distribution of shear stress in a circular shaft under pure torsion. (b) Pure shear state of stress and principal directions x and y. (c) Mohr's circle.

stress. By performing the integration using the linear distribution property of τ, we get

$$\tau = \frac{Tr}{J} \tag{4.29}$$

where J is the polar moment of area, which is defined as

$$J = \int_A r^2 \, dA \tag{4.30}$$

Note that in the foregoing equations, r is any radius within the shaft cross-section and τ is the shear stress at that location. For the rest of the discussion, assume that r denotes the outer (maximum) radius and that τ is the corresponding shear stress τ_{max} (i.e., use $r \equiv r_{max}$ and $\tau \equiv \tau_{max}$ to avoid the use of subscripts).

In a small square (two-dimensional) element on the cylindrical outer surface of the shaft, with one side parallel to the shaft axis and another side along the shaft circumference, the state of stress is pure shear, as shown in figure 4.20b. Mohr's circle of the state of stress on the outer surface of the shaft is indicated in figure 4.20c. It follows that on the outer surface of the shaft, for an element as shown in figure 4.20b, at 45° to the shaft axis, the state of stress is tension or compression without shear. Accordingly, x and y in figure 4.20b are directions of principal stress on the shaft surface. Note that the principal stress (tension or compression) σ is given by (figure 4.20c)

$$\sigma = \tau \tag{4.31}$$

Now, to determine strain along the two principal directions x and y, we use stress–strain constitutive relations for a plane stress problem; thus,

$$\epsilon_x = \frac{1}{E}(\sigma_x - \nu\sigma_y) \tag{4.32}$$

$$\epsilon_y = \frac{1}{E}(\sigma_y - \nu\sigma_x) \tag{4.33}$$

where E = Young's modulus of elasticity
$\quad\quad \nu$ = Poisson's ratio

Using the fact that

$$\sigma_x = -\sigma_y = \sigma \tag{4.34}$$

we get, by solving equations 4.32 and 4.33,

$$\epsilon_x = -\epsilon_y = \epsilon = \frac{(1 + \nu)}{E}\sigma \tag{4.35}$$

Now, in view of equations 4.31 and 4.29, we have

$$\epsilon = \frac{(1 + \nu)r}{EJ}T \tag{4.36}$$

or, using the fact that shear modulus G is given by

$$E = 2(1 + \nu)G \tag{4.37}$$

we can write

$$\epsilon = \frac{r}{2GJ}T \tag{4.38}$$

It follows from either equation 4.36 or equation 4.38 that torque T can be determined by measuring direct strain ϵ on the shaft surface along a principal stress direction (i.e., at 45° to the shaft axis). This is the basis of torque sensing using strain measurements. Using the general bridge equation 4.18 along with 4.19 in equation 4.38, we can obtain torque T from bridge output δv_o:

$$T = \frac{8GJ}{kS_s r} \frac{\delta v_o}{v_{\text{ref}}} \tag{4.39}$$

where S_s is the gage factor (or sensitivity) of the strain gages. The bridge constant k depends on the number of active strain gages used. Strain gages are assumed to be mounted along a principal direction. Three possible configurations are shown in figure 4.21. In configurations (a) and (b) only two strain gages are used, and the bridge constant $k = 2$. Note that both axial loads and bending are compensated with the given configurations because resistance in both gages will be changed by the same amount (same sign and same magnitude) that cancels out up to first order, for the bridge circuit connection shown in figure 4.21. Configuration (c) has two pairs of gages, mounted on the two opposite surfaces of the shaft. The bridge constant is doubled in this configuration, and here again, the sensor clearly self-compensates for axial and bending loads up to first order $[O(\delta R)]$.

Design Considerations

Two conflicting requirements in the design of a torsion element for torque sensing are sensitivity and bandwidth. The element has to be sufficiently flexible in order to get an acceptable level of sensor sensitivity (i.e., a sufficiently large output signal).

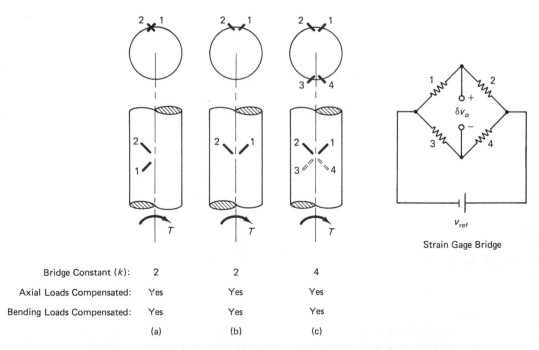

Bridge Constant (k):	2	2	4
Axial Loads Compensated:	Yes	Yes	Yes
Bending Loads Compensated:	Yes	Yes	Yes
	(a)	(b)	(c)

Figure 4.21. Strain gage configurations for a circular shaft torque sensor.

According to equation 4.38, this requires a small torsional rigidity GJ, so as to produce a large strain for a given torque. Unfortunately, since the torsion-sensing element is connected in series between a drive element and a driven element, high flexibility of the torsion element will result in low overall stiffness in the system. Specifically, with reference to figure 4.22, the overall stiffness K_{old} prior to connecting the torsion element is given by

$$\frac{1}{K_{old}} = \frac{1}{K_m} + \frac{1}{K_L} \tag{4.40}$$

and the stiffness K_{new} after connecting the torsion member is given by

$$\frac{1}{K_{new}} = \frac{1}{K_m} + \frac{1}{K_L} + \frac{1}{K_s} \tag{4.41}$$

where K_m is the equivalent stiffness of the drive unit, K_L is the equivalent stiffness of the load, and K_s is the stiffness of the torque-sensing element. Since $1/K_{new} > 1/K_{old}$, we have $K_{new} < K_{old}$. This reduction in stiffness is associated with a reduction in natural frequency and bandwidth, resulting in slower response to control commands in the overall system. Furthermore, a reduction in stiffness causes a reduction in the loop gain, and steady-state error in some motion variables can increase, which will demand more effort from the controller to achieve the required level of accuracy. One aspect of the design of the torsion element is to guarantee that the element stiffness is small enough to provide adequate sensitivity but large enough to maintain adequate bandwidth and system gain. In situations where K_s cannot be increased adequately without seriously jeopardizing the sensor sensitivity, system bandwidth can be improved by decreasing either the load inertia or the drive unit (motor) inertia.

Example 4.8

Consider a rigid load that has polar moment of inertia J_L driven by a motor with a rigid rotor that has inertia J_m. A torsion member of stiffness K_s is connected between the rotor and the load, as shown in figure 4.23a, in order to measure the torque transmitted to the load. Determine the transfer function between the motor torque T_m and the twist angle θ of the torsion member. What is the torsional natural frequency ω_n of the system? Discuss why the system bandwidth depends on ω_n. Show that the bandwidth can be improved by increasing K_s, by decreasing J_m, or by decreasing J_L. Mention some advantages and disadvantages of introducing a gear box at the motor output.

Solution From the free-body diagrams shown in figure 4.23b, the equations of motion can be written:

$$\text{For motor:} \quad J_m \ddot{\theta}_m = T_m - K_s(\theta_m - \theta_L) \tag{i}$$

$$\text{For load:} \quad J_L \ddot{\theta}_L = K_s(\theta_m - \theta_L) \tag{ii}$$

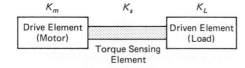

Figure 4.22. Stiffness degradation due to the flexibility of the torque-sensing element.

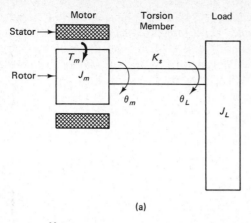

(a)

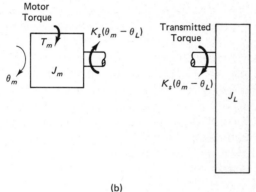

(b)

Figure 4.23. An example of bandwidth analysis of a system with a torque sensor: (a) system model; (b) free-body diagrams.

Note that θ_m is the motor rotation and θ_L is the load rotation. Divide equation (i) by J_m, divide equation (ii) by J_L, and subtract the second equation from the first; thus,

$$\ddot{\theta}_m - \ddot{\theta}_L = \frac{T_m}{J_m} - \frac{K_s}{J_m}(\theta_m - \theta_L) - \frac{K_s}{J_L}(\theta_m - \theta_L)$$

This equation can be expressed in terms of the twist angle:

$$\theta = \theta_m - \theta_L \qquad\qquad\qquad \text{(iii)}$$

$$\ddot{\theta} + K_s\left(\frac{1}{J_m} + \frac{1}{J_L}\right)\theta = \frac{T_m}{J_m} \qquad\qquad \text{(iv)}$$

Hence, the transfer function $G(s)$ between input T_m and output θ is obtained by introducing the Laplace variable s in place of the time derivative d/dt. Specifically,

$$G(s) = \frac{1/J_m}{s^2 + K_s(1/J_m + 1/J_L)} \qquad\qquad \text{(v)}$$

The characteristic equation of the twisting system is

$$s^2 + K_s\left(\frac{1}{J_m} + \frac{1}{J_L}\right) = 0 \qquad\qquad \text{(vi)}$$

It follows that the torsional (twisting) natural frequency ω_n is given by

$$\omega_n = \sqrt{K_s\left(\frac{1}{J_m} + \frac{1}{J_L}\right)} \tag{vii}$$

In addition to this natural frequency, there is a zero natural frequency in the overall system, which corresponds to rotation of the entire system as a rigid body without any twisting in the torsion member (the rigid-body mode). Note that both natural frequencies will be obtained if the output is taken as either θ_m or θ_L, not the twist angle θ. When the output is taken as the twist angle θ, the response is measured relative to the rigid-body mode; hence, the zero-frequency term disappears from the characteristic equation.

The transfer function given by equation (v) may be written as

$$G(s) = \frac{1/J_m}{s^2 + \omega_n^2} \tag{viii}$$

In the frequency domain, $s = j\omega$, and the resulting frequency transfer function is

$$G(j\omega) = \frac{1/J_m}{\omega_n^2 - \omega^2} \tag{ix}$$

It follows that if ω is small in comparison to ω_n, the transfer function can be approximated by

$$G(j\omega) = \frac{1/J_m}{\omega_n^2} \tag{x}$$

which is a static relationship, implying an instantaneous response without any dynamic delay. Since system bandwidth represents the excitation frequency range ω within which the system responds sufficiently fast, it follows that system bandwidth improves when ω_n is increased. Hence, ω_n is a measure of system bandwidth.

Now, observe from equation (vii) that ω_n (and the system bandwidth) increases when K_s is increased, when J_m is decreased, or when J_L is decreased. If a gear box is added to the system, the equivalent inertia increases and the equivalent stiffness decreases. This reduces the system bandwidth, resulting in a slower response. Another disadvantage of a gear box is the backlash and friction that enter the system. The main advantage, however, is that torque transmitted to the load is amplified through speed reduction between motor and load. However, high torques and low speeds can be achieved by using torque motors without employing any speed reducers or by using backlash-free transmissions such as harmonic drives and traction (friction) drives.

The design of a torsion element for torque sensing can be viewed as the selection of the polar moment of area J of the element to meet the following four requirements:

1. The strain capacity limit specified by the strain gage manufacturer is not exceeded.
2. A specified upper limit on nonlinearity for the strain gage is not exceeded for linear operation.
3. Sensor sensitivity is acceptable in terms of the output signal level of the difference amplifier in the bridge circuit.

4. The overall stiffness (bandwidth, steady-state error, etc.) of the system is acceptable.

Let us develop design criteria for each of these requirements.

Strain capacity of the gage.   The maximum strain handled by a strain gage element is limited by factors such as strength, creep problems associated with the bonding material, and hysteresis. This limit ϵ_{max} is specified by the strain gage manufacturer. For a typical semiconductor gage, the maximum strain limit is on the order of 3,000 $\mu\epsilon$. If the maximum torque that is required to be handled by the sensor is T_{max}, we have, from equation 4.38,

$$\frac{r}{2GJ}T_{max} < \epsilon_{max}$$

or

$$J > \frac{r}{2G}\frac{T_{max}}{\epsilon_{max}}$$

Now, introducing a safety factor ϕ, we can express the following design criterion for the torsion element.

$$J = \frac{\phi r}{2G}\frac{T_{max}}{\epsilon_{max}} \tag{4.42}$$

with ϵ_{max} being specified.

Strain gage nonlinearity limit.   For large strains, the characteristic equation of a strain gage will be increasingly nonlinear. This is particularly true for semiconductor gages. If we assume the quadratic equation 4.20, the percentage nonlinearity N_p is given by equation 4.22. For a specified nonlinearity, an upper limit for strain can be determined using this result; thus,

$$\epsilon_{max} = \frac{N_p S_1}{50 S_2} \tag{4.43}$$

The corresponding J is determined using equation 4.42; thus,

$$J = \frac{25\phi r S_2}{GS_1}\frac{T_{max}}{N_p} \tag{4.44}$$

with N_p being specified.

Sensitivity requirement.   The output signal from the strain gage bridge is provided by a differential amplifier that detects the voltages at the two output nodes of the bridge (A and B in figure 4.9b), takes the difference, and amplifies it by a gain K_a. This output signal is supplied to an analog-to-digital converter (ADC) that provides a digital signal to the computer for performing further processing and control. The signal level of the amplifier output has to be sufficiently high so that the signal-to-noise ratio (SNR) is adequate. Otherwise, serious noise problems will result. Typically, a maximum voltage on the order of ± 10 V is desired.

Amplifier output v_o is given by

$$v_o = K_a \, \delta v_o \tag{4.45}$$

where δv_o is the bridge output before amplification. It follows that the desired signal level can be obtained by simply increasing the amplifier gain. There are limits to this approach, however. In particular, a large gain will increase the susceptibility of the amplifier to saturation and instability problems, such as drift and errors due to parameter changes. Hence, sensitivity has to be improved as much as possible through mechanical considerations.

By substituting equation 4.39 into 4.45, we get the signal level requirement

$$v_o \leq \frac{K_a k S_s r v_{\text{ref}}}{8GJ} T_{\text{max}}$$

where v_o is the specified lower limit on the output signal from the bridge amplifier. Then

$$J \leq \frac{K_a k S_s r v_{\text{ref}}}{8G} \frac{T_{\text{max}}}{v_o}$$

Hence, the limiting design value for J is given by

$$J = \frac{K_a k S_s r v_{\text{ref}}}{8G} \frac{T_{\text{max}}}{v_o} \tag{4.46}$$

with v_o being specified

Stiffness requirement. The lower limit of the overall stiffness of the system is constrained by factors such as speed of response (represented by system bandwidth) and steady-state error (represented by system gain). The polar moment of area J should be chosen such that the stiffness of the torsion element does not fall below a specified limit K.

First, let us obtain an expression for the torsional stiffness of a circular shaft. For a shaft of length L and radius r, a twist angle of θ corresponds to a shear strain of

$$\gamma = \frac{r\theta}{L} \tag{4.47}$$

on the outer surface. Accordingly, shear stress is given by

$$\tau = \frac{Gr\theta}{L} \tag{4.48}$$

Now in view of equation 4.29, the torsional stiffness of the shaft is given by

$$K_s = \frac{T}{\theta} = \frac{GJ}{L} \tag{4.49}$$

Note that the stiffness can be increased by increasing GJ. However, this decreases sensor sensitivity because, in view of equation 4.38, measured direct strain ϵ decreases for a given torque when GJ is increased. Note that there are two other parameters—outer radius r and length L of the torsion element—that we can manipulate. Although, for a solid shaft, J increases (to the fourth power) with r, it is possible to manipulate J and r independently, with practical limitations, for hollow shafts. For this reason, hollow elements are commonly used as torque-sensing elements. With these design freedoms, for a given value of GJ, we can increase r to increase the sensitivity of the strain gage bridge without changing the system stiffness, and we can decrease L to increase the system stiffness without affecting the bridge sensitivity.

Assuming that the shortest possible length L is used in the sensor, for a specified stiffness limit K we should have

$$\frac{GJ}{L} \geq K$$

The limiting design value for J is given by

$$J = \frac{L}{G}K \tag{4.50}$$

with K being specified.

The governing formulas for the polar moment of area J of the torque sensor, based on the four criteria discussed earlier, are summarized in table 4.1.

Example 4.9

A joint of a direct-drive robotic arm is sketched in figure 4.24. Note that the rotor of the drive motor is an integral part of the driven link, without the use of gears or any other speed reducers. Also, the motor stator is an integral part of the drive link. A

TABLE 4.1   DESIGN CRITERIA FOR A STRAIN GAGE TORQUE-SENSING ELEMENT

Criterion	Specification	Governing formula for the second moment of area J
Strain capacity of strain gage element	ϵ_{max}	$\dfrac{\phi r}{2G} \cdot \dfrac{T_{max}}{\epsilon_{max}}$
Strain gage nonlinearity	N_p	$\dfrac{25\phi r S_2}{GS_1} \cdot \dfrac{T_{max}}{N_p}$
Sensor sensitivity	v_o	$\dfrac{K_a k S_s r v_{ref}}{8G} \cdot \dfrac{T_{max}}{v_o}$
Sensor stiffness (system bandwidth and gain)	K	$\dfrac{L}{G} \cdot K$

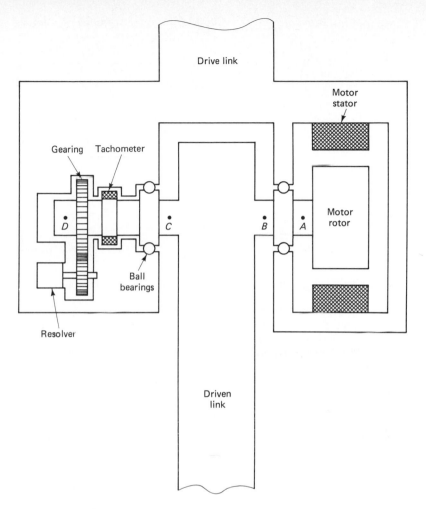

Figure 4.24. A joint of a direct-drive robotic arm.

tachometer measures the joint speed (relative), and a resolver measures the joint rotation (relative). Gearing is used to improve the performance of the resolver (see chapters 3 and 5). Neglecting mechanical loading from sensors but including bearing friction, sketch the torque distribution along the joint axis. Suggest a location (or locations) for measuring the net torque transmitted to the driven link using a strain gage torque sensor.

Solution   For simplicity, assume point torques. Denoting the motor torque by T_m, the total rotor inertia torque and frictional torque in the motor by T_I, and the frictional torques at the two bearings by T_{f1} and T_{f2}, the torque distribution can be sketched as shown in figure 4.25. The net torque transmitted to the driven link is T_L. Locations available to install strain gages include A, B, C, and D. Note that T_L is given by the difference between the torque at B and the torque at C. Hence, strain gage torque sensors should be mounted at B and C and the difference of the readings should be taken for accurate measurement of T_L. Since bearing friction is small for most practical purposes,

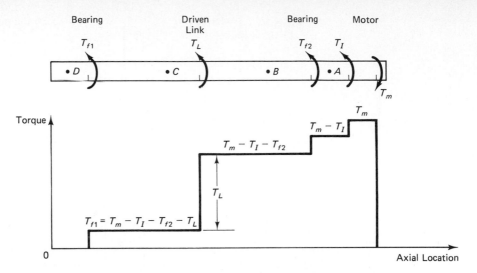

Figure 4.25. Torque distribution along the axis of a direct-drive manipulator joint.

a single torque sensor located at B will provide reasonably accurate results. The motor torque T_m is also approximately equal to the transmitted torque when bearing friction and motor loading effects (inertia and friction) are negligible. This is the reason behind using motor current (field or armature) to measure joint torque in some robotic applications.

Example 4.10

Consider the design of a tubular torsion element. Using the notation of table 4.1, the following design specifications are given: $\epsilon_{max} = 3,000 \ \mu\epsilon$; $N_p = 5\%$; $v_o = 10$ V; and for a system bandwidth of 50 Hz, $K = 2.5 \times 10^4$ lbf.in./rad. A bridge with four active strain gages is used to measure torque in the torsion element. The following parameter values are provided:

1. For strain gages:

$$S_s = S_1 = 115, \ S_2 = 3500$$

2. For the torsion element

$$\text{Outer radius } r = 1''$$

$$\text{Shear modulus } G = 5 \times 10^6 \ \text{lbf/in.}^2$$

$$\text{Length } L = 1''$$

3. For the bridge circuitry:

$$v_{ref} = 20 \text{ V} \quad \text{and} \quad K_a = 100$$

The maximum torque that is expected is

$$T_{max} = 100 \text{ lbf.in.}$$

Using these values, design a torsion element for the sensor. Compute the specifications of the designed sensor.

Solution Let us assume a safety factor of $\phi = 1$. We can compute the polar moment of area J using each of the four criteria given in table 4.1:

1. For $\epsilon_{max} = 3{,}000~\mu\epsilon$:

$$J = \frac{1 \times 1 \times 100}{2 \times 5 \times 10^6 \times 3 \times 10^{-3}} \text{ in.}^4 = 3.33 \times 10^{-3} \text{ in.}^4$$

2. For $N_p = 5$:

$$J = \frac{25 \times 1 \times 1 \times 3500 \times 100}{5 \times 10^6 \times 115 \times 5} \text{ in.}^4 = 3.04 \times 10^{-3} \text{ in.}^4$$

3. For $v_o = 10$ V:

$$J = \frac{100 \times 4 \times 115 \times 1 \times 20 \times 100}{8 \times 5 \times 10^6 \times 10} \text{ in.}^4 = 0.23 \text{ in.}^4$$

4. For $K = 2.5 \times 10^4$ lbf.in./rad:

$$J = \frac{1 \times 2.5 \times 10^4}{5 \times 10^6} \text{ in.}^4 = 5 \times 10^{-3} \text{ in.}^4$$

It follows that for an acceptable sensor, we should satisfy

$$J \geq (3.35 \times 10^{-3}) \text{ and } (3.04 \times 10^{-3}) \text{ and } (5 \times 10^{-3}) \text{ and } J \leq 0.23 \text{ in.}^4$$

We pick $J = 0.23$ in.^{4} so that the tube thickness is sufficiently large to transmit load without buckling or yielding. Since, for a tubular shaft,

$$J = \frac{\pi}{2}(r_o^4 - r_i^4)$$

where r_o is the outer radius and r_i is the inner radius, we have

$$0.23 = \frac{\pi}{2}(1^4 - r_i^4)$$

or

$$r_i = 0.86 \text{ in.}$$

Now, with the chosen value for J:

$$\epsilon_{max} = \frac{0.23}{3.33 \times 10^{-3}} \times 3{,}000~\mu\epsilon = 2.07 \times 10^5~\mu\epsilon$$

$$N_p = \frac{3.04 \times 10^{-3}}{0.23} \times 5\% = 0.07\%$$

$$v_o = 10 \text{ V}$$

$$K = \frac{0.23}{5 \times 10^{-3}} \times 2.5 \times 10^4 = 1.15 \times 10^6 \text{ lbf.in./rad}$$

Since natural frequency is proportional to the square root of stiffness, for a given inertia, we note that a bandwidth of

$$50 \sqrt{\frac{1.15 \times 10^6}{2.5 \times 10^4}} = 339 \text{ Hz}$$

is possible with this design.

Although the manner in which strain gages are configured on a torque sensor can be exploited to compensate for cross-sensitivity effects arising from factors such as tensile and bending loads, it is advisable to use a torque-sensing element that inherently possesses low sensitivity to these factors that cause error in a torque measurement. The tubular torsion element discussed in this section is convenient for analytical purposes because of the simplicity of the associated expressions for design parameters. Unfortunately, such an element is not very rigid to bending and tensile loads. Alternative shapes and structural arrangements have to be considered if inherent rigidity (insensitivity) to cross-loads is needed. Furthermore, a tubular element has the same strain at all locations on the element surface. This does not give us a choice with respect to mounting locations of strain gages in order to maximize the torque sensor sensitivity. Another disadvantage of the basic tubular element is that the surface is curved; therefore, much care is needed in mounting fragile semiconductor gages, which would be easily damaged even with slight bending. Hence, a sensor element that has flat surfaces to mount the strain gages would be desirable.

A torque-sensing element that has the foregoing desirable characteristics (i.e., inherent insensitivity to cross-loading, nonuniform strain distribution on the surface, and availability of flat surfaces to mount strain gages) is shown in figure 4.26. Note that two sensing elements are connected radially between the drive unit and the driven member. The sensing elements undergo bending to transmit a torque between the driver and the driven member. Bending strains are measured at locations of high sensitivity and are taken to be proportional to the transmitted torque. Analytical determination of the calibration constant is not easy for such complex sensing elements, but experimental determination is straightforward. Note that the strain gage torque sensors measure the direction as well as the magnitude of the torque transmitted through it.

Deflection Torque Sensors

Instead of measuring strain in the sensor element, the actual deflection (twisting or bending) can be measured and used to determine torque, through a suitable calibration constant. For a circular shaft (solid or hollow) torsion element, the governing relationship is given by equation 4.49, which can be written in the form

$$T = \frac{GJ}{L} \theta \qquad (4.51)$$

The calibration constant GJ/L has to be small in order to achieve high sensitivity. This means that the element stiffness should be low. This limits the bandwidth (which measures speed of response) and gain (which determines steady-state error) of the overall system. The twist angle θ is very small (e.g., a fraction of a degree) in

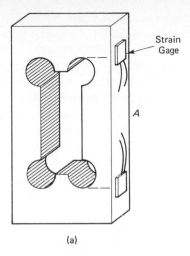

(a)

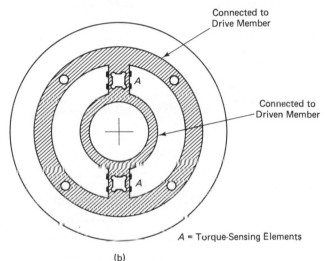

A = Torque-Sensing Elements

(b)

Figure 4.26. A bending element for torque sensing: (a) shape of the sensing element; (b) element location.

systems with high bandwidth. This requires very accurate measurement of θ in order to determine torque T.

Two types of displacement torque sensors will be described here. One sensor directly measures the angle of twist, and the other sensor uses the change in magnetic induction associated with sensor deformation.

Direct-deflection torque sensor. Direct measurement of the twist angle between two axial locations in the torsion member, using an angular displacement sensor, can be used to determine torque. The difficulty in this case is that under dynamic conditions, relative deflection has to be measured while the torsion element is rotating. One type of displacement sensor that could be used here is a synchro transformer (see chapter 3). The two rotors of the synchro are mounted at the two ends of the torsion member. The synchro output gives the relative angle of rotation of the two rotors. Another type of displacement sensor that could be used is shown in figure 4.27a. Two ferromagnetic gear wheels are splined at two axial locations of the torsion element. Two stationary proximity probes of the magnetic induction type (self-induction or mutual induction) are placed radially, facing the gear teeth, at the two locations. As the shaft rotates, the gear teeth change the flux linkage of the proximity sensor coils. The resulting output signals of the two probes are pulse sequences, shaped somewhat like sine waves. The phase shift of one signal with respect to the other determines the relative angular deflection of one gear wheel with respect to the other, assuming that the two probes are synchronized under no-torque conditions. Both the magnitude and the direction of the transmitted torque are determined using this method. A 360° phase shift corresponds to a relative deflection by an integer multiple of the gear pitch. It follows that deflections less than half the pitch can be measured without ambiguity. Assuming that the output signals of the two probes are sine waves (narrow-band filtering can be used to achieve this), the phase shift ϕ is proportional to the angular twist θ. If the gear wheel has n teeth, a primary phase shift of 2π corresponds to a twist angle of $2\pi/n$ radians. Hence,

$$\theta = \frac{\phi}{n}$$

and from equation 4.51, we get

$$T = \frac{GJ\phi}{Ln} \qquad (4.52)$$

where G = shear modulus of the torsion element
$\quad J$ = polar moment of area of the torsion element
$\quad \phi$ = phase shift between the two proximity probe signals
$\quad L$ = axial separation of the proximity probes
$\quad n$ = number of teeth in each gear wheel

Note that the proximity probes are noncontact devices, unlike DC strain gage sensors. Also, note that eddy current proximity probes and Hall effect proximity probes (see chapter 5) could be used instead of magnetic induction probes in this method of torque sensing.

Variable-reluctance torque sensor.   A torque sensor that is based on sensor element deformation and that does not require a contacting commutator is shown in figure 4.27b. This is a variable-reluctance device that operates like a differential transformer (RVDT or LVDT). The torque-sensing element is a ferromagnetic tube that has two sets of slits, typically oriented along the two principal stress directions of the tube (45°) under torsion. When a torque is applied to the torsion element, one set of gaps closes and the other set opens as a result of the principal stresses normal to the slit axes. Primary and secondary coils are placed around the slitted tube, and they remain stationary. One segment of the secondary coil is placed around one set of slits, and the second segment is placed around the other (perpendicular) set. The primary coil is excited by an AC supply, and the induced voltage v_o in the secondary coil is measured. As the tube deforms, it changes the magnetic reluctance in the flux

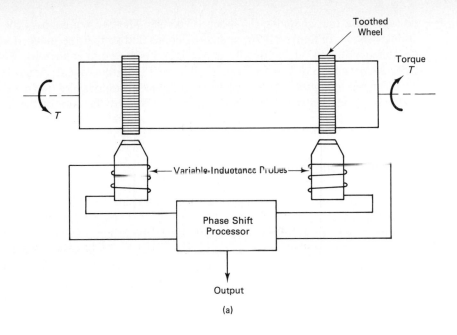

(a)

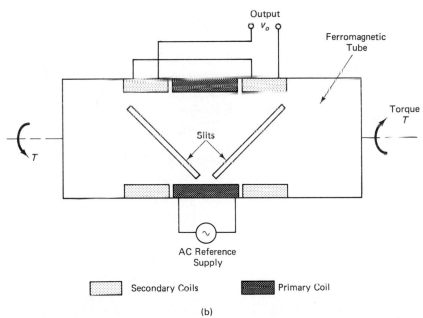

(b)

Figure 4.27. Deflection torque sensors: (a) a direct-defelection torque sensor; (b) a variable-reluctance torque sensor.

linkage path, thus changing the induced voltage. The two segments of the secondary coil, as shown in figure 4.27b, should be connected so that the induced voltages are absolutely additive (algebraically subtractive), because one voltage increases and the other decreases, to obtain the best sensitivity. The output signal should be demodulated (by removing the carrier frequency component) to measure transient torques effectively. Note that the direction of torque is given by the sign of the demodulated signal.

Reaction Torque Sensors

The foregoing methods of torque sensing use a sensing element that is connected between the drive member and the driven member. A major drawback of such an arrangement is that the sensing element modifies the original system in an undesirable manner, particularly by decreasing the system stiffness and adding inertia. Not only will the overall bandwidth of the system decrease, but the original torque will also be changed (mechanical loading) because of the inclusion of an auxiliary sensing element. Furthermore, under dynamic conditions, the sensing element will be in motion, thereby making torque measurement more difficult. The reaction method of torque sensing eliminates these problems to a large degree. This method can be used to measure torque in a rotating machine. The supporting structure (or housing) of the rotating machine (e.g., motor, pump, compressor, turbine, generator) is cradled by releasing fixtures, and the effort necessary to keep the structure from moving is measured. A schematic representation of the method is shown in figure 4.28a. Ideally, a lever arm is mounted on the cradled housing, and the force required to fix the housing is measured using a force sensor (load cell). The reaction torque on the housing is given by

$$T_R = F_R \cdot L \tag{4.53}$$

where F_R = reaction force measured using load cell
L = lever arm length

Alternatively, strain gages or other types of force sensors could be mounted directly at the fixture locations (e.g., at the mounting bolts) of the housing to measure the reaction forces without cradling the housing. Then the reaction torque is determined with a knowledge of the distance of the fixture locations from the shaft axis.

The reaction-torque method of torque sensing is widely used in dynamometers (reaction dynamometers) that determine transmitted power in rotating machinery through torque and shaft speed measurements. A major drawback of reaction-type torque sensors can be explained using figure 4.28b. A motor with rotor inertia J, which rotates at angular acceleration $\ddot{\theta}$, is shown. By Newton's third law (action = reaction), the electromagnetic torque generated at the rotor of the motor T_m will be reacted back onto the stator and housing. By applying Newton's second law to the motor rotor, we get

$$J\ddot{\theta} = T_m - T_{f1} - T_{f2} - T_L$$

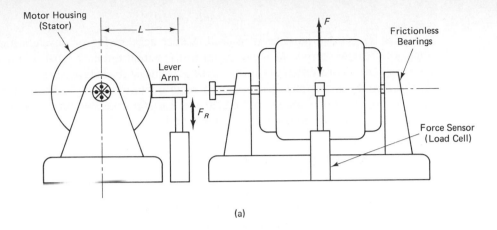

(a)

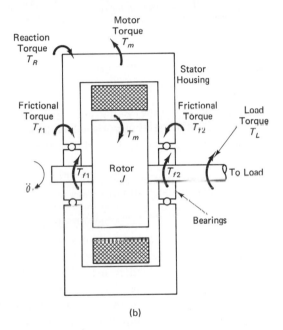

(b)

Figure 4.28. (a) Schematic representation of a reaction torque sensor setup (reaction dynamometer). (b) The relationship between reaction torque and load torque.

where T_{f1} and T_{f2} denote the frictional torques at the two bearings and T_L is the torque transmitted to the driven load. Thus,

$$T_L = T_m - T_{f1} - T_{f2} - J\ddot{\theta} \tag{i}$$

and since the motor housing is stationary, we have

$$T_R + T_{f1} + T_{f2} - T_m = 0 \tag{ii}$$

Subtracting equation (ii) from (i) gives

$$T_L = T_R - J\ddot{\theta} \tag{4.54}$$

Note that T_L is what must be measured. Under accelerating or decelerating conditions, the reaction torque T_R is not equal to the actual torque T_L that is transmitted. One method of compensating for this error is to measure shaft acceleration, compute inertia torque, and adjust the measured reaction torque using this inertia torque. Note that the frictional torque in the bearings does not enter the final equation (equation 4.54). This is an advantage of this method.

Motor Current Torque Sensors

Torque in an electric motor is generated as a result of the electromagnetic interaction between the armature winding of the motor and the field winding (see chapter 7 for details). For a DC motor, the armature winding is located on the rotor and the field winding is on the stator. The motor torque T_m in a DC motor is given by

$$T_m = ki_f i_a \tag{4.55}$$

where i_f = field current
   i_a = armature current
   k = torque constant of the motor

It is seen from equation 4.55 that the motor torque can be determined by measuring either i_a or i_f while the other is kept constant at a known value. In particular, note that i_f is assumed constant in armature control and i_a is assumed constant in field control.

In the past, DC motors were predominantly used in complex control applications. Although AC synchronous motors were limited mainly to constant-speed applications in the past, they are finding numerous uses in variable-speed applications (e.g., robotic manipulators) because of rapid advances in solid-state drives. Today, AC motor drive systems employing thyristors—such as *silicon-controlled rectifiers* (SCRs), which are unidirectional solid-state switches, and *triacs,* which are bidirectional solid-state switches—are widely used. Thyristor circuits are used to vary the frequency of the supply voltage, thereby controlling the motor speed. Torque in AC motors also can be determined by monitoring the motor current. For example, consider the three-phase synchronous motor shown schematically in figure 4.29. The armature winding of a conventional synchronous motor is carried by the stator (in contrast to DC motors). Suppose that the currents in the three phases (armature currents) are denoted by i_1, i_2, and i_3, respectively. The DC field current in the rotor winding is denoted by i_f. The motor torque T_m can be expressed as

$$T_m = ki_f\left[i_1 \sin\theta + i_2 \sin\left(\theta - \frac{2\pi}{3}\right) + i_3 \sin\left(\theta - \frac{4\pi}{3}\right) \right] \tag{4.56}$$

where θ denotes the angular rotation of the rotor and k is the torque constant of the synchronous motor. Note that since i_f is assumed fixed, the motor torque can be determined by measuring the phase currents. For the special case of a "balanced" three-phase supply, we have

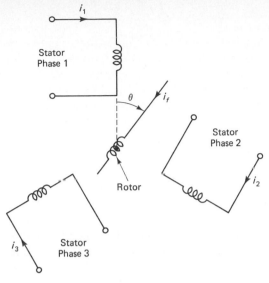

Stator
Phase 1

θ i_f

Stator
Phase 2

i_2

Rotor

i_3

Stator
Phase 3

Figure 4.29. Schematic representation of a three-phase synchronous motor.

$$i_1 = i_a \sin \omega t \tag{4.57}$$

$$i_2 = i_a \sin \left(\omega t - \frac{2\pi}{3} \right) \tag{4.58}$$

$$i_3 = i_a \sin \left(\omega t - \frac{4\pi}{3} \right) \tag{4.59}$$

where ω denotes the line frequency (frequency of the current in each supply phase) and i_a is the amplitude of the phase current. Substituting equations 4.57 through 4.59 into 4.56, and simplifying using well-known trigonometric identities, we get

$$T_m = 1.5ki_f i_a \cos (\theta - \omega t) \tag{4.60}$$

We know that the angular speed of a three-phase synchronous motor with one pole pair per phase is equal to the line frequency ω (see chapter 7). Accordingly, we have

$$\theta = \theta_0 + \omega t \tag{4.61}$$

where θ_0 denotes the angular position of the rotor at $t = 0$. It follows that with a balanced three-phase supply, the torque of a synchronous motor is given by

$$T_m = 1.5ki_f i_a \cos \theta_0 \tag{4.62}$$

This expression is quite similar to the one for a DC motor, as given by equation 4.55.

It should be understood that motor current provides only an estimate for motor torque. The actual torque that is transmitted through the motor shaft (the load torque) is different from the motor torque generated at the stator–rotor interface of the motor. This difference is necessary for overcoming the inertia torque of the moving parts of the motor unit (particularly rotor inertia) and frictional torque (particularly bearing friction).

4.6 FORCE SENSORS

Force sensors are useful in numerous applications. For example, cutting forces generated by a machine tool may be monitored to detect an impending failure and to diagnose the causes of failure, in controlling the machine tool, and in evaluating the product quality. In vehicle testing, force sensors are used to monitor impact forces. Robotic handling and assembly tasks are controlled by measuring the forces generated at the end effector. Direct measurement of forces is useful in nonlinear feedback control of mechanical systems.

Force sensors that employ strain gage elements or piezoelectric (quartz) crystals with built-in microelectronics are common. For example, thin-film and foil sensors that employ the strain gage principle for measuring forces and pressures are commercially available. Several such sensors are shown in figure 4.30. Both impulsive forces and slowly varying forces can be monitored using these sensors. Some types of force sensors are based on measuring a deflection caused by the force. Relatively high deflections (typically several mils) would be necessary for this technique to be feasible. Commercially available sensors range from sensitive devices that can detect forces on the order of thousandth of a newton to heavy-duty load cells that can handle very large forces (e.g., 10 tons). Since the techniques of torque sensing can be extended in a straightforward manner to force sensing, further discussion of the topic is not undertaken here. Instead, we will discuss a nonlinear feedback control method that employs force sensing.

Nonlinear Feedback Control

Force and acceleration measurements can be used to linearize the behavior of a complex mechanical system, thereby making it amenable to linear control techniques. This approach is commonly known as nonlinear feedback control. Many researchers have proposed this method for controlling robotic manipulators, which are highly nonlinear, and coupled mechanical dynamic systems. This section presents a general discussion of the method. Specific applications can be found in the literature on robotic manipulator control.

Consider a nonlinear, coupled, mechanical system represented by the dynamic model

$$\mathbf{M}(\mathbf{y})\ddot{\mathbf{y}} + \mathbf{b}(\mathbf{y}, \dot{\mathbf{y}}, t) = \mathbf{f} \qquad (4.63)$$

where $\mathbf{M}(\mathbf{y})$ = mass matrix of the system

$\mathbf{y}$ = response vector (a vector of generalized coordinates, giving the displacement in each degree of freedom)

$\mathbf{f}$ = forcing vector (a vector of forces along the degrees of freedom of the system)

Note that the vector $\mathbf{b}$ represents all the remaining forces present along each degree of freedom. For example, $\mathbf{b}$ includes dissipation forces (coulomb friction, structural damping, etc.), gravity forces, nonlinear inertia forces such as centrifugal and coriolis forces, and backlash effects due to loose parts and gears. The inertia

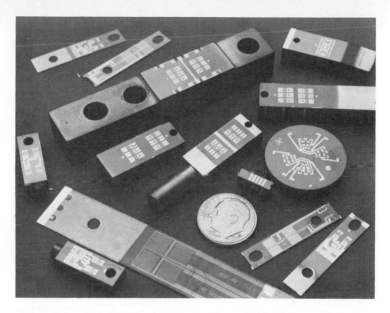

Figure 4.30. Thin-film and foil force sensors (Courtesy of Transducers, Inc.).

matrix $\mathbf{M}$ depends nonlinearly on $\mathbf{y}$ in general; these nonlinearities are kinematic in nature and usually are trigonometric functions. The vector $\mathbf{b}$ depends nonlinearly on the velocity vector $\dot{\mathbf{y}}$ as well as $\mathbf{y}$. Furthermore, its parameters might be time-variant; hence, the time variable t is explicitly indicated as an argument. We know that, in practice, it is much easier to model the inertia matrix (or even to measure it) than to model and accurately determine the $\mathbf{b}$ vector. Now, suppose that we have a good estimate for $\mathbf{M}$. Then, by measuring the force vector $\mathbf{f}$ and the acceleration vector $\ddot{\mathbf{y}}$, we can estimate $\mathbf{b}$. Note that, in many cases, determination of $\mathbf{M}$ will require measurement of the displacement vector $\mathbf{y}$ as well.

Suppose that we make an estimate $\hat{\mathbf{b}}$ for the unknown forces $\mathbf{b}$ in equation 4.63, and the driving force vector $\mathbf{f}$ is generated according to

$$\mathbf{f} = \hat{\mathbf{b}} + \mathbf{u} \tag{4.64}$$

where $\mathbf{u}$ is an external input vector. By substituting equation 4.64 in 4.63, we note that if our estimate $\hat{\mathbf{b}}$ is accurate, then we have

$$\mathbf{M}(\mathbf{y})\ddot{\mathbf{y}} = \mathbf{u} \tag{4.65}$$

This is obviously a much simpler system, and it should be easier to control. Still, the system given by equation 4.65 is nonlinear. Now suppose that we measure the displacement vector $\mathbf{y}$, determine (possibly compute) the inertia matrix $\mathbf{M}(\mathbf{y})$, which is nonlinear in $\mathbf{y}$, and form the input vector $\mathbf{u}$ according to the relation

$$\mathbf{u} = \hat{\mathbf{M}}(\mathbf{y})\mathbf{K}[\mathbf{y}_d - \mathbf{y}] \tag{4.66}$$

where $\hat{\mathbf{M}}(\mathbf{y})$ = computed (estimated) mass matrix
$\quad\quad \mathbf{K}$ = a constant matrix of control gains
$\quad\quad \mathbf{y}_d$ = desired (reference) value for the displacement vector

By substituting equation 4.66 in 4.65, we observe that if the estimated mass matrix is sufficiently accurate, and since $\mathbf{M}$ is a positive definite matrix, the resulting control system is given by

$$\ddot{\mathbf{y}} = \mathbf{K}[\mathbf{y}_d - \mathbf{y}] \tag{4.67}$$

Note that this is a "linear" system. Furthermore, if $\mathbf{K}$ is chosen as a diagonal matrix, we get an uncoupled system. The overall control system is shown by the block diagram in figure 4.31. The control problem now is to select the constant gains $\mathbf{K}$ so that $\mathbf{y}$ is sufficiently close to the desired $\mathbf{y}_d$. If the closeness achieved is not adequate, we can improve the control system—for example, by adding a velocity feedback loop. In any event, since equation 4.67 is linear and uncoupled, the associated control problem will be much simpler. Much past experience is available, and the associated theory is well developed and quite exact for this type of linear control problem. This is a significant advantage of the nonlinear feedback control method presented here. Equation 4.67 is exact if and only if our estimates for $\mathbf{b}$ and $\mathbf{M}$ are exact. As always, some error will be present. This error will be small, however, if the measurements are made accurately and if the model for the inertia matrix is sufficiently accurate. Then the linear equation 4.67 can still be used, with possible errors being represented by an unknown disturbance. Since this disturbance input is small, the associate control problem is still relatively simple. The major difficulty in this method is the time delay of the control action; if $\mathbf{b}$ is estimated using one measurement sample, nonlinear feedback can be quick but inaccurate. On the other hand, accurate estimation of $\mathbf{b}$ introduces large time delays in the control action.

Stability Problems in Force Feedback

In force feedback control, the stability of the control system can depend on the location of the force sensors used. In particular, in robotic manipulator applications, it has been experienced that with some locations and configurations of a force-sensing

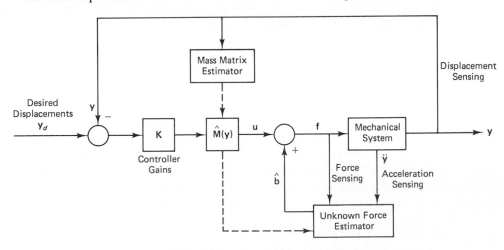

Figure 4.31. Nonlinear feedback control using force acceleration, and displacement measurements.

wrist at the robot end effector, dynamic instabilities were present in the manipulator response for some (large) values of control gains in the force feedback loop. These instabilities were found to be limit-cycle-type motions in most cases. Generally, it is known that when the force sensors are more remotely located with respect to the drive actuators of a mechanical system, the system is more likely to exhibit instabilities under force feedback control. Hence, it is desirable to make force measurements very close to the actuator locations when force feedback is used.

4.7 TACTILE SENSING

Tactile sensing is usually interpreted as *touch sensing*, but tactile sensing is different from a simple touch where very few discrete force measurements are made. In tactile sensing, a force "distribution" is measured, using a closely spaced array of force sensors and usually exploiting the skinlike properties of the sensor array.

Tactile sensing is particularly important in two types of operations: (1) grasping and (2) object identification. In grasping, the object has to be held in a stable manner without being allowed to slip and without being damaged. Object identification includes recognizing or determining the shape, location, and orientation of an object as well as detecting or identifying surface properties (e.g., density) and defects. Ideally, these tasks would require two types of sensing:

1. Continuous-variable sensing of contact forces
2. Sensing of the surface deformation profile

These two types of data are generally related through the constitutive relations (e.g., stress–strain relations) of the tactile sensor touch surface or of the object that is being grasped. As a result, either the almost-continuous-variable sensing of tactile forces or the sensing of the tactile deflection profile, separately, is often termed tactile sensing. Note that learning experience is also an important part of tactile sensing. For example, picking up a fragile object such as an egg and picking up an object that has the same shape but is made of a flexible material are not identical processes; they require some learning through touch, particularly when vision capability is not available.

Tactile Sensor Requirements

Significant advances in tactile sensing are taking place in the robotics area. Applications, which are very general and numerous, include automated inspection of surface profiles and joints for defects, material handling or parts transfer (e.g., pick and place), parts assembly (e.g., parts mating), parts identification and gaging in manufacturing applications (e.g., determining the size and shape of a turbine blade picked from a bin), and fine-manipulation tasks (e.g., production of arts and crafts, robotic engraving, and robotic microsurgery). Note that some of these applications might need only simple touch (force-torque) sensing if the parts being grasped are properly oriented and if adequate information about the process is already available.

Naturally, the frequently expressed design objective for tactile sensing devices has been to mimic the capabilities of human fingers. Specifically, the tactile sensor should have a compliant covering with skinlike properties, along with enough degrees of freedom for flexibility and dexterity, adequate sensitivity and resolution for information acquisition, adequate robustness and stability to accomplish various tasks, and some local intelligence for identification and learning purposes. Although the spatial resolution of a human fingertip is about 2 mm, much finer spatial resolutions (less than 1 mm) can be realized if information through other senses (e.g., vision), prior experience, and intelligence are used simultaneously during the touch. The force resolution (or sensitivity) of a human fingertip is on the order of 1 gm. Also, human fingers can predict "impending slip" during grasping, so that corrective actions can be taken before the object actually slips. At an elementary level, this requires a knowledge of shear stress distribution and friction properties at the common surface of the object and the hand. Additional information and an "intelligent" processing capability are also needed to predict slip accurately and to take corrective actions to prevent slipping. These are, of course, ideal goals for a tactile sensor, but they are not unrealistic in the long run. Typical specifications for an industrial tactile sensor are as follows:

1. Spatial resolution of about 2 mm
2. Force resolution (sensitivity) of about 2 gm
3. Force capacity (maximum touch force) of about 1 kg
4. Response time of 5 ms or less
5. Low hysteresis (low energy dissipation)
6. Durability under harsh working conditions
7. Insensitivity to change in environmental conditions (temperature, dust, humidity, vibration, etc.)
8. Capability to detect and even predict slip

Although the technology of tactile sensing has not peaked yet, and the widespread use of tactile sensors in industrial applications is still to come, several types of tactile sensors that meet and even exceed the foregoing specifications are commercially available. In future developments of these sensors, two separate groups of issues need to be addressed:

1. Ways to improve the mechanical characteristics and design of a tactile sensor so that accurate data with high resolution can be acquired quickly using the sensor
2. Ways to improve signal analysis and processing capabilities so that useful information can be extracted from the data acquired through tactile sensing

Under the second category, we also have to consider techniques for using tactile information in the feedback control of processes. In this context, the development of control algorithms, rules, and inference techniques for intelligent controllers that use tactile information has to be addressed.

Construction and Operation of Tactile Sensors

The touch surface of a tactile sensor is usually made of an elastomeric pad or flexible membrane. The contact force distribution in a tactile sensor is usually measured using an array of force sensors located under the flexible membrane. The deflection profile is determined using a matrix of proximity sensors or deflection sensors. Arrays of piezoelectric sensors and metallic or semiconductor strain gages (piezoresistive sensors) in sufficient density (number of elements per unit area) may be used for the measurement of the tactile force distribution. Electromagnetic and capacitive sensors may be used in obtaining the deflection profile. The principles of operation of these types of sensors have been discussed in this chapter and in the preceding chapter. In particular, semiconductor elements are poor in mechanical strength but have good sensitivity. The skinlike membrane itself can be made from a conductive elastomer (e.g., graphite-leaded neoprene rubber) whose resistance changes can be monitored and used in determining the force and deflection distribution. Common problems with conductive elastomers are electrical noise, nonlinearity, hysteresis, low sensitivity, drift, low bandwidth, and poor material strength. Optical tactile sensors use light-sensitive elements (photodetectors) to sense the intensity of light (or laser beams) reflected from the tactile surface. Since the light intensity depends on the distance from the tactile surface to the photosensor, the deflection profile can be determined. Optical methods have the advantages of being free from electromagnetic noise and safe in explosive environments, but they can have errors due to stray light reaching the sensor, variation in intensity of the light source, and changes in properties of the light-propagation medium (e.g., dirt, humidity, and smoke). Note that since force and deflection are related through a constitutive law for the tactile sensor (touch pad), only one type of measurement, not both force and deflection, is needed in tactile sensing.

A schematic representation of an optical tactile sensor (built at the Man-Machine Systems Laboratory at MIT) is shown in figure 4.32. If a beam of light (or laser) is projected onto a reflecting surface, the intensity of light reflected back and received by a light receiver depends on the distance (proximity) of the reflecting surface. For example, in figure 4.32a, more light is received by the light receiver when the reflecting surface is at Position 2 than when it is at Position 1. But if the reflecting surface actually touches the light source, light will be completely blocked off, and no light will reach the receiver. Hence, in general, the proximity-intensity curve for an optical proximity sensor will be nonlinear and will have the shape shown in figure 4.32a. Using this (calibration) curve, we can determine the position (x) once the intensity of the light received at the photosensor is known. This principle, which was explained in chapter 3, is the principle of operation of many optical tactile sensors. In the system shown in figure 4.32b, the flexible tactile element consists of a thin, light-reflecting surface embedded within an outer layer (touch pad) of high-strength rubber and an inner layer of transparent rubber. Optical fibers are uniformly and rigidly mounted across this inner layer of rubber so that light can be projected directly onto the reflecting surface.

The light source, the beam splitter, and the solid-state (charge-coupled device, or CCD) camera form an integral unit that can be moved laterally in known steps to

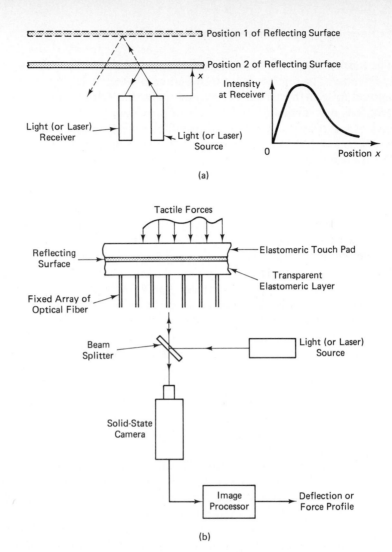

Figure 4.32. (a) The principle of an optical proximity sensor. (b) Schematic representation of a fiber optic tactile sensor.

scan the entire array of optical fiber if a single image frame of the camera does not cover the entire array. The splitter plate reflects part of the light from the light source onto a bundle of optical fiber. This light is reflected by the reflecting surface and is received by the solid-state camera. Since the intensity of the light received by the camera depends on the proximity of the reflecting surface, the density image detected by the camera will determine the deflection profile of the tactile surface. Using appropriate constitutive relations for the tactile sensor, the tactile force distribution can be determined as well. The image processor computes the deflection profile and the associated tactile force distribution in this manner. Note that the sensor resolution can be improved at the expense of the thickness of the elastomer layer, which determines the robustness of the sensor.

In the described fiber optic tactile sensor (figure 4.32), the optical fibers serve as the medium through which light or laser rays are transmitted to the tactile site. Alternatively, the light source and receiver can be located at the tactile site itself. The principle of operation of this type of tactile sensor is shown in figure 4.33. When the elastomeric touch pad is pressed at a sensitive point, the pin connected to the pad at that point moves (in the x direction), thereby obstructing the light received by the photodiode from the light-emitting diode (LED). Hence, the output signal of the photodiode measures the pin movement. One type of piezoresistive tactile sensor uses an array of semiconductor strain gages mounted under the touch pad on a rigid base. In this manner, the force distribution on the touch pad is measured directly. Ultrasonic tactile sensors are based, for example, on pulse-echo ranging (see chapter 3). In this method, the tactile surface is two membranes separated by an air gap. The time taken for an ultrasonic pulse to travel through the gap and be reflected back onto a receiver depends, in particular, on the thickness of the air gap. Since this time interval changes with deformation of the tactile surface, it can serve as a measure of the deformation of the tactile surface at a given location. Other possibilities for tactile sensors include the use of chemical effects that might be present when an object is touched and the influence of grasping on the natural frequencies of an array of sensing elements.

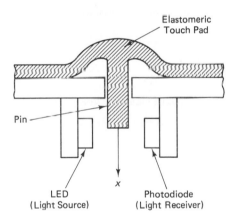

Figure 4.33. An optical tactile sensor with localized light sources and photosensors.

Example 4.11

A strain gage tactile sensor has been developed by the Eaton Corporation in Troy, Michigan. The concept behind it can be used to determine the size and location of a point-contact force, which is useful, for example, in parts-mating applications. Consider a square plate of length a, simply supported by frictionless hinges at its four corners on strain gage load cells, as shown in figure 4.34a. Show that the magnitude, direction, and location of a point force P applied normally to the plate can be determined using the readings of the four load cells. If a particular parts-mating process limits the tolerance on the measurement error of the force location to δr, determine the tolerance δF on the load cell error. Note that typical values for a and P are 2 in. and 200 lb, respectively.

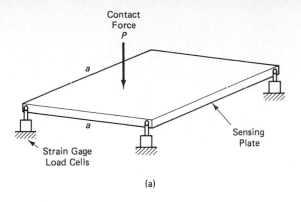

Contact
Force
P

a

a

Sensing
Plate

Strain Gage
Load Cells

(a)

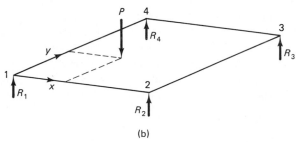

(b)

Figure 4.34. (a) Schematic represenation of a strain gage point-contact sensor. (b) Free-body diagram.

Solution   Consider the free-body diagram shown in figure 4.34b. The location of force P is given by the coordinates (x, y) in the Cartesian coordinate system (x, y, z), with origin located at 1, as shown. The load cell reading at location i is denoted by R_i. Equilibrium in the z direction gives the force balance

$$P = R_1 + R_2 + R_3 + R_4 \qquad \text{(i)}$$

Equilibrium about the y-axis gives the moment balance

$$Px = R_2a + R_3a \qquad \text{(ii)}$$

or

$$x = \frac{a}{P}(R_2 + R_3) \qquad \text{(iii)}$$

Similarly, equilibrium about the x-axis gives

$$y = \frac{a}{P}(R_3 + R_4) \qquad \text{(iv)}$$

It follows from equations (i), (iii), and (iv) that the force P (direction as well as magnitude) and its location (x, y) are completely determined by the load cell readings.

Now take the differentials of equations (i) and (ii):

$$\delta P = \delta R_1 + \delta R_2 + \delta R_3 + \delta R_4$$

$$P\,\delta x + x\,\delta P = a\,\delta R_2 + a\,\delta R_3$$

Direct substitution gives

$$\delta x = \frac{a}{P}(\delta R_2 + \delta R_3) - \frac{x}{P}(\delta R_1 + \delta R_2 + \delta R_3 + \delta R_4)$$

Note that x lies between 0 and a, and each δR_i can vary up to $\pm\delta F$. Hence, the largest error in x is given by $(2a/P)\,\delta F$. This is limited to δr. Hence, we have

$$\delta r = \frac{2a}{P}\,\delta F$$

or

$$\delta F = \frac{P}{2a}\,\delta r$$

which gives the tolerance on the force error. The same result is obtained by considering y instead of x.

Sensor density or resolution, dynamic range, response time or bandwidth, strength and physical robustness, size, stability (dynamic robustness), linearity, flexibility, and localized intelligence (including data processing and learning) are important factors that require consideration in the analysis, design, or selection of a tactile sensor. The specifications chosen will depend on the particular application. Typical values are 100 sensor elements spaced at 1 mm, a dynamic range of 60 dB, and a bandwidth of over 100 Hz (a response time of 10 ms or less). Because of the large number of sensor elements, signal conditioning and processing for tactile sensors present enormous difficulties. For instance, in piezoelectric tactile sensors, it is usually impractical to use a separate charge amplifier (or a voltage amplifier) for each piezoelectric element, even when built-in microamplifiers are available. Instead, analog multiplexing could be employed, along with a few high-performance signal amplifiers. The sensor signals could then be serially transferred for digital processing. The obvious disadvantages here are the increase in data acquisition time and the resulting reduction in sensor bandwidth.

Passive Compliance

Tactile sensing is used for the active control of processes. This is particularly useful in robotic applications that call for fine manipulation (e.g., microsurgery, assembly of delicate instruments, and robotic artwork). Heavy-duty industrial manipulators are often not suitable for fine manipulation because of errors that arise from such factors as backlash, friction, drift, errors in control hardware and algorithms, and generally poor dexterity. This situation can be improved to a great extent by using a heavy-duty robot for gross manipulations and using a microminiature robot (or hand or gripper or finger manipulator) to serve as an end effector for fine-manipulation purposes. The end effector will use tactile sensing and localized control to realize required levels of accuracy in fine manipulation. This approach to accurate manipulation is expensive, of course, primarily because of the sophisticated instrumentation and local processing needed at the end effector. A more cost-effective approach is to use passive remote-center compliance (RCC) at the end effector. With this method, passive devices (linear or nonlinear springs) are added in the end effector design (typically, at the wrist) so that some compliance (flexibility) is present. Errors in manipulation (e.g., jamming during parts mating) will generate forces and moments that will deflect the end effector so as to correct the situation. Active compliance, which employs local sensors to adaptively change the end effector compliance, is used in operations where passive compliance alone is not adequate. Impedance control is useful here.

PROBLEMS

4.1. A signature verification pen has been patented by IBM Corporation. The purpose of the pen is to detect whether the user is forging someone else's signature. The instrumented pen has analog sensors. Sensor signals are conditioned using microcircuitry built into the pen and sampled into a digital computer at the rate of 80 samples/second using an ADC. Typically, about 1,000 data samples are collected per signature. Prior to the pen's use, authentic signatures are collected offline and stored in a reference data base. When a signature and the corresponding identification code are supplied to the computer for verification, a program in the processor retrieves the authentic signature from the data base by referring to the identification code and then compares the two sets of data for authenticity. This process takes about three seconds. Discuss the types of sensors that could be used in the pen. Estimate the total time required for a signal verification. What are the advantages and disadvantages of this method in comparison to having the user punch in an identification code alone or provide the signature without the identification code?

4.2. Describe three practical applications that employ force sensing, torque sensing, or tactile sensing. Using a practical example, explain the difference between force feedback control and feedforward force control.

4.3. Under what conditions can displacement control be treated as force control? Describe a situation in which this is not feasible.

4.4. Consider the joint of a robotic manipulator, shown schematically in figure P4.4. Torque sensors are mounted at locations 1, 2, and 3. If the electromagnetic torque gen-

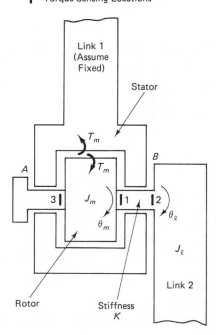

▮ = Torque-Sensing Locations

Figure P4.4. Schematic representation of a manipulator joint.

erated at the motor rotor is T_m, write equations for the torque transmitted to link 2, the frictional torque at bearing A, the frictional torque at bearing B, and the reaction torque on link 1 in terms of the measured torques, the inertia torque of the rotor, and T_m.

4.5. A model for a machining operation is shown in figure P4.5. The cutting force is denoted by f, and the cutting tool with its fixtures is modeled by a spring (stiffness k), a viscous damper (damping constant b), and a mass m. The actuator (hydraulic) with its controller is represented by an active stiffness g. Obtain a transfer relation between the actuator input u and the cutting force f. Discuss a control strategy for countering effects due to random variations in the cutting force. Note that this is important for controlling the product quality.

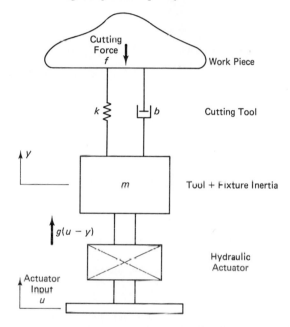

Figure P4.5. A model for a machining operation.

4.6. A strain gage sensor to measure the torque T_m generated by a motor is shown schematically in figure P4.6. The motor is floated on frictionless bearings. A uniform rectangular lever arm is rigidly attached to the motor housing, and its projected end is restrained by a pin joint. Four identical strain gages are mounted on the lever arm, as shown. Three of the strain gages are at point A, which is located at a distance a from the motor shaft, and the fourth strain gage is at point B, which is located at a distance $2a$ from the motor shaft. The pin joint is at a distance ℓ from the motor shaft. Strain gages 2, 3, and 4 are on the top surface of the lever arm, and gage 1 is on the bottom surface. Obtain an expression for T_m in terms of the bridge output δv_o and the following additional parameters:

S_s = gage factor (strain gage sensitivity)

v_{ref} = supply voltage to the bridge

b = width of the lever arm cross-section

h = height of the lever arm cross-section

E = Young's modulus of the lever arm

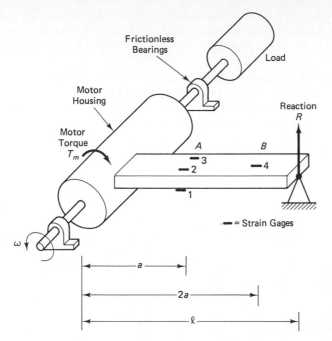

Figure P4.6. A strain gage sensor for measuring motor torque.

Verify that the bridge sensitivity does not depend on ℓ. Describe means to improve the bridge sensitivity. Explain why the sensor reading is only an approximation to the torque transmitted to the load. Give a relation to determine the net normal reaction force at the bearings, using the bridge output.

4.7. The sensitivity S_s of a strain gage consists of two parts: the contribution from the change in resistivity of the material and the direct contribution due to the change in shape of the strain gage when deformed. Show that the second part may be approximated by $(1 + 2\nu)$, where ν denotes the Poisson's ratio of the strain gage material.

4.8. A bridge with two active strain gages is being used to measure bending moment M (figure P4.8a) and torque T (figure P4.8b) in a machine part. Using sketches, suggest the orientations of the two gages mounted on the machine part and the corresponding bridge connections in each case in order to obtain the best sensitivity from the bridge. What is the value of the bridge constant in each case?

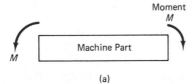

(a)

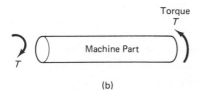

(b)

Figure P4.8. Sensing elements: (a) bending member; (b) torsion member.

4.9. Compare the potentiometer (ballast) circuit with the Wheatstone bridge circuit for strain gage measurements with respect to the following considerations:
(a) Sensitivity to the measured strain
(b) Error due to ambient effects (e.g., temperature changes)
(c) Signal-to-noise ratio of the output voltage
(d) Circuit complexity and cost
(e) Linearity

4.10. In the strain gage bridge shown in figure 4.9b, suppose that the load current i is not negligible. Derive an expression for the output voltage v_o in terms of R_1, R_2, R_3, R_4, R_L, and v_{ref}. Initially, the bridge was balanced, with the resistances in the four arms being equal. Then one of the resistances (say, R_1) was increased by 1 percent. Plot to scale the ratio (actual output from the bridge)/(output under open-circuit, or infinite-load-impedance, conditions) as a function of the nondimensionalized load resistance R_L/R in the range 0.1 to 10.0, where R denotes the resistance in each arm of the bridge initially.

4.11. What is meant by the term *bridge sensitivity* in strain gage measurements? Describe methods of increasing bridge sensitivity. Assuming the load resistance to be very high in comparison with the arm resistances in the strain gage bridge shown in figure 4.9b, obtain an expression for the power dissipation p in terms of bridge resistances and supply voltage. Discuss how the limitation on power dissipation can affect bridge sensitivity.

4.12. Consider the strain gage bridge shown in figure 4.9b. Initially, the bridge is balanced, with $R_1 = R_2 = R$. (*Note: R_3 may not be equal to R_1.*) Then R_1 is changed by δR. Assuming the load current to be negligible, derive an expression for the percentage error due to neglecting the second-order and higher-order terms in δR. If $\delta R/R = 0.05$, estimate this nonlinearity error.

4.13. Discuss the advantages and disadvantages of the following techniques in the context of measuring transient signals.
(a) DC bridge circuits versus AC bridge circuits
(b) Slip ring and brush commutators versus AC transformer commutators
(c) Strain gage torque sensors versus variable-inductance torque sensors
(d) Piezoelectric accelerometers versus strain gage accelerometers
(e) Tachometer velocity transducers versus piezoelectric velocity transducers

4.14. For a semiconductor strain gage characterized by the quadratic strain–resistance relationship

$$\frac{\delta R}{R} = S_1 \epsilon + S_2 \epsilon^2$$

obtain an expression for the equivalent gage factor (sensitivity) S_s using the least squares error linear approximation. Assume that only positive strains up to ϵ_{max} are measured with the gage. Derive an expression for the percentage nonlinearity. Taking $S_1 = 117$, $S_2 = 3,600$, and $\epsilon_{max} = 0.01$ strain, compute S_s and the percentage nonlinearity.

4.15. Consider a balanced bridge circuit with four identical strain gages that have resistance R. Suppose that one gage is active and its resistance is increased by δR. Also, because of a temperature rise, the resistance of each of the four gages increased by $\delta R'$. Show that up to first order, the temperature effects are compensated. When higher-order terms are included, show that the magnitude of the bridge output error due to temperature effects may be approximated by

$$|\delta v_o'| = \frac{v_{\text{ref}}}{4R^2}|\delta R||\delta R'|$$

where v_{ref} is the supply voltage to the bridge.

4.16. Semiconductor strain gages are made from P-type material (e.g., silicon crystals doped with boron) as well as N-type material (e.g., silicon crystals doped with phosphorous). The sensitivity (gage factor) S_s is positive for P-type strain gages and negative for N-type strain gages. The temperature coefficient of resistance (α) is positive for both types of gages, and the temperature coefficient of gage factor (β) is negative for both types. In terms of these features of semiconductor strain gages, discuss how a combination of P-type gages and N-type gages may be used effectively to obtain a temperature-compensated strain gage bridge circuit.

4.17. Briefly describe how strain gages may be used to measure

(a) Force   (d) Pressure

(b) Displacement   (e) Temperature

(c) Acceleration

Show that if a compensating resistance R_c is connected in series with the supply voltage v_{ref} to a strain gage bridge that has four identical members, each with resistance R, the output equation is given by

$$\frac{\delta v_o}{v_{\text{ref}}} = \frac{R}{(R + R_c)}\frac{kS_s}{4}\epsilon$$

in the usual rotation.

A foil-gage load cell uses a simple (one-dimensional) tensile member to measure force. Suppose that k and S_s are insensitive to temperature change. If the temperature coefficient of R is α_1, that of the series compensating resistance R_c is α_2, and that of the Young's modulus of the tensile member is $(-\beta)$, determine an expression for R_c that would result in automatic (self) compensation for temperature effects. Under what conditions is this arrangement realizable?

4.18. Draw a block diagram for a single joint of a robot, identifying inputs and outputs. Using the diagram, explain the advantages of torque sensing in comparison to displacement and velocity sensing at the joint. What are the disadvantages of torque sensing?

4.19. A single joint of a direct-drive manipulator has a motor with rotor inertia J_m. In the neighborhood of a particular orientation (or configuration), this motor drives an inertia load that has an equivalent moment of inertia J_L. The motor rotor is coupled to the driven link through a torque-sensing shaft of torsional stiffness K.

(a) Obtain a differential equation for the twist angle θ of the shaft, using the motor torque T_m as the input.

(b) What is the transfer function between T_m and θ?

(c) Give an expression for the twisting natural frequency of the joint.

(d) If the "dynamic" frequency response should be no greater than 1.02 times the static response, obtain an expression for the mechanical bandwidth of the joint in terms of J_m, J_L, and K.

4.20. In example 4.8, obtain the transfer functions relating input T_m and outputs θ_m and θ_L. What are the system natural frequencies, and what modes of motion do they represent?

4.21. Figure P4.21 shows a schematic diagram of a measuring device.

(a) Identify the various components in this device.

(b) Describe the operation of the device, explaining the function of each component and identifying the nature of the measurand and the output of the device.

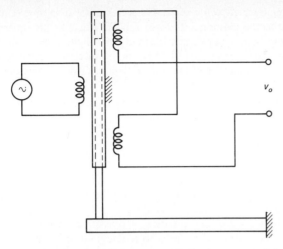

Figure P4.21. An analog sensor.

(c) List the advantages and disadvantages of the device.

(d) Describe a possible application of this device.

4.22. Consider a nonlinear device represented by the static relationship

$$y = y(u)$$

This may be approximated by the linear relationship

$$y = ku + y_o$$

where $y_o = y(u_o)$ is the operating point about which the operation is linearized. Show that

$$k = \frac{\partial y}{\partial u}(u_o)$$

Note that if the device is actually linear, k would be a constant that is independent of u, given by

$$k = \frac{y(u) - y_o}{u}$$

But in the nonlinear case, if u changes, k in the linearized relationship changes. Discuss the implications of these observations on the accuracy of strain gage measurements under large variations in strain. (*Hint:* Consider the case when y = bridge output, u = measured strain, and k = bridge constant.)

4.23. Discuss the advantages and disadvantages of torque sensing by the motor current method. Show that for a synchronous motor with a balanced three-phase supply, the electromagnetic torque generated at the rotor–stator interface is given by

$$T_m = ki_f i_a \cos(\theta - \omega t)$$

where i_f = DC current in the rotor (field) winding
  i_a = amplitude of the supply current to each phase in the stator (armature)
  θ = angle of rotation
  ω = frequency (angular) of the AC supply
  t = time
  k = motor torque constant

4.24. Describe an application in which tactile sensing is preferred over sensing of a few point forces. A piezoelectric tactile sensor has 25 force-sensing elements per square centimeter. Each sensor element in the sensor can withstand a maximum load of 40 N and can detect load changes on the order of 0.01 N. What is the force resolution of the tactile sensor? What is the spatial resolution of the sensor? What is the dynamic range in decibels?

4.25. Discuss factors that limit the lower frequency and upper frequency limits of measurements obtained from the following devices:
   (a) Strain gage
   (b) Rotating shaft torque sensor
   (c) Reaction torque sensor

4.26. Briefly describe a situation in which tension in a moving belt or cable has to be measured under transient conditions. What are some of the difficulties associated with measuring tension in a moving member? A strain gage tension sensor for a belt-drive system is shown in figure P4.26. Two identical active strain gages, G_1 and G_2, are mounted at the root of a cantilever element with rectangular cross-section, as shown. A light, frictionless pulley is mounted at the free end of the cantilever element. The belt makes a 90° turn when passing over this idler pulley.

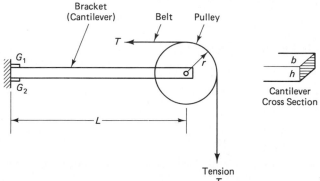

Figure P4.26. A strain gage tension sensor.

   (a) Using a circuit diagram, show the Wheatstone bridge connections necessary for the strain gages G_1 and G_2 so that strains due to axial forces in the cantilever member have no effect on the bridge output (i.e., effects of axial loads are compensated) and the sensitivity to bending loads is maximized.
   (b) Obtain an equation relating the belt tension T and the bridge output δv_o in terms of the following additional parameters:

   S_s = gage factor (sensitivity) of each strain gage
   E = Young's modulus of the cantilever element
   L = length of the cantilever element
   b = width of the cantilever cross-section
   h = height of the cantilever cross-section

   Note that the radius of the pulley does not enter this equation.

4.27. Explain why the strain gage bridge equation 4.14 is not valid for large changes in resistance. Consider equation 4.6 with R_1 being the only active gage and $R_3 = R_4$. Obtain

Torque, Force, and Tactile Sensors   Chap. 4

an expression for R_1 in terms of R_2, v_o, and v_{ref}. Show that when $R_1 = R_2$, we get $v_o = 0$, the balanced bridge, as required. Note that the equation for R_1, assuming that v_o is measured using a high-impedance voltmeter, can be used to detect large resistance changes in R_1. Now suppose that the active gage R_1 is connected to the bridge using a long, twisted wire pair, with each wire having a resistance of R_c. The bridge circuit has to be modified as in figure P4.27 in this case. Using the expression for R_1 obtained earlier, show that the equation of the modified bridge is given by

$$R_1 = R_2\left[\frac{v_{ref} + 2v_o}{v_{ref} - 2v_o}\right] + 4R_c\frac{v_o}{[v_{ref} - 2v_o]}$$

Obtain an expression for the fractional error in the R_1 measurement due to cable resistance R_c. Show that this can be decreased by increasing R_2 and v_{ref}.

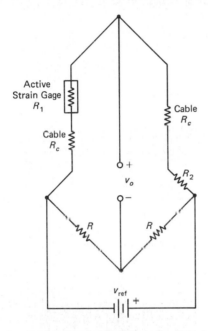

Figure P4.27. The influence of cable resistance on strain gage bridge measurements.

4.28. The read-write head in a disk drive of a digital computer should float at a constant but small height (say 15 μin.) above the disk surface. Because of aerodynamics resulting from the surface roughness and the surface deformations of disk, the head can be excited into vibrations that could cause head–disk contacts. These contacts, which are called head–disk interferences (HDIs), are clearly undesirable. They can occur at very high frequencies (say, 100 kHz). The purpose of a glide test is to detect HDIs and to determine the nature of these interferences. Glide testing can be used to determine the effect of parameters such as the flying height of the head and the speed of the disk and to qualify (certify the quality of) disk drive units. Discuss the basic instrumentation needed in glide testing. In particular, describe the types of sensors that could be used and their advantages and disadvantages.

4.29. What are the typical requirements for an industrial tactile sensor? Explain how a tactile sensor differs from a simple touch sensor. Define spatial resolution and force resolution (or sensitivity) of a tactile sensor. The spatial resolution of your fingertip can be determined by a simple experiment using two pins and a helper. Close your eyes. Instruct the helper to apply one pin or both pins randomly to your fingertip so that you will feel the pressure of the tip of the pins. You should respond by telling the helper whether you feel both pins or only one pin. If you feel both pins, the helper should decrease the spacing of the two pins in the next round of tests. The test should be repeated in this manner by successively decreasing the spacing between the pins until you feel only one pin when both pins are actually applied. Then measure the distance between the two pins in millimeters. The largest spacing between the two pins that will result in this incorrect sensation corresponds to the spatial resolution of your fingertip. Repeat this experiment for all your fingers, repeating the test several times on each finger. Compute the average and the standard deviation. Then perform the test on other subjects (preferably of the opposite sex). Discuss your results. Do you notice large variations in the results?

4.30. Torque, force, and tactile sensing can be very useful in many applications, particularly in the manufacturing industry. For each of the following applications, discuss the types of sensors that would be useful for performing the task properly.
   (a) Controlling the operation of inserting printed circuit boards in card cages using a robotic end effector
   (b) Controlling a robotic end effector that screws a threaded part into a hole
   (c) Failure prediction and diagnosis of a drilling operation
   (d) Gripping a fragile and delicate object by a robotic hand without damaging the object
   (d) Gripping a metal part using a two-fingered gripper
   (f) Quickly identifying and picking a complex part from a bin containing many different parts

4.31. The dexterity of a device is usually defined as the ratio (number of degrees of freedom in the device)/(motion resolution of the device). We will call this *motion dexterity*. Suppose that we define *force dexterity* as follows:

$$\text{force dexterity} = \frac{\text{number of df}}{\text{force resolution}}$$

Discuss the implications of these two types of dexterity, giving examples for situations in which both mean the same thing and situations in which the two terms mean different things. Explain how force dexterity of a device (say, an end effector) can be improved by using tactile sensors. Discuss dexterity requirements for the following tasks, clearly indicating whether motion dexterity or force dexterity is preferred in each case:
   (a) Gripping a hammer and driving a nail with it
   (b) Threading a needle
   (c) Seam tracking of a complex part in robotic arc welding
   (d) Finishing the surface of a complex metal part using robotic grinding

4.32. Discuss why nonlinear feedback control could be very useful in controlling complex mechanical systems with nonlinear and coupled dynamics. What are the shortcomings of nonlinear feedback control? Consider the two-link manipulator that carries a point load (weight W) at the end effector, as shown in figure P4.32. Its dynamics can be expressed as

$$\mathbf{I\ddot{q} + b = \tau}$$

where $\mathbf{q}$ = vector of (relative) joint rotations q_1 and q_2
   $\mathbf{\tau}$ = vector of drive torques τ_1 and τ_2 at the two joints, corresponding to the coordinates q_1 and q_2

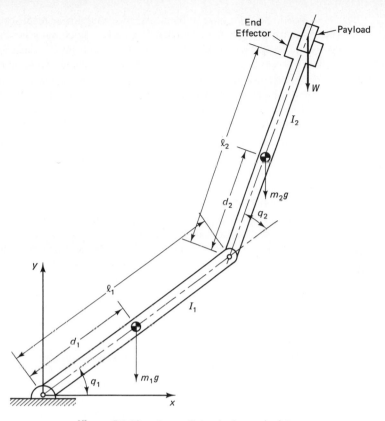

Figure P4.32. A two-link robotic manipulator.

$\mathbf{I}$ = second-order inertia matrix

$$= \begin{bmatrix} I_{11} & I_{12} \\ I_{21} & I_{22} \end{bmatrix}$$

$\mathbf{b}$ = vector of joint-friction, gravitational, centrifugal, and coriolis torques (components are b_1 and b_2)

Neglecting joint friction, and with zero payload ($W = 0$), show that

$$I_{11} = m_1 d_1^2 + I_1 + I_2 + m_2(\ell_1^2 + d_2^2 + 2\ell_1 d_2 \cos q_2)$$

$$I_{12} = I_{21} = I_2 + m_2 d_2^2 + m_2 \ell_1 d_2 \cos q_2$$

$$I_{22} = I_2 + m_2 d_2^2$$

$$b_1 = m_1 g d_1 \cos q_1 + m_2 g[d_1 \cos q_1 + d_2 \cos (q_1 + q_2)]$$
$$\qquad - m_2 \ell_1 d_2 \dot{q}_2^2 \sin q_2 - 2m_2 \ell_1 d_2 \dot{q}_1 \dot{q}_2 \sin q_2$$

$$b_2 = m_2 g d_2 \cos (q_1 + q_2) + m_2 \ell_1 d_2 \dot{q}_1^2 \sin q_2$$

where I_1, I_2 = moments of inertia of the links about their centroids
m_1, m_2 = masses of the links

The geometric parameters ℓ_1, ℓ_2, d_1, and d_2 are as defined in figure P4.32.

What variables have to be measured for nonlinear feedback control? Noting that the elements of **b** are more complex (even after neglecting joint friction, backlash, and payload) than the elements of **I**, justify using nonlinear feedback control for this system instead of using a control method based on an accurate dynamic model.

4.33. Describe four advantages and four disadvantages of a semiconductor strain gage weight sensor. A weight sensor is used in a robotic wrist. What would be the purpose of this sensor? How can the information obtained from the weight sensor be used in controlling the robotic manipulator?

4.34. Discuss whether there is any relationship between the dexterity and the stiffness of a manipulator hand. The stiffness of a robotic hand can be improved during gripping operations by temporarily decreasing the number of degrees of freedom of the hand using suitable fixtures. What purpose does this serve? Using schematic diagrams, explain the operation of such a jig hand.

4.35. Force switches are used in applications where only a force limit, rather than a continuous force signal, has to be detected. Examples include detecting closure force (torque) in valve closing, detecting fit in parts assembly, and product filling in containers by weight. Expensive force sensors are not needed in such applications because a continuous history of a force signal is not needed. With sketches, describe the construction of a simple spring-loaded force switch.

4.36. Using equation 4.14, show that if the resistance elements R_1 and R_2 have the same temperature coefficient of resistance and if R_3 and R_4 have the same temperature coefficient of resistance, the temperature effects are compensated up to first order. A strain gage accelerometer uses two semiconductor strain gages, one integral with the cantilever element near the fixed end (root) and the other mounted at an unstrained location in the accelerometer housing. Describe the operation of the accelerometer. What is the purpose of the second strain gage?

REFERENCES

ASADA, H., YOUCEF-TOUMI, K., and LIM, S. K. "Joint Torque Measurement of a Direct-Drive Arm." In *Proceedings of the 23rd Conference on Decision and Control*. IEEE. Las Vegas, December 1984, pp. 1332–37.

BOWMAN, R. "Diffused Semiconductor Strain Gages." *Measurements and Control Journal* 106: 214–18, September 1984.

BOYER, D. "Glide Test Calibration for Rigid Disk Magnetic Media." *Sensors* 3(9): 80–88, September 1986.

CRANDALL, S. H., DAHL, N. C., and LARDNER, T. J. *An Introduction to the Mechanics of Solids*. McGraw-Hill, New York, 1972.

DALLY, J. W., RILEY, W. F., and MCCONNELL, K. G. *Instrumentation for Engineering Measurements*. Wiley, New York, 1984.

DARIO, P., and DEROSSI, D. "Tactile Sensing and the Gripping Challenge." *IEEE Spectrum* 22(8): 46–52, August 1985.

DESILVA, C. W. "Torque Measurement." *Measurements and Control Journal* 127: 193–94, February 1988.

DESILVA, C. W., PRICE, T. E., and KANADE, T. "Torque Sensor for Direct-Drive Manipulators." *Journal of Engineering for Industry, Trans. ASME* 109(2): 122–27, 1987.

DOEBELIN, E. O. *Measurement Systems Application and Design,* 3d ed. McGraw-Hill, New York, 1983.

HEWIT, J. R., and BURDESS, J. S. "Fast Dynamic Coupled Control for Robotics Using Active Force Control." *Mechanism and Machine Theory* 16(5): 535–42, 1981.

HOGAN, N. "Application of Impedance Control to Automated Deburring." *Proceedings of the Fourth Yale Workshop on Applications of Adaptive Systems Theory.* Center for Systems Science, Yale University, May 1985, pp. 203–207.

JUDD, B. "Interfacing Sensors to Measurement Equipment." *Sensors* 3(9): 25–33, September 1986.

MATSCH, L. W. *Electromagnetic and Electromechanical Machines.* Intext, Scranton, Pa., 1972.

MCALPINE, G. A. "Tactile Sensing." *Sensors* 3(4): 7–16, April 1986.

5

Digital Transducers

5.1 INTRODUCTION

Any transducer that presents information as discrete samples and that does not introduce a *quantization error* when the reading is represented in the digital form may be classified as a digital transducer. A digital processor plays the role of controller in a digital control system. This facilitates complex processing of measured signals and other known quantities in order to obtain control signals for the actuators that drive the plant of the control system. If the measured signals are in analog form, an analog-to-digital conversion (ADC) stage is necessary prior to digital processing. There are several other shortcomings of analog signals in comparison to digital signals, as outlined in chapter 1. These considerations help build a case in favor of direct digital measuring devices for digital control systems.

Digital measuring devices (or digital transducers, as they are commonly known) generate discrete output signals such as pulse trains or encoded data that can be directly read by a control processor. Nevertheless, the sensor stage of digital measuring devices is usually quite similar to that of their analog counterparts. There are digital measuring devices that incorporate microprocessors to perform numerical manipulations and conditioning locally and provide output signals in either digital or analog form. These measuring systems are particularly useful when the required variable is not directly measurable but could be computed using one or more measured outputs (e.g., power = force × speed). Although a microprocessor is an integral part of the measuring device in this case, it performs not a measuring task but, rather, a conditioning task. For our purposes, we shall consider the two tasks separately.

The objective of this chapter is to study the operation and utilization of several types of direct digital transducers. Our discussion will be limited to motion transducers. Note, however, that by using a suitable auxiliary front-end sensor, other measurands—such as force, torque, and pressure—may be converted into a motion and subsequently measured using a motion transducer. For example, altitude (or pressure) measurements in aircraft and aerospace applications are made using a pressure-sensing front end, such as a bellows or diaphragm device, in conjunction with an optical encoder to measure the resulting displacement. Motion, as manifested in physical systems, is typically continuous in time. Therefore, we cannot speak of digital motion sensors in general. Actually, it is the transducer stage that generates the discrete output signal in a digital motion measuring device. Commercially available *direct digital transducers* are not as numerous as analog sensors, but what is available has found extensive application.

When the output of a digital transducer is a pulse signal, a counter is used either to count the pulses or to count clock cycles over one pulse duration. The count is first represented as a digital word according to some code; then it is read by a data acquisition and control computer. If, on the other hand, the output of digital transducer is automatically available in a coded form (e.g., binary, binary-coded decimal, ASCII), it can be directly read by a computer. In the latter case, the coded signal is normally generated by a parallel set of pulse signals; the word depends on the pattern of the generated pulses.

5.2 SHAFT ENCODERS

Any transducer that generates a coded reading of a measurement can be termed an encoder. Shaft encoders are digital transducers that are used for measuring angular displacements and angular velocities. Applications of these devices include motion measurement in performance monitoring and control of robotic manipulators, machine tools, digital tape-transport mechanisms, servo plotters and printers, satellite mirror positioning systems, and rotating machinery such as motors, pumps, compressors, turbines, and generators. High resolution (depending on the word size of the encoder output and the number of pulses per revolution of the encoder), high accuracy (particularly due to noise immunity of digital signals and superior construction), and relative ease of adaption in digital control systems (because transducer output is digital), with associated reduction in system cost and improvement of system reliability, are some of the relative advantages of digital transducers over their analog counterparts.

Encoder Types

Shaft encoders can be classified into two categories, depending on the nature and the method of interpretation of the transducer output: (1) incremental encoders and (2) absolute encoders. The output of an incremental encoder is a pulse signal that is generated when the transducer disk rotates as a result of the motion that is being measured. By counting the pulses or by timing the pulse width using a clock signal, both angular displacement and angular velocity can be determined. Displacement, however, is obtained with respect to some reference point on the disk, as indicated by a reference pulse (index pulse) generated at that location on the disk. The index pulse count determines the number of full revolutions.

An absolute encoder (or whole-word encoder) has many pulse tracks on its transducer disk. When the disk of an absolute encoder rotates, several pulse trains—equal in number to the tracks on the disk—are generated simultaneously. At a given instant, the magnitude of each pulse signal will have one of two signal levels (i.e., a binary state), as determined by a level detector. This signal level corresponds to a binary digit (0 or 1). Hence, the set of pulse trains gives an encoded binary number at any instant. The pulse windows on the tracks can be organized into some pattern (code) so that each of these binary numbers corresponds to the angular position of the encoder disk at the time when the particular binary number is detected. Furthermore, pulse voltage can be made compatible with some form of digital logic (e.g., transistor-to-transistor logic, or TTL). Consequently, the direct digital readout of an angular position is possible, thereby expediting digital data acquisition and processing. Absolute encoders are commonly used to measure fractions of a revolution. However, complete revolutions can be measured using an additional track that generates an index pulse, as in the case of incremental encoder.

The same signal generation (and pick-off) mechanism may be used in both types of transducers. Four techniques of transducer signal generation can be identified:

1. Optical (photosensor) method
2. Sliding contact (electrical conducting) method
3. Magnetic saturation (reluctance) method
4. Proximity sensor method

For a given type of encoder (incremental or absolute), the method of signal interpretation is identical for all four types of signal generation. Thus, we shall describe the principle of signal generation for all four mechanisms, but we will consider only the optical encoder in the context of signal interpretation and processing.

The optical encoder uses an opaque disk (code disk) that has one or more circular tracks, with some arrangement of identical transparent windows (slits) in each track. A parallel beam of light (e.g., from a set of light-emitting diodes or a tungsten lamp) is projected to all tracks from one side of the disk. The transmitted light is picked off using a bank of photosensors on the other side of the disk that typically has one sensor for each track. This arrangement is shown in figure 5.1a, which indicates just one track and one pick-off sensor. The light sensor could be a silicon photodiode, a phototransistor, or a photovoltaic cell. Since the light from the source is interrupted by the opaque areas of the track, the output signal from the probe is a series of voltage pulses. This signal can be interpreted to obtain the angular position and angular velocity of the disk. Note that in the standard terminology, the sensor element of such a measuring device is the encoder disk that is coupled to the rotating object (directly or through a gear mechanism). The transducer stage is the conversion of disk motion into the pulse signals. The opaque background of transparent windows (the window pattern) on an encoder disk is produced by contact printing

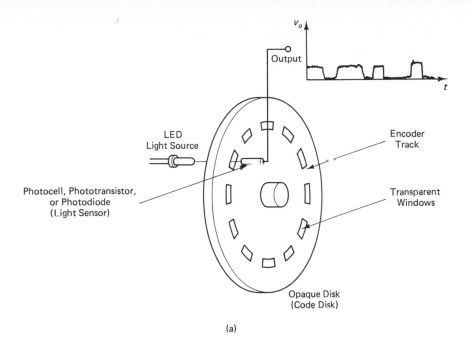

(a)

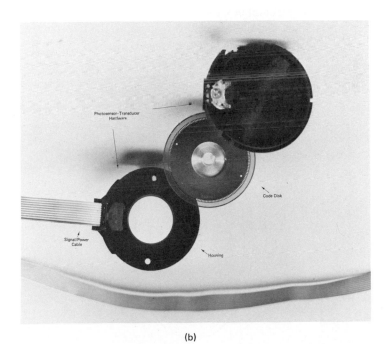

(b)

Figure 5.1. (a) Schematic representation of an optical encoder. (b) Components of an incremental optical encoder (courtesy of Hewlett-Packard Conmpany).

techniques. The precision of this production procedure is a major factor that determines the accuracy of optical encoders. Note that a transparent disk with opaque spots will work equally well as the encoder disk of an optical encoder. The code disk, housing, and signal/power cable of a commercially available incremental optical encoder are shown in figure 5.1b.

In a sliding contact encoder, the transducer disk is made of an electrically insulating material (see figure 5.2). Circular tracks on the disk are formed by implanting a pattern of conducting areas. These conducting regions correspond to the transparent windows on an optical encoder disk. All conducting areas are connected to a common slip ring on the encoder shaft. A constant voltage v_{ref} is applied to the slip ring using a brush mechanism. A sliding contact such as a brush touches each track, and as the disk rotates, a voltage pulse signal is picked off by it (see figure 5.2). The pulse pattern depends on the conducting–nonconducting pattern on each track as well as the nature of rotation of the disk. The signal interpretation is done as it is for optical encoders. The advantages of sliding contact encoders include high sensitivity (depending on the supply voltage) and simplicity of construction (low cost). The disadvantages include the familiar drawbacks of contacting and commutating devices (e.g., friction, wear, brush bounce due to vibration, and signal glitches and metal oxidation due to electrical arcing). A transducer's accuracy is very much dependent upon the precision of the conducting patterns of the encoder disk. One method of generating the conducting pattern on the disk is electroplating.

Magnetic encoders have high-strength magnetic areas imprinted on the encoder disk using techniques such as etching, stamping, or recording (similar to tape recording). These magnetic areas correspond to the transparent windows on an opti-

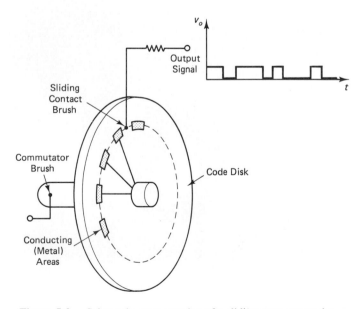

Figure 5.2. Schematic representation of a sliding contact encoder.

cal encoder disk. The signal pick-off device is a microtransformer that has primary and secondary windings on a circular ferromagnetic core. This pick-off sensor resembles a core storage element in older mainframe computers. The encoder arrangement is illustrated schematically in figure 5.3. A high-frequency (typically 100 kHz) primary voltage induces a voltage in the secondary winding of the sensing element at the same frequency, operating as a transformer. A magnetic field of sufficient strength can saturate the core, however, thereby significantly increasing the reluctance and dropping the induced voltage. By demodulating the induced voltage, a pulse signal is obtained. This signal can be interpreted in the usual manner. Note that a pulse peak corresponds to a nonmagnetic area and a pulse valley corresponds to a magnetic area on each track. Magnetic encoders have noncontacting pick-off sensors, which is an advantage. They are more costly than the contacting devices, however, primarily because of the cost of transformer elements and demodulating circuitry for generating the output signal.

Proximity sensor encoders use a proximity sensor as the signal pick-off element. Any type of proximity sensor may be used—for example, a magnetic induction probe or an eddy current probe, as discussed in chapter 3. In the magnetic induction probe, for example, the disk is made of ferromagnetic material. The encoder tracks have raised spots of the same material (see figure 5.4), serving a purpose analogous to that of the windows on an optical encoder disk. As a raised spot approaches the probe, the flux linkage increases as a result of the associated decrease in reluctance, thereby raising the induced voltage level. The output voltage is a pulse-modulated signal at the frequency of the supply (primary) voltage of the proximity sensor. This is then demodulated, and the resulting pulse signal is interpreted. In principle, this device operates like a conventional digital tachometer. If an eddy current probe is used, pulse areas in the track are plated with a conducting material. A flat plate may be used in this case, because the nonconducting areas on the disk do not generate eddy currents.

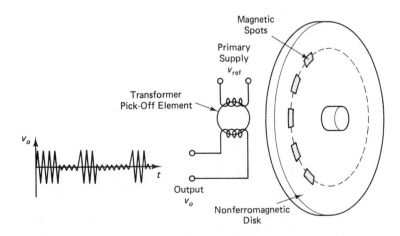

Figure 5.3. Schematic representation of a magnetic encoder.

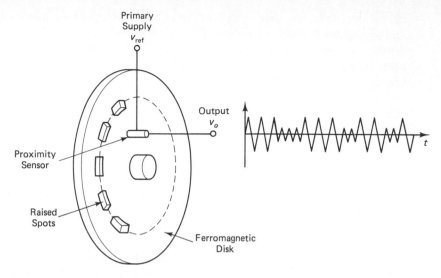

Figure 5.4. Schematic representation of a proximity probe encoder.

Note that an incremental encoder disk requires only one primary track that has equally spaced and identical window (pick-off) areas. The window area is equal to the area of the interwindow gap. Usually, a reference track that has just one window is also present in order to generate a pulse (known as the index pulse) to initiate pulse counting for angular position measurement and to detect complete revolutions. In contrast, absolute encoder disks have several rows of tracks, equal in number to the bit size of the output data word. Furthermore, the track windows are not equally spaced but are arranged in a specific pattern on each track so as to obtain a binary code (or a gray code) for the output data from the transducer. It follows that absolute encoders need at least as many signal pick-off sensors as there are tracks, whereas incremental encoders need one pick-off sensor to detect the magnitude of rotation and an additional sensor at a quarter-pitch separation (pitch = center-to-center distance between adjacent windows) to identify the direction of rotation. Some designs of incremental encoders have two identical tracks, one a quarter-pitch offset from the other, and the two pick-off sensors are placed radially without offset. A pick-off sensor for receiving a reference pulse is also used in some designs of incremental encoders (three-track incremental encoders).

In many control applications, encoders are built into the plant itself, rather than being externally fitted onto a rotating shaft. For instance, in a robot arm, the encoder might be an integral part of the joint motor and may be located within its housing. This reduces coupling errors (e.g., errors due to backlash, shaft flexibility, and resonances added by the transducer and fixtures), installation errors (e.g., eccentricity), and overall cost.

Since the signal interpretation techniques are quite similar for the various types of encoder signal generation techniques, we shall limit further discussion to optical encoders. These are the predominantly employed types of shaft encoders in practical applications. Signal interpretation depends on whether the particular optical encoder is an incremental device or an absolute device.

5.3 INCREMENTAL OPTICAL ENCODERS

There are two possible configurations for an incremental encoder disk: (1) the offset sensor configuration and (2) the offset track configuration. The first configuration is shown schematically in figure 5.5. The disk has a single circular track with identical and equally spaced transparent windows. The area of the opaque region between adjacent windows is equal to the window area. Two photodiode sensors (pick-offs 1 and 2 in figure 5.5) are positioned facing the track a quarter-pitch (half the window length) apart. The ideal forms of their output signals (v_1 and v_2) after passing them through pulse-shaping circuitry are shown in figure 5.6a and 5.6b for the two directions of rotation.

In the second configuration of incremental encoders, two identical tracks are used, one offset from the other by a quarter-pitch. In this case, one pick-off sensor is positioned facing each track—on a radial line, without any circumferential offset—unlike the previous configuration. The output signals from the two sensors are the same as before, however (figure 5.6).

In both configurations, an additional track with a lone window and associated sensor is also usually available. This track generates a reference pulse (index pulse) per revolution of the disk (see figure 5.6c). This pulse is used to initiate the counting

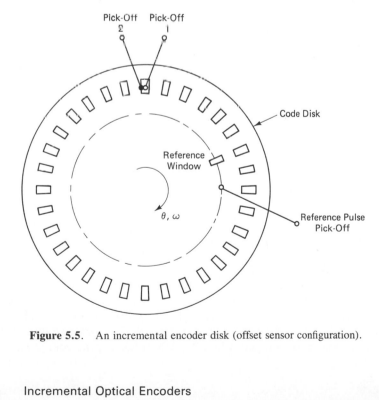

Figure 5.5. An incremental encoder disk (offset sensor configuration).

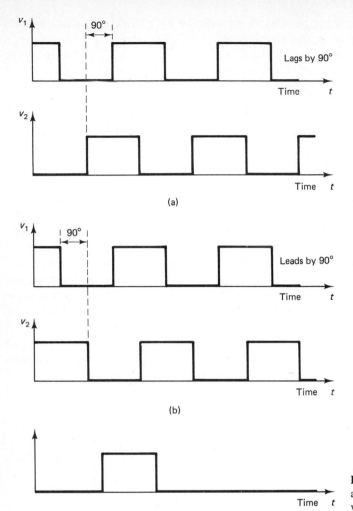

Figure 5.6. Shaped pulse signals from an incremental encoder: (a) for clockwise rotation; (b) for counterclockwise rotation; (c) reference pulse signal.

operation. Furthermore, the index pulse count gives the number of complete revolutions, which is required in absolute angular rotation measurements. Note that the pulse width and pulse-to-pulse period (encoder cycle) are constant in each sensor output when the disk rotates at constant angular velocity. When the disk accelerates, the pulse width decreases continuously; when the disk decelerates, the pulse width increases.

Direction of Rotation

The quarter-pitch offset in sensor location or track position is used to determine the direction of rotation of the disk. For example, figure 5.6a shows idealized sensor outputs (v_1 and v_2) when the disk rotates in the clockwise direction; and figure 5.6b shows the sensor outputs when the disk rotates in the counterclockwise direction. It is clear from these two figures that in clockwise rotation, v_1 lags v_2 by a quarter of a cycle (i.e., a phase lag of 90°); and in counterclockwise rotation, v_1 leads v_2 by a quarter of a cycle. Hence, the direction of rotation is obtained by determining the phase difference of the two output signals, using phase-detection circuitry.

One method for determining the phase difference is to time the pulses using a high-frequency clock signal. For example, if the counting (timing) operation is initiated when the v_1 signal begins to rise, and if n_1 = number of clock cycles (time) until v_2 begins to rise and n_2 = number of clock cycles until v_1 begins to rise again, then $n_1 > n_2 - n_1$ corresponds to clockwise rotation and $n_1 < n_2 - n_1$ corresponds to counterclockwise rotation. This should be clear from figures 5.6a and 5.6b.

Construction Features

The actual internal hardware of commercial encoders is not as simple as what is suggested by figure 5.5. (see figure 5.1b). A more detailed schematic diagram of the signal generation mechanism of an optical incremental encoder is shown in figure 5.7. The light generated by the light-emitting diode (LED) is collimated (forming parallel rays) using a lens. This pencil of parallel light passes through a window of the rotating code disk. The grating (masking) disk is stationary and has a track of windows identical to that in the code disk. A significant amount of light passes through the grating window only if it is aligned with a window of the code disk. Because of the presence of the grating disk, more than one window of the code disk may be illuminated by the same LED, thereby improving the intensity of light received by the photosensor but not introducing any error caused by the diameter of the pencil of light being larger than the window length. When the windows of the code disk face the opaque areas of the grating disk, virtually no light is received by the photosensor. Hence, as the code disk moves, alternating light and dark spots (a moiré pattern) are seen by the photosensor. Note that the grating disk helps increase

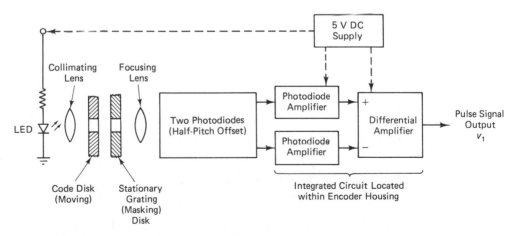

Figure 5.7. Internal hardware of an optical incremental encoder (for a single output pulse signal).

the output signal level significantly. But the supply voltage fluctuations also directly influence the light level received by the photosensor. If the sensitivity of the photosensor is not high enough, a low light level might be interpreted as no light, which would result in measurement error. Such errors due to instabilities and changes in the supply voltage can be eliminated by using two photosensors, one placed half a pitch away from the other along the window track. This arrangement should not be confused with the quarter-of-a-pitch offset arrangement that is required for direction detection. This arrangement is for contrast detection. The sensor facing the opaque region of the masking disk will always read a low signal. The other sensor will read either a high signal or a low signal, depending on whether it faces a window or an opaque region of the code disk. The two signals from these two sensors are amplified separately and fed into a differential amplifier. If the output is high, we have a pulse. In this manner, a stable and accurate output pulse signal can be obtained even under unstable voltage supply conditions. The signal amplifiers are integrated circuit devices and are housed within the encoder itself. Additional pulse-shaping circuitry may also be present. The power supply has to be provided separately as an external component. The voltage level and pulse width of the output pulse signal are logic-compatible (e.g., transistor-to-transistor logic, or TTL) so that they may be read directly using a digital board. The schematic diagram in figure 5.7 shows the generation of only one (v_1) of the two quadrature pulse signals. The other pulse signal (v_2) is generated using identical hardware but at a quarter of a pitch offset. The index pulse (reference pulse) signal is also generated in a similar manner. The cable of the encoder (usually a ribbon cable) has a multipin connector (see figure 5.1b). Three of the pins provide the three output pulse signals. Another pin carries the DC supply voltage (typically 5 V) from the power supply into the encoder. Note that the only moving part in the system shown in figure 5.7 is the code disk.

Displacement and Velocity Computation

A digital processor computes angular displacements and velocities using the digital data read into it from encoders, along with other pertinent parameters. To compute the angular position θ, suppose that the maximum count possible is M pulses and the range of the encoder is $\pm\theta_{max}$. Then the angle corresponding to a count of n pulses is

$$\theta = \frac{n}{M}\,\theta_{max} \tag{5.1}$$

Note that if the data size is r bits, allowing for a sign bit,

$$M = 2^{r-1} \tag{5.2}$$

where zero count is also included. Strictly speaking,

$$M = 2^{r-1} - 1$$

if zero count is not included. Note that if θ_{max} is 2π and $\theta_{min} = 0$, for example, then θ_{max} and θ_{min} will correspond to the same position of the code disk. To avoid this ambiguity, we use

$$\theta_{min} = \frac{\theta_{max}}{2^{r-1}}$$

Then equation 5.2 leads to the conventional definition for digital resolution:

$$\frac{(\theta_{max} - \theta_{min})}{(2^{r-1} - 1)}$$

Two methods are available for determining velocities using an incremental encoder: (1) the pulse-counting method and (2) the pulse-timing method. In the first method, the pulse count over the sampling period of the digital processor is measured and is used to calculate the angular velocity. For a given sampling period, there is a lower speed limit below which this method is not very accurate. In the second method, the time for one encoder cycle is measured using a high-frequency clock signal. This method is particularly suitable for measuring low speeds accurately.

To compute the angular velocity ω using the first method, suppose that the count during a sampling period T is n pulses. Hence, the average time for one pulse is T/n. If there are N windows on the disk, the average time for one revolution is NT/n. Hence,

$$\omega = \frac{2\pi n}{NT} \tag{5.3}$$

For the second method of velocity computation, suppose that the clock frequency is f Hz. If m cycles of the clock signal are counted during an encoder period (interval between two adjacent windows), the time for that encoder cycle (i.e., the time to rotate through one encoder pitch) is given by m/f. With a total of N windows on the track, the average time for one revolution of the disk is Nm/f. Hence,

$$\omega = \frac{2\pi f}{Nm} \tag{5.4}$$

Note that a single incremental encoder can serve as both position sensor and speed sensor. Hence, a position loop and a speed loop in a control system can be closed using a single encoder, without having to use a conventional (analog) speed sensor such as a tachometer. The speed resolution of the encoder (depending on the method of speed computation—pulse counting or pulse timing) can be chosen to meet the accuracy requirements for the speed control loop. A further advantage of using an encoder rather than a conventional (analog) motion sensor is that an analog-to-digital converter (ADC) would be unnecessary. For example, the pulses generated by the encoder could be used as *interrupts* for the control computer. These interrupts are then directly counted (by an up/down counter or indexer) and timed (by a clock) within the control computer, thereby providing position and speed readings. Another way of interfacing an encoder to the control computer (without the need for an ADC) will be explained next.

Data Acquisition Hardware

A method for interfacing an incremental encoder to a digital processor (digital controller) is shown schematically in figure 5.8. The pulse signals are fed into an up/down counter that has circuitry to detect pulses (for example, by rising-edge detection or by level detection) and logic circuitry to determine the direction and to code the count. A pulse in one direction (say, clockwise will increment the count by one (an upcount), and a pulse in the opposite direction will decrement the count by one (a downcount). The coded count may be directly read by the processor through its input/output (I/O) board without the need for an ADC. The count is transferred to a latch buffer so that the measurement is read from the buffer rather than from the counter itself. This arrangement provides an efficient means of data acquisition because the counting process can continue without interruption while the count is being read by the processor from the latch buffer. The processor identifies various components in the measurement system using addresses, and this information is communicated to the individual components through the address bus. The start, end, and nature of an action (e.g., data read, clear the counter, clear the buffer) are communicated to various devices by the processor through the control bus. The processor can command an action to a component in one direction of the bus, and the component can respond with a message (e.g., job completed) in the opposite direction. The data (e.g., the count) are transmitted through the data bus. While the processor reads (samples) data from the buffer, the control signals guarantee that no data are transferred to that buffer from the counter. It is clear that the data acquisition consists of handshake operations between the processor and the auxiliary components. More than one encoder may be addressed, controlled, and read by the

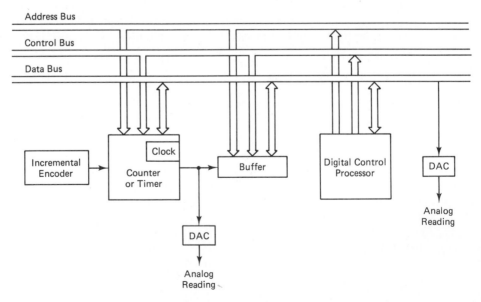

Figure 5.8. A data acquisition system for an incremental encoder.

same three buses. The buses are typically multicore cables carrying signals in parallel logic. Slower communication in serial logic is also common.

In measuring position using an incremental encoder, the counter may be continuously monitored through a digital-to-analog converter (DAC in figure 5.8). However, the count is read by the processor only at every sampling instant. Since a cumulative count is required in displacement measurement, the buffer is not cleared once the count is read in by the processor.

For velocity measurement by the pulse-counting method, the buffer is read at intervals of T, which is also the counting-cycle time. The counter is cleared every time a count is transferred to the buffer, so that a new count can begin. With this method, a new reading is available at every sampling period.

In the pulse-timing method of velocity computation, the counter is actually a timer. The encoder cycle is timed using a clock (internal or external), and the count is passed on to the buffer. The counter is then cleared and the next timing cycle is started. The buffer is read by the processor periodically. With this method, a new reading is available at every encoder cycle. Note that under transient velocities, the encoder-cycle time is variable and is not directly related to the sampling period. Nevertheless, it is desirable to make the sampling period smaller than the encoder-cycle time, in general, so that no count is missed by the processor.

More efficient use of the digital processor may be achieved by using an interrupt routine. With this method, the counter (or buffer) sends an interrupt request to the processor when a new count is ready. The processor then temporarily suspends the current operation and reads in the new data. Note that the processor does not continuously wait for a reading in this case.

Displacement Resolution

The resolution of an encoder represents the smallest change in measurement that can be measured realistically. Since an encoder can be used to measure both displacement and velocity, we can identify a resolution for each case. Displacement resolution is governed by the number of windows N in the code disk and the digital size (number of bits) r of the buffer (counter output). The physical resolution is determined by N. If only one pulse signal is used (i.e., no direction sensing), and if the rising edges of the pulses are detected (i.e., full cycles of the encoder are counted), the physical resolution is given by $(360/N)°$. But if both pulse signals (quadrature signals) are available and the capability to detect rising and falling edges of a pulse is also present, four counts can be made per encoder cycle, thereby improving the resolution by a factor of four. Hence, the physical resolution is given by

$$\Delta\theta_p = \frac{360°}{4N} \tag{5.5}$$

To understand this, note in figure 5.6a (or figure 5.6b) that when the two signals v_1 and v_2 are added, the resulting signal has a transition at every quarter of the encoder cycle. This is illustrated in figure 5.9. By detecting each transition (through edge detection or level detection), four pulses can be counted within every main cy-

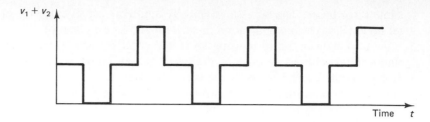

Figure 5.9. Quadrature signal addition to improve physical resolution.

cle. It should be mentioned that each signal (v_1 or v_2) separately has a resolution of half a pitch, provided that transitions (rising edges and falling edges) are detected and counted instead of pulses being counted. Accordingly a disk with 10,000 windows has a resolution of $0.018°$ if only one pulse signal is used (and both transitions, rise and fall, are detected). When both signals (with a phase shift of a quarter of a cycle) are used, the resolution improves to $0.009°$. This resolution is achieved directly from the mechanics of the transducer; no interpolation is involved. It assumes, however, that the pulses are nearly ideal and, in particular, that the transitions are perfect. In practice, this cannot be achieved if the pulse signals are noisy; pulse shaping might be necessary.

Assuming that the maximum angle measured is $360°$ (or $\pm 180°$), the digital resolution is given by (see equations 5.1 and 5.2)

$$\Delta\theta_d = \frac{180°}{2^{r-1}} = \frac{360°}{2^r} \tag{5.6}$$

This should be clear, because a digital word containing r bits can represent 2^r different values (unsigned). As mentioned earlier, we have used $360°/2^r$ instead of $360°/(2^r - 1)$ in equation 5.6 for digital resolution, and this is further supported by the fact that $0°$ and $360°$ represent the same position of the code disk. An ambiguity does not arise if we take the minimum value of θ to be $360°/2^r$, not zero. Then, by definition, the digital resolution is given by

$$\frac{(360° - 360°/2^r)}{(2^r - 1)}$$

This result is exactly the same as what is given by equation 5.6.

The larger of the two resolutions in equations 5.5. and 5.6 governs the displacement resolution of the encoder.

Example 5.1

For an ideal design of an incremental encoder, obtain an equation relating the parameters d, w, and r, where

d = diameter of encoder disk
w = number of windows per unit diameter of disk
r = word size (bits) of angle measurements

Assume that quadrature signals are available. If $r = 12$ and $w = 1,000/\text{in.}$, determine a suitable disk diameter.

Solution   In this problem, we are required to assign a word size for the resolution available from the number of windows. The position resolution by physical constraint (assuming that quadrature signals are available) is

$$\Delta\theta_p = \frac{1}{4}\left(\frac{360}{wd}\right)^\circ$$

The resolution available from the digital word size of the buffer is

$$\Delta\theta_d = \left(\frac{360}{2^r}\right)^\circ$$

Note that $\Delta\theta_p = \Delta\theta_d$ provides an ideal design. Hence,

$$\frac{1}{4}\frac{360}{wd} = \frac{360}{2^r}$$

Simplifying, we have

$$wd = 2^{r-2}$$

with $r = 12$ and $w = 1,000/\text{in.}$,

$$d = \left(\frac{2^{12-2}}{1,000}\right) \text{ in.} = 1.024 \text{ in.}$$

The physical resolution of an encoder can be improved by using step-up gearing so that one rotation of the moving object that is being monitored corresponds to several rotations of the code disk of the encoder. This improvement is directly proportional to the gear ratio. Backlash in the gearing mechanism introduces a new error, however. For best results, this backlash error should be several times smaller than the resolution with no backlash. Note that the digital resolution is not improved by gearing, because the maximum angle of rotation of the moving object (say, 360°) still corresponds to the buffer size, and the change in the least significant bit (LSB) of the buffer corresponds to the same change in the angle of rotation of the moving object. In fact, the overall displacement resolution can be harmed in this case if excessive backlash is present.

Example 5.2

By using high-precision techniques to imprint window tracks on the code disk, it is possible to attain the window density of 1,000 windows/in. of diameter. Consider a 3,000-window disk. Suppose that step-up gearing is used to improve resolution and the gear ratio is 10. If the word size of the output buffer is 16 bits, examine the displacement resolution of this device.

Solution   First consider the case in which gearing is not present. With quadrature signals, the physical resolution is

$$\Delta\theta_p = \frac{360°}{4 \times 3,000} = 0.03°$$

Now, for a range of measurement given by $\pm 180°$, a 16-bit output provides a digital resolution of

$$\Delta\theta_d = \frac{180°}{2^{15}} = 0.005°$$

Hence, in the absence of gearing, the overall displacement resolution is $0.03°$. On the other hand, with a gear ratio of 10, and neglecting gear backlash, the physical resolution improves to $0.003°$, but the digital resolution remains unchanged at best. Hence, the overall displacement resolution has improved to $0.005°$ as a result of gearing.

In summary, the displacement resolution of an incremental encoder depends on the following factors:

1. Number of windows on the code track (or disk diameter)
2. Gear ratio
3. Word size of the measurement buffer

The angular resolution of an encoder can be further improved, through interpolation, by adding equally spaced pulses in between every pair of pulses generated by the encoder circuit. These auxiliary pulses are not true measurements, and they can be interpreted as a linear interpolation scheme between true pulses. One method of accomplishing this interpolation is by using the two pick-off signals that are generated by the encoder (quadrature signals). These signals are nearly sinusoidal prior to shaping (say, by level detection). They can be filtered to obtain two sine signals that are $90°$ out of phase (i.e., a sine signal and a cosine signal). By weighted combination of these two signals, a series of sine signals can be generated such that each signal lags the preceding signal by any integer fraction of $360°$. By level detection or edge detection (rising and falling edges), these sine signals can be converted into square wave signals. Then, by logical combination of the square waves, an integer number of pulses can be generated within each encoder cycle. These are the interpolation pulses that are added to improve the encoder resolution. In practice, about twenty interpolation pulses can be added between adjacent main pulses by this method.

Velocity Resolution

An incremental encoder is also a velocity-measuring device. The velocity resolution of an incremental encoder depends on the method that is employed to determine velocity. Since the pulse-counting method and the pulse-timing method are both based on counting, the resolution corresponds to the change in angular velocity that results from changing (incrementing or decrementing) the count by one. If the pulse-counting method is employed, it is clear from equation 5.3 that a unity change in the count n corresponds to a speed change of

$$\Delta\omega_c = \frac{2\pi}{NT} \qquad (5.7)$$

where N is the number of windows in the code track and T is the sampling period. Equation 5.7 gives the velocity resolution by this method. Note that this resolution is independent of the angular velocity itself. The resolution improves, however, with the number of windows and the sampling period. But under transient conditions, the accuracy of a velocity reading decreases with increasing T (the sampling frequency has to be at least double the highest frequency of interest in the velocity signal). Hence, the sampling period should not be increased indiscriminately.

If the pulse-timing method is employed, the velocity resolution is given by (see equation 5.4)

$$\Delta\omega_t = \frac{2\pi f}{Nm} - \frac{2\pi f}{N(m+1)} = \frac{2\pi f}{Nm(m+1)} \tag{5.8}$$

where f is the clock frequency. For large m, $(m+1)$ can be approximated by m. Then, by substituting equation 5.4 in 5.8, we get

$$\Delta\omega_t = \frac{N\omega^2}{2\pi f} \tag{5.9}$$

Note that in this case, the resolution degrades quadratically with speed. This observation confirms the previous suggestion that the pulse-timing method is appropriate for low speeds. For a given speed, the resolution degrades with increasing N. The resolution can be improved, however, by increasing the clock frequency. Gearing up has a detrimental effect on the speed resolution in the pulse-timing method, but it has a favorable effect in the pulse-counting method (see problem 5.6). In summary, the speed resolution of an incremental encoder depends on the following factors:

1. Number of windows N
2. Sampling period T
3. Clock frequency f
4. Speed ω
5. Gear ratio

5.4 ABSOLUTE OPTICAL ENCODERS

Absolute encoders directly generate coded data to represent angular positions using a series of pulse signals. No pulse counting is involved in this case. A simplified code pattern on an absolute encoder disk that utilizes the direct binary code is shown in figure 5.10a. The number of tracks (n) in this case is 4, but in practice n is on the order of 14. The disk is divided into 2^n sectors. Each partitioned area of the matrix thus formed corresponds to a bit of data. For example, a transparent area may correspond to binary 1 and an opaque area to binary 0. Each track has a pick-off sensor similar to those used in incremental encoders. The set of n pick-off sensors is arranged on a radial line and facing the tracks on one side of the disk. A light source (e.g., light-emitting diode or LED) illuminates the other side of the disk. As the disk rotates, the bank of pick-off sensors generates a set of pulse signals. At a given in-

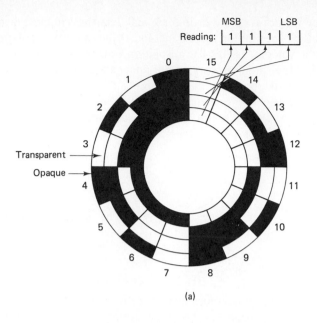

(a)

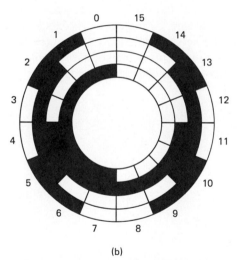

(b)

Figure 5.10. Schematic diagram of an absolute encoder disk pattern: (a) binary code; (b) gray code.

stant, the combination of signal levels will provide a coded data word that uniquely determines the position of the disk at that time, with a resolution given by the sector angle. In figure 5.10a, the word size of the data is 4 bits. This can represent decimal numbers from 0 to 15, as given by the sixteen sectors of the disk. In each sector, the outermost element is the least significant bit (LSB) and the innermost element is the most significant bit (MSB). The direct binary representation of the disk sectors (position) is given in table 5.1. The angular resolution for this simplified example is $(360/2^4)°$, or $22.5°$. If $n = 14$, the angular resolution improves to $(360/2^{14})°$, or $0.022°$.

TABLE 5.1 SECTOR CODING FOR THE
ABSOLUTE ENCODER EXAMPLE

Sector number	Straight binary code (MSB → LSB)	Gray code
0	0 0 0 0	0 1 1 1
1	0 0 0 1	0 1 1 0
2	0 0 1 0	0 1 0 0
3	0 0 1 1	0 1 0 1
4	0 1 0 0	0 0 0 1
5	0 1 0 1	0 0 0 0
6	0 1 1 0	0 0 1 0
7	0 1 1 1	0 0 1 1
8	1 0 0 0	1 0 1 1
9	1 0 0 1	1 0 1 0
10	1 0 1 0	1 0 0 0
11	1 0 1 1	1 0 0 1
12	1 1 0 0	1 1 0 1
13	1 1 0 1	1 1 0 0
14	1 1 1 0	1 1 1 0
15	1 1 1 1	1 1 1 1

Gray Coding

There is a data interpretation problem associated with the straight binary code in absolute encoders. Notice in table 5.1 that in straight binary, the transition from one sector to the adjacent sector may need more than one switching of bits of the binary data. For example, the transition from 0011 to 0100 or from 1011 to 1100 requires three bit switchings, and the transition from 0111 to 1000 or from 1111 to 0000 requires four bit switchings. If the pick-off sensors are not properly aligned along a radius of the encoder disk, or if excessive manufacturing error tolerances were allowed in imprinting the code pattern on the disk, or if environmental effects have resulted in large irregularities in the sector matrix, then bit switching will not take place simultaneously. This results in ambiguous readings during the transition period. For example, in changing from 0011 to 0100, if the LSB switches first, the reading becomes 0010. In decimal form, this incorrectly indicates that the rotation was from angle 3 to angle 2, whereas, it was actually a rotation from angle 3 to angle 4. Such ambiguities can be avoided by using a gray code, as shown in figure 5.10b for this example. The coded representation of the sectors is given in table 5.1. Note that in this case, each adjacent transition involves only one bit switching. A disadvantage of utilizing a gray code is that it requires additional logic to convert the gray-coded number to the corresponding binary number.

As with incremental encoders, the resolution of an absolute encoder can be improved by interpolation using auxiliary pulses. This requires an interpolation track and two pick-off sensors placed a quarter-pitch apart. This is equivalent to having an incremental encoder and an absolute encoder in the same unit. The resolution is limited by the word size of the output data. Step-up gear mechanisms can also be employed to improve encoder resolution.

Absolute encoders can be used for angular velocity measurement as well. For this, either the pulse-timing method or the angle-measurement method may be used. With the first method, the interval between two consecutive readings is strobed (or timed) using a high-frequency strobe (clock) signal, as in the case of an incremental encoder. Typical strobing frequency is 1 MHz. The start and stop of strobing are triggered by the coded data from the encoder. The clock cycles are counted by a counter, as in the case of an incremental encoder, and the count is reset (cleared) after each counting cycle. The angular speed can be computed using these data, as shown earlier for an incremental encoder. With the second method, the change in angle is measured from one sample to the next, and the angular speed is computed as the ratio (angle change)/(sampling period).

Because the code matrix on the disk is more complex in an absolute encoder, and because more light sensors are required, an absolute encoder can be nearly twice as expensive as an incremental encoder. An absolute encoder does not require digital counters and buffers, however, unless data interpolation is done using an auxiliary track or pulse-timing is used for velocity calculation. Also, an absolute encoder has the advantage that if a reading is missed, it will not affect the next reading, whereas, a missed pulse in an incremental encoder would carry an error to the subsequent readings until the counter is cleared. Furthermore, incremental encoders have to be powered throughout operation of the system. Thus, a power failure can introduce an error unless the reading is reinitialized (calibrated). An absolute encoder must be powered *and* monitored only when a reading is taken.

5.5 ENCODER ERROR

Errors in shaft encoder readings can come from several factors. The primary sources of these errors are as follows:

1. Quantization error (due to digital word size limitations)
2. Assembly error (eccentricity, etc.)
3. Coupling error (gear backlash, belt slippage, loose fit, etc.)
4. Structural limitations (disk deformation and shaft deformation due to loading)
5. Manufacturing tolerances (errors from inaccurately imprinted code patterns, inexact positioning of the pick-off sensors, limitations and irregularities in signal generation and sensing components, etc.)
6. Ambient effects (vibration, temperature, light noise, humidity, dirt, smoke, etc.)

These factors can result in erroneous displacement and velocity readings and inexact direction detection.

One form of error in an encoder reading is the hysteresis. For a given position of the moving object, if the encoder reading depends on the direction of motion, the measurement has a hysteresis error. In that case, if the object rotates from position A to position B and back to position A, for example, the initial and the final readings of the encoder will not match. The causes of hysteresis include backlash in gear couplings, loose fits, mechanical deformation in the code disk and shaft, delays in electronic circuitry (electrical time constants), and noisy pulse signals that make the detection of pulses (say, by level detection or edge detection) less accurate.

The raw pulse signal from an optical encoder is somewhat irregular, primarily because of noise in the signal generation circuitry, including the noise created by imperfect light sources and photosensors. Noisy pulses have imperfect edges. As a result, pulse detection through edge detection can result in errors such as multiple triggering for the same edge of a pulse. This can be avoided by including a Schmitt trigger (a logic circuit with electronic hysteresis) in the edge-detection circuit, so that slight irregularities in the pulse edges will not cause erroneous triggering, provided that the noise level is within the hysteresis band of the trigger. A disadvantage of this method, however, is that hysteresis will be present even when the encoder itself is perfect. Virtually noise-free pulses can be generated if two photosensors are used to detect adjacent transparent and opaque areas on a track simultaneously and a separate circuit (a comparator) is used to create a pulse that depends on the sign of the voltage difference of the two sensor signals. (We described this method earlier. A schematic diagram of this arrangement is given in figure 5.7.)

Eccentricity Error

Eccentricity (denoted by e) of an encoder is defined as the distance between the center of rotation C of the code disk and the geometric center G of the circular code track. Nonzero eccentricity causes a measurement error known as the *eccentricity error*. The primary contributions to eccentricity are

1. Shaft eccentricity (e_s)
2. Assembly eccentricity (e_a)
3. Track eccentricity (e_t)
4. Radial play (e_p)

Shaft eccentricity results if the rotating shaft on which the code disk is mounted is imperfect, so that its axis of rotation does not coincide with its geometric axis. Assembly eccentricity is caused if the code disk is improperly mounted on the shaft, so that the center of the code disk does not fall on the shaft axis. Track eccentricity comes from irregularities in the code track imprinting process, so that the center of the track circle does not coincide with the nominal geometric center of the disk. Radial play is caused by any looseness in the assembly in the radial direction. All four of these parameters are random variables that have mean values μ_s, μ_a, μ_t, and μ_p, respectively, and standard deviations σ_s, σ_a, σ_t, and σ_p, respectively. A very conservative upper bound for the mean value of the overall eccentricity is given by the sum of the individual mean values, each value being considered positive. A more reasonable estimate is provided by the *root-mean-square (rms)* value, as given by

$$\mu = \sqrt{\mu_s^2 + \mu_a^2 + \mu_t^2 + \mu_p^2} \tag{5.10}$$

Furthermore, assuming that the individual eccentricities are independent random variables, the standard deviation of the overall eccentricity is given by

$$\sigma = \sqrt{\sigma_s^2 + \sigma_a^2 + \sigma_t^2 + \sigma_p^2} \tag{5.11}$$

Knowing the mean value μ and the standard deviation σ of the overall eccentricity, it is possible to obtain a reasonable estimate for the maximum eccentricity that can occur. It is reasonable to assume that the eccentricity has a Gaussian (or normal) distribution, as shown in figure 5.11. The probability that the eccentricity lies between two given values is obtained by the area under the probability density curve within these two values (points) on the *x*-axis (also see chapter 2). In particular, for the normal distribution, the probability that the eccentricity lies within $\mu - 2\sigma$ and $\mu + 2\sigma$ is 95.5 percent, and the probability that the eccentricity falls within $\mu - 3\sigma$ and $\mu + 3\sigma$ is 99.7 percent. We can say, for example, that at a confidence level of 99.7 percent, the net eccentricity will not exceed $\mu + 3\sigma$.

Example 5.3

The mean values and the standard deviations of the four primary contributions to eccentricity in a shaft encoder are as follows (in millimeters):

Shaft eccentricity = (0.1, 0.01)
Assembly eccentricity = (0.2, 0.05)
Track eccentricity = (0.05, 0.001)
Radial play = (0.1, 0.02)

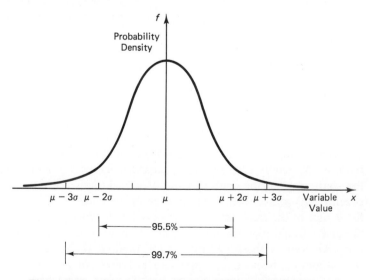

Figure 5.11. Gaussian (normal) probability density function.

Estimate the overall eccentricity at a confidence level of 96 percent.

Solution The mean value of the overall eccentricity may be estimated as the rms value of the individual means; thus, from equation 5.10,

$$\mu = \sqrt{0.1^2 + 0.2^2 + 0.05^2 + 0.1^2} = 0.25 \text{ mm}$$

Using equation 5.11, the standard deviation of the overall eccentricity is estimated as

$$\sigma = \sqrt{0.01^2 + 0.05^2 + 0.001^2 + 0.02^2} = 0.055 \text{ mm}$$

Now, assuming Gaussian distribution, an estimate for the overall eccentricity at a confidence level of 96 percent is given by

$$\hat{e} = 0.25 + 2 \times 0.055 = 0.36 \text{ mm}$$

Once the overall eccentricity is estimated in the foregoing manner, the corresponding measurement error can be determined. Suppose that the true angle of rotation is θ and the corresponding measurement is θ_m. The eccentricity error is given by

$$\Delta\theta = \theta_m - \theta \tag{5.12}$$

The maximum error can be shown to exist when the line of eccentricity (CG) is symmetrically located within the angle of rotation, as shown in figure 5.12. For this configuration, the sine rule for triangles gives

$$\frac{\sin (\Delta\theta/2)}{e} = \frac{\sin (\theta/2)}{r}$$

where r denotes the code track radius, which can be taken as the disk radius for most practical purposes. Hence, the eccentricity error is given by

$$\Delta\theta = 2 \sin^{-1} \left(\frac{e}{r} \sin \frac{\theta}{2} \right) \tag{5.13}$$

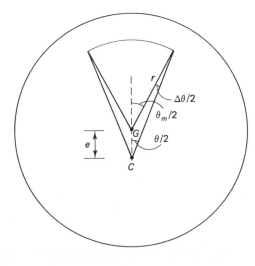

Figure 5.12. Nomenclature for eccentricity error (C = center of rotation, G = geometric center of the code track).

Example 5.4

Show analytically that the eccentricity error of an encoder disk does not enter measurements of complete revolutions of the disk.

Solution   It is intuitively clear that the eccentricity error should not enter measurements of complete revolutions, and this can be shown analytically by using equation 5.13. In this case, $\theta = 2\pi$. Accordingly, $\Delta\theta = 0$. For multiple revolutions, the eccentricity error is periodic with period 2π.

For small angles, the sine of an angle is approximately equal to the angle itself, in radians. Hence, for small $\Delta\theta$, the eccentricity error may be expressed as

$$\Delta\theta = \frac{2e}{r} \sin \frac{\theta}{2} \tag{5.14}$$

Furthermore, for small angles of rotation, the fractional eccentricity error is given by

$$\frac{\Delta\theta}{\theta} = \frac{e}{r} \tag{5.15}$$

which is, in fact, the worst fractional error. As the angle of rotation increases, the fractional error decreases (as shown in figure 5.13), reaching the zero value for a full revolution. From the point of view of gross error, the worst value occurs when $\theta = \pi$, which corresponds to half a revolution. From equation 5.13, it is clear that the maximum gross error due to eccentricity is given by

$$\Delta\theta_{max} = 2 \sin^{-1} \frac{e}{r} \tag{5.16}$$

If this value is less than half the resolution of the encoder, the eccentricity error becomes inconsequential. For all practical purposes, since e is much less than r, we may use the following expression for the maximum eccentricity error:

$$\Delta\theta_{max} = \frac{2e}{r} \tag{5.17}$$

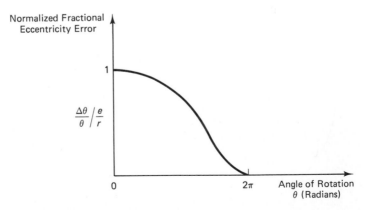

Figure 5.13.   Fractional eccentricity error variation with the angle of rotation.

Digital Transducers   Chap. 5

Example 5.5

Suppose that in example 5.3, the radius of the code disk is 5 cm. Estimate the maximum error due to eccentricity. If each track has 1,000 windows, determine whether the eccentricity error is significant.

Solution With the given level of confidence, we have calculated the overall eccentricity to be 0.36 mm. Now, from equation 5.16 or 5.17, the maximum angular error is

$$\Delta\theta_{max} = \frac{2 \times 0.36}{50} = 0.014 \text{ rad} = 0.83°$$

Assuming that quadrature signals are used to improve the encoder resolution, we have

$$\text{resolution} = \frac{360°}{4 \times 1,000} = 0.09°$$

Note that the maximum error due to eccentricity is more than ten times the encoder resolution. Hence, eccentricity will significantly affect the accuracy of the encoder.

Eccentricity also affects the phase angle between the quadrature signals of an incremental encoder if a single track and two pick-off sensors (with circumferential offset) are used. This error can be reduced using the two-track arrangement, with the two sensors positioned along a radial line, so that eccentricity affects the two outputs equally.

5.6 DIGITAL RESOLVERS

Digital resolvers, or mutual induction encoders, operate somewhat like analog resolvers, using the principle of mutual induction. They are known commercially as Inductosyns (Ferrand Controls, Valhalla, N.Y.). A digital resolver has two disks facing each other (but not touching), one (the stator) stationary and the other (the rotor) coupled to the rotating object. The rotor has a fine electric conductor foil imprinted on it, as shown schematically in figure 5.14. The printed pattern is a closely spaced set of radial pulses, all of which are connected to a high-frequency AC supply of voltage v_{ref}. The stator disk has two separate printed patterns that are identical to the rotor pattern, but one pattern on the stator is shifted by a quarter-pitch from the other pattern. The primary voltage in the rotor circuit induces voltages in the two secondary (stator) foils at the same frequency. As the rotor turns, the level of the induced voltage changes, depending on the relative position of the foil patterns on the two disks. When the foil pulse patterns coincide, the induced voltage is maximum (positive or negative), and when the rotor foil pattern has a half-pitch offset from the stator foil pattern, the induced voltage in adjacent parts cancel each other, producing a zero output. If the speed of rotation is constant, the output voltages v_1 and v_2 in the two foils of the stator become signals that have a carrier frequency (supply frequency) component modulated by periodic and nearly sinusoidal signals with a

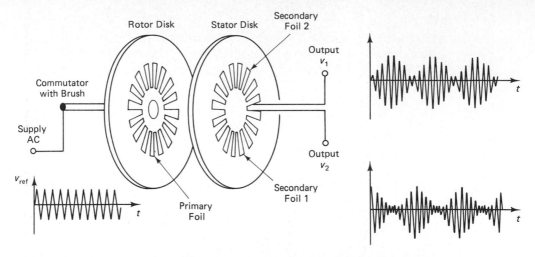

Figure 5.14. Schematic diagram of a digital resolver.

phase shift of 90°. The modulation signals can be extracted by demodulation and converted into pulse signals. When the speed is not constant, pulse width will vary with time. As in the case of an incremental encoder, angular displacement and angular velocity are determined by counting or timing the pulses. The direction of rotation is determined by the phase difference in the two modulating output signals. (In one direction, the phase shift is 90°; in the other direction, it is −90°.) Very fine resolutions are obtainable from a digital resolver; it is usually not necessary to use step-up gear systems or other techniques to improve resolution. Resolutions up to 0.0005° can be obtained from these transducers, but they are usually more expensive than optical encoders.

5.7 DIGITAL TACHOMETERS

Since shaft encoders are also used for measuring angular velocities, they can be considered tachometers. In classic terminology, a digital tachometer is a device that employs a toothed wheel to measure angular velocities. A schematic diagram of one such device is shown in figure 5.15. This is a magnetic induction tachometer of the variable-reluctance type. The teeth on the wheel are made of ferromagnetic material. The two magnetic induction (and variable-reluctance) proximity probes are placed facing the teeth radially, a quarter-pitch apart. When the toothed wheel rotates, the two probes generate output signals that are 90° out of phase. One signal leads the other in one direction of rotation and lags the other in the opposite direction of rotation. In this manner, directional readings are obtained. The speed is computed either by counting pulses over a sampling period or by timing the pulse width, as in the case of an incremental encoder.

Alternative types of digital tachometers use eddy current proximity probes or capacitive proximity probes (see chapter 3). In the case of an eddy current tachome-

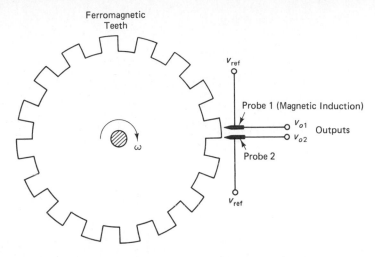

Figure 5.15. Schematic diagram of a pulse tachometer.

ter, the teeth of the pulsing wheel are made of or plated with electricity-conducting material, and the probe emits a radio-frequency magnetic field. In the case of a capacitive tachometer, the toothed wheel forms one plate of the capacitor; the other plate is the probe and is kept stationary. As the wheel turns, the capacitor gap width fluctuates. If the capacitor is excited by an AC voltage of high frequency (typically 1 MHz), a near-pulse-modulated signal at that carrier frequency is obtained. This can be detected by a suitable capacitance bridge circuit. By demodulating the output signal, the modulating-pulse signal can be extracted. This pulse signal is used in the angular velocity computation.

The advantages of these digital (pulse) tachometers over optical encoders include simplicity, robustness, and low cost. The disadvantages include poor resolution (determined by the number of teeth, the speed of rotation, and the word size used for data transmission), and mechanical errors due to loading, hysteresis, and manufacturing irregularities.

5.8 HALL EFFECT SENSORS

Consider a semiconductor element subject to a DC voltage v_{ref}. If a magnetic field is applied perpendicular to the direction of this voltage, a voltage v_o will be generated in the third orthogonal direction within the semiconductor element. This is known as the Hall effect (observed by E. H. Hall in 1879). A schematic representation of a Hall effect sensor is shown in figure 5.16.

A Hall effect sensor may be used for motion sensing in many ways—for example, as an analog proximity sensor, a digital limit switch, or a digital shaft encoder. Since the output voltage v_o increases as the distance from the magnetic source to the semiconductor element decreases, the output signal v_o can be used as a measure of proximity. Alternatively, a certain threshold level of output voltage v_o can be used

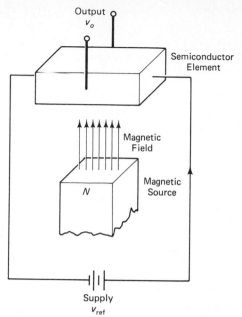

Figure 5.16. Schematic representation of a Hall effect sensor.

to activate a digital switch or to create a digital output, hence forming a digital limit switch.

A more practical arrangement would be to have the semiconductor element and the magnetic source fixed relative to one another in a single package. By moving a ferromagnetic member into the air gap between the magnetic source and the semiconductor element, the flux linkage can be altered. This changes v_o. This arrangement is suitable both as an analog proximity sensor and as a limit switch. Furthermore, if a toothed ferromagnetic wheel is used to change v_o, we have a shaft encoder or a digital tachometer (see figure 5.17).

The longitudinal arrangement of a proximity sensor, in which the moving element approaches head-on toward the sensor, is not suitable when there is a danger of overshooting the target, since it will damage the sensor. A more desirable configuration is the lateral arrangement, in which the moving member slides by the

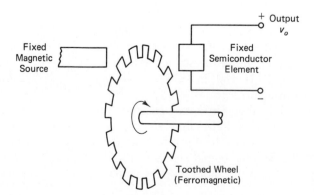

Figure 5.17. Schematic diagram of a Hall effect shaft encoder or digital tachometer.

Digital Transducers   Chap. 5

sensing face of the sensor. The sensitivity will be lower, however, with this lateral arrangement.

The relationship between the output voltage v_o and the distance x of a Hall effect sensor measured from the moving member is nonlinear. Linear Hall effect sensors use calibration to linearize their output.

5.9 MEASUREMENT OF TRANSLATORY MOTIONS

Digital rectilinear transducers are useful in many applications. Typical applications include x-y positioning tables, machine tools, valve actuators, read-write heads in disk drive systems, and robotic manipulators (e.g., at prismatic joints) and robot hands. The principles used in angular motion transducers described so far in this book can be used in measuring rectilinear motions as well. In rectilinear encoders, for example, rectangular flat plates moving rectilinearly, instead of rotating disks, are used with the same types of signal generation and interpretation mechanisms.

Cable Extension Sensors

In many applications, rectilinear motion is produced from a rotary motion (say, of a motor) through a suitable transmission device, such as rack and pinion or lead screw and nut. In these cases, rectilinear motion can be determined by measuring the associated rotary motion, assuming that errors due to backlash, flexibility, and so forth, in the transmission device can be neglected. Another way to measure rectilinear motions using a rotary sensor is to use a modified sensor that has the capability to convert a rectilinear motion into a rotary motion within the sensor itself. An example would be the *cable extension method* of sensing rectilinear motions. This method is particularly suitable for measuring motions that have large excursions. The cable extension method uses an angular motion sensor with a spool rigidly coupled to the rotating part of the sensor (e.g., the encoder disk) and a cable that wraps around the spool. The other end of the cable is attached to the object whose rectilinear motion is to be sensed. The housing of the rotary sensor is firmly mounted on a stationary platform, so that the cable can extend in the direction of motion. When the object moves, the cable extends, causing the spool to rotate. This angular motion is measured by the rotary sensor. With proper calibration, this device can give rectilinear measurements directly. As the object moves toward the sensor, the cable has to retract without slack. This is guaranteed by using a device such as a spring motor to wind the cable back. The disadvantages of the cable extension method include mechanical loading of the moving object, time delay in measurements, and errors caused by the cable, including irregularities, slack, and tensile deformation.

Moiré Fringe Displacement Sensors

Another device for measuring rectilinear motions employs the moiré fringe technique. The operation of this motion sensor is somewhat analogous to that of a digital resolver, and the signal generation method is similar to that in an optical encoder.

Thus, this device can be viewed as an optical encoder with improved sensitivity and resolution. A transparent plate with a series of opaque lines arranged in parallel in the transverse direction forms the stationary plate (grating plate) of the transducer. This is called the mask plate. A second transparent plate, with an identical set of ruled lines, forms the moving plate. The lines on both plates are evenly spaced, and the line width is equal to the spacing between adjacent lines. A light source is placed on the moving plate side, and light transmitted through the common area of the two plates is detected on the other side using one or more photosensors. When the lines on the two plates coincide, the maximum amount of light will pass through the common area of the two plates. When the lines on one plate fall on the transparent spaces of the other plate, virtually no light will pass through the plates. Accordingly, as one plate moves relative to the other, a pulse train is generated by the photosensor, and it can be used to determine rectilinear displacement and velocity, as in the case of an incremental encoder. Moiré fringes are the shadow patterns formed in this manner. They can also be detected and observed by photographic means. With this technique, very small resolutions (e.g., 0.0002 in.) can be realized. Note that the method provides improved sensitivity over a basic optical encoder because light passing through many gratings is received by the same photosensor. Also, finer line spacing (in conjunction with wider light sensors) can be used in this method, thereby providing increased resolution.

The moiré device is used to measure rigid-body movements of one plate of the sensor with respect to the other, and it can be used to detect deformations (e.g., elastic deformations) of one plate with respect to the other in the direction orthogonal to the grating lines. In this case, depending on the nature of the plate deformation, some transparent lines of one plate will be completely covered by the opaque lines of the other plate, and some other transparent lines of the first plate will have coinciding transparent lines on the second plate. Thus, the observed image will have dark lines (moiré fringes) corresponding to the regions with clear/opaque overlaps of the two plates and bright lines corresponding to the regions with clear/clear overlaps of the two plates. Hence, the moiré fringe pattern will provide the deformation pattern of one plate with respect to the other.

Example 5.6

Suppose that each plate of a moiré fringe deformation sensor has a line pitch of 0.01 mm. A tensile load is applied to one plate in the direction perpendicular to the lines. Five moiré fringes are observed in 10 cm of the moiré image under tension. What is the tensile strain in the plate?

Solution There is one moiré fringe in every $10/5 = 2$ cm of the plate. Hence, extension of a 2 cm portion of the plate = 0.01 mm, and

$$\text{tensile strain} = \frac{0.01 \text{ mm}}{2 \times 10 \text{ mm}} = 0.0005\epsilon = 500\mu\epsilon$$

In the foregoing example, we have assumed that the strain distribution (or deformation) of the plate is uniform. Under nonuniform strain distributions, the observed moiré fringe pattern generally will not be parallel straight lines.

5.10 LIMIT SWITCHES

Limit switches are sensors used in detecting limits of mechanical motions. A limit of a movement can be detected by using a simple contact mechanism to close a circuit or trigger a pulse. Hence, the information provided by a limit switch takes only two states (on/off, present/absent, go/no-go, etc.); it can be represented by one bit. In this sense, a limit switch is considered a digital transducer. Additional logic is needed if the direction of contact is also needed. Limit switches are available for both rectilinear and angular motions.

A microswitch is a solid-state switch that can be used as a limit switch. Microswitches are commonly used in counting operations—for example, to keep a count of completed products in a factory warehouse.

Although a purely mechanical device consisting of linkages, gears, ratchet wheels and pawls, and so forth, can serve as a limit switch, electrical and solid-state switches are usually preferred for such reasons as accuracy, durability, a low activating force (practically zero) requirement, low cost, and small size. Any proximity sensor could serve as the sensing element of a limit switch. The proximity sensor signal is then used in the required manner—for example, to activate a counter, a mechanical switch, or a relay circuit, or simply as an input to a control computer.

PROBLEMS

5.1. Identify active transducers among the following types of shaft encoders, and justify your claims. Also, discuss the relative merits and drawbacks of the four types of encoders.
(a) Optical encoders
(b) Sliding contact encoders
(c) Magnetic encoders
(d) Proximity sensor encoders

5.2. Explain why the speed resolution of a shaft encoder depends on the speed itself. What are some of the other factors that affect speed resolution? The speed of a DC motor was increased from 50 rpm to 500 rpm. How would the speed resolution change if the speed was measured using an incremental encoder
(a) By the pulse-counting method?
(b) By the pulse-timing method?

5.3. Discuss construction features and operation of an optical encoder for measuring *rectilinear* displacements and velocities.

5.4. What is hysteresis in an optical encoder? List several causes of hysteresis and discuss ways to minimize hysteresis.

5.5. Describe methods of improving resolution in an encoder. An incremental encoder disk has 5,000 windows. The word size of the output data is 12 bits. What is the angular resolution of the device? Assume that quadrature signals are available but that no interpolation is used.

5.6. A shaft encoder that has N window per track is connected to a shaft through a gear system with gear ratio p. Derive formulas for calculating angular velocity of the shaft by
(a) The pulse-counting method
(b) The pulse-timing method

What is the speed resolution in each case? What effect does step-up gearing have on the speed resolution?

5.7. An optical encoder has n windows/inch diameter (in each track). What is the eccentricity tolerance e below which readings are not affected by eccentricity error?

5.8. Show that in the single-track, two-sensor design of an incremental encoder, the phase angle error (in quadrature signals) due to eccentricity is inversely proportional to the second power of the radius of the code disk for a given window density. Suggest a way to reduce this error.

5.9. Encoders that can provide 50,000 counts/turn with ± 1 count accuracy are commercially available. What is the resolution of such an encoder? Describe the physical construction of an encoder that has this resolution.

5.10. A particular type of multiplexer can handle ninety-six sensors. Each sensor generates a pulse signal with variable pulse width. The multiplexer scans the incoming pulse sequences, one at a time, and passes the information onto a control computer.
(a) What is the main objective of using a multiplexer?
(b) What type of sensors could be used with this multiplexer?

5.11. Suppose that a feedback control sytem is expected to provide an accuracy within $\pm \Delta y$ in a response variable y. Explain why the sensor that measures y should have a resolution of $\pm (\Delta y / 2)$ or better for this accuracy to be possible. An x-y table has a travel of 2 m. The feedback control system is expected to provide an accuracy of ± 1 mm. An optical encoder is used to measure position for feedback in each direction (x and y). What is the minimum bit size that is required for each encoder output buffer? If the motion sensor used is an absolute encoder, how many tracks and how many sectors should be present on the encoder disk?

5.12. Discuss the advantages of solid-state limit switches over mechanical limit switches. Solid-state limit switches are used in many applications, particularly in the aircraft and aerospace industries. One such application is in landing gear control, to detect up, down, and locked conditions of the landing gear. High reliability is of utmost importance in such applications. Mean time between failure (MTBF) of over 100,000 hours is possible with solid-state limit switches. Using your engineering judgment, give an MTBF value for a mechanical limit switch.

5.13. Explain how resolution of a shaft encoder could be improved by pulse interpolation. Suppose that a pulse generated from an incremental encoder can be approximated by

$$v = v_o \left(1 + \sin \frac{2\pi\theta}{\Delta\theta} \right)$$

where θ denotes the angular position of the encoder window with respect to the photosensor position. Let us consider rotations of a half-pitch or smaller (i.e., $0 \le \theta \le \Delta\theta/2$, where $\Delta\theta$ is the window pitch angle). By using this sinusoidal approximation for a pulse, show that we can improve the resolution of an encoder indefinitely simply by measuring the shape of each pulse at clock cycle intervals using a high-frequency clock signal.

5.14. What is a Hall effect tachometer? Discuss the advantages and disadvantages of a Hall effect motion sensor in comparison to an optical motion sensor (e.g., an optical encoder).

5.15. The pulses generated by the coding disk of an incremental optical encoder are approximately triangular in shape. Explain the reason for this. Describe a method for converting these triangular pulses into sharp rectangular pulses.

5.16. A brand of autofocusing camera uses a microprocessor-based feedback control system consisting of a charge-coupled device (CCD) imaging system, a microprocessor, a drive motor, and an optical encoder. The purpose of the control system is to focus the camera automatically, based on the image of the subject as sensed by a matrix of CCDs (a set of metal oxide semiconductor field-effect transistors, or MOSFETs). The light rays from the subject that pass through the lens will fall onto the CCD matrix. This will generate a matrix of charge signals, which are shifted one at a time, row by row, into an output buffer and passed on to the microprocessor after conditioning the resulting video signal. The CCD image obtained by sampling the video signal is analyzed by the microprocessor to determine whether the camera is focused properly. If not, the lens is moved by the motor so as to achieve focusing. Draw a schematic diagram for the autofocusing control system and explain the function of each component in the control system, including the encoder.

5.17. A Schmitt trigger is a semiconductor device that can function as a level detector or a switching element with hysteresis. The presence of hysteresis can be used, for example, to eliminate chattering during switching caused by noise in the switching signal. The input/output characteristic of a Schmitt trigger is shown in figure P5.17a. If the input signal is as shown in figure P5.17b, determine the output signal.

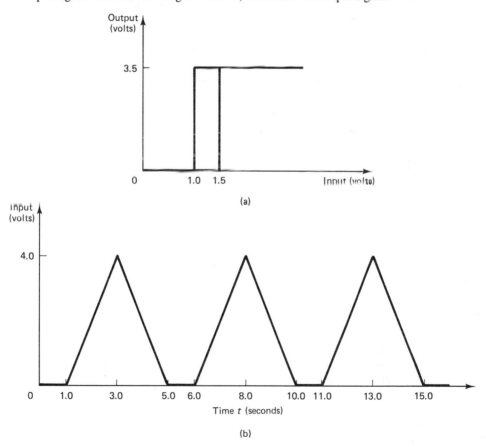

Figure P5.17. (a) The input/output characteristic of a Schmitt trigger. (b) A triangular input signal

5.18. Displacement sensing and speed sensing are essential for a position servo. If a digital controller is employed to generate the servo signal, one option would be to use an analog displacement sensor and an analog speed sensor, along with analog-to-digital converters (ADCs) to produce the necessary digital feedback signals. Alternatively, an incremental encoder could be used to provide both displacement and speed feedbacks. In this case, ADCs are not needed. Encoder pulses will provide interrupts to the digital controller. Displacement is obtained by counting the interrupts. The speed is obtained by timing the interrupts. In some applications, analog speed signals are needed. Explain how an incremental encoder and a frequency-to-voltage converter (FVC) may be used to generate an analog speed signal.

5.19. Consider the two quadrature pulse signals (say, *A* and *B*) from an incremental encoder. Using sketches of these signals, show that in one direction of rotation, signal *B* is at a high level during the up-transition of signal *A*, and in the opposite direction of rotation, signal *B* is at a low level during the up-transition of signal A. Note that the direction of motion can be determined in this manner by using level detection of one signal during the up-transition of the other signal.

REFERENCES

BARNEY, G. C. *Intelligent Instrumentation*. Prentice-Hall, Englewood Cliffs, N.J., 1985.

CERNI, R. H., and FOSTER, L. E. *Instrumentation for Engineering Measurement*. Wiley, New York, 1962.

DESILVA, C. W. "Motion Sensors in Industrial Robots." *Mechanical Engineering* 107(6): 40–51, June 1985.

————. "Counters/Frequency Tachometers." *Measurements and Control Journal* 116: 201, April 1986.

FRANKLIN, G. F., and POWELL, J. D. *Digital Control of Dynamic Systems*. Addison-Wesley, Reading, Mass., 1980.

WOOLVET, G. A. *Transducers in Digital Systems*. Peter Peregrinus, London, 1979.

6

Stepper Motors

6.1 INTRODUCTION

The actuator is the device that mechanically drives a control system. There are many classifications of actuators. Those that directly operate a process (load, plant) are termed *process actuators*. Joint motors in a robotic manipulator are good examples of process actuators. In process control applications in particular, actuators are often used to operate controller components (final control elements), such as servovalves, as well. Actuators in this category are termed *control actuators*. Actuators that automatically use response signals from a process in feedback to correct the operation of the process are termed *servoactuators*. In particular, motors that use position, speed, and perhaps load torque measurements, and armature current or field current in feedback are termed *servomotors*. Factors such as power, motion resolution, repeatability, and operating bandwidth requirements for an actuator can differ significantly, depending on the particular control system and the specific function of the actuator within the system. Proper selection of actuators for a particular application is of utmost importance in the instrumentation and design of control systems.

Most actuators used in control applications are continuous-drive devices. Examples are DC torque motors, induction motors, hydraulic and pneumatic motors, and piston-cylinder drives (rams). Continuous-drive actuators will be discussed in chapter 7. Stepper motors are incremental-drive actuators. It is reasonable to treat them as digital actuators. Unlike continuous-drive actuators, stepper motors are driven in fixed angular steps (increments). Each step of rotation is the response of the motor rotor to an input pulse (or a digital command). In this manner, the stepwise rotation of the rotor can be synchronized with pulses in a command-pulse train, assuming, of course, that no steps are missed, thereby making the motor respond faithfully to the input signal (pulse sequence) in an open-loop manner. It should be remembered that, like a conventional continuous-drive motor, the stepper motor is also an electromagnetic actuator, in that it converts electromagnetic energy into mechanical energy to perform mechanical work.

The terms *stepper motor*, *stepping motor*, and *step motor* are synonymous and are often used interchangeably. Actuators that can be classified as stepper motors

253

have been in use for more than fifty years, but only after the incorporation of solid-state circuitry and logic devices in their drive systems have stepper motors emerged as cost-effective alternatives for DC servomotors in high-speed motion-control applications. Many kinds of actuators fall into the stepper motor category, but only those that are widely used in industry are discussed in this chapter. Note that even if the mechanism by which the incremental motion is generated differs from one type of stepper motor to the next, the same control techniques can be used in the associated control systems, making a general treatment of stepper motors possible, at least from the control point of view.

6.2 PRINCIPLE OF OPERATION

One common feature in any stepper motor is that the stator of the motor contains several pairs of field windings that can be switched on to produce electromagnetic pole pairs (N and S). The polarities can be reversed in two ways:

1. By reversing the direction of current in the winding
2. By using two pairs of windings (*bifilar windings*) for each pole pair, one pair giving one set of poles when energized and the other pair giving the opposite polarities

Note that in the case of bifilar windings, a relatively simple on/off switching mechanism is adequate for reversing polarities, thereby providing simplified drive circuitry. The drive circuitry for unifilar windings is more complex; in particular, current reversal circuitry is needed. Twice the normal number of windings would be required in bifilar-wound motors, most of which are inactive at a given time. This increases the motor size for a given torque rating. This drawback can be counteracted to some extent by decreasing the wire diameter, which results in an added advantage: The reduced wire diameter provides increased resistance for a given length of wire, which increases damping and decreases the electrical time constant of the motor, thus providing a better (fast but less oscillatory) single-step response. A further advantage of a bifilar stepper is its smaller levels of induced voltages by self-induction and mutual induction, because the current reversals are absent. For this reason, the effective (dynamic) torque at a given speed (stepping rate) is usually larger for bifilar stepper motors than for their unifilar counterparts, particularly at high speeds (see figure 6.1.) At very low stepping rates, however, dissipation effects will dominate induced-voltage effects, thereby providing higher torques with unifilar (standard) windings at these low stepping rates, as shown in figure 6.1.

Most classifications of stepper motors are based on the nature of the motor rotor. One such classification distinguishes variable-reluctance (VR) stepper motors, which have soft-iron rotors, from permanent-magnet (PM) stepper motors, which have magnetized rotors. The two types of motors operate in a somewhat similar manner. Hybrid motors possess characteristics of both VR steppers and PM steppers. A disadvantage of VR stepper motors is that since the rotor is not magnetized, the holding torque is zero when the stator windings are not energized (power off).

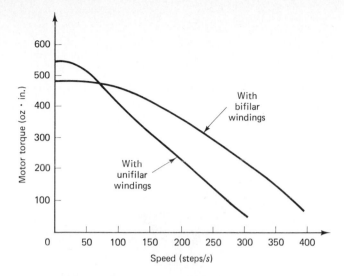

Figure 6.1. The effect of bifilar windings on motor torque.

Hence, there is no capability to hold the load at a given position under power-off conditions unless mechanical brakes are employed. Permanent-magnet stepper motors that have electromagnetic rotors, as opposed to permanently magnetized rotors, are also available. In this case, magnetizing coils, activated by a unidirectional DC source, are placed on the stator to magnetize the rotor.

A photograph of the internal components of a two-stack stepping motor is shown in figure 6.2a. The principle of operation of a permanent-magnet stepper motor is explained by the simple schematic diagram shown in figure 6.2b. The stator has two sets of windings (two *phases*), placed at 90°. This arrangement has four *salient poles* in the stator, each pole being geometrically separated by a 90° angle from the adjacent one. The rotor is a two-pole permanent magnet. Each phase can take one of the three states 1, 0, and −1, which are defined as follows:

State 1: current in a specified direction
State −1: current in the opposite direction
State 0: no current

By switching the currents in the two phases in an appropriate sequence, either a clockwise (CW) rotation or a counterclockwise (CCW) rotation can be produced. The CW rotation sequence is shown in figure 6.3. Note that S_i denotes the state of the ith phase. The *step angle* for this motor is 45°. At the end of each step, the rotor assumes the *minimum reluctance position* that corresponds to that particular magnetic polarity pattern in the stator. (Reluctance measures the magnetic resistance in a flux path). This is a *stable equilibrium configuration* and is known as the *detent position* for that step. When the stator currents (states) are switched for the next step, the minimum reluctance position changes (rotates by the step angle) and the rotor assumes the corresponding stable equilibrium position; the rotor turns through a single step (45° in this example). Table 6.1 gives the stepping sequences necessary for a complete clockwise rotation and a complete counterclockwise rotation. Note that a

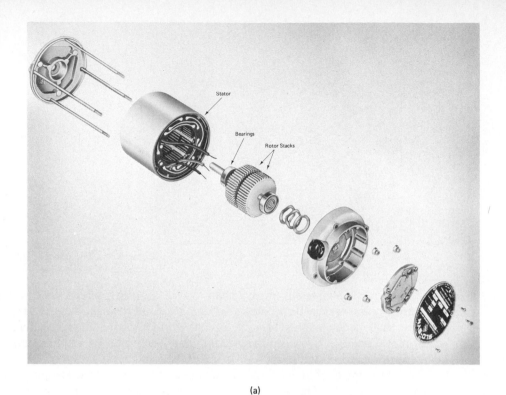

(a)

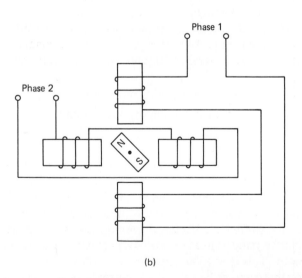

(b)

Figure 6.2. (a) Constructional details of a two-stack stepping motor (courtesy of the Superior Electric Company). (b) Schematic diagram of a two-phase permanent magnet stepper motor.

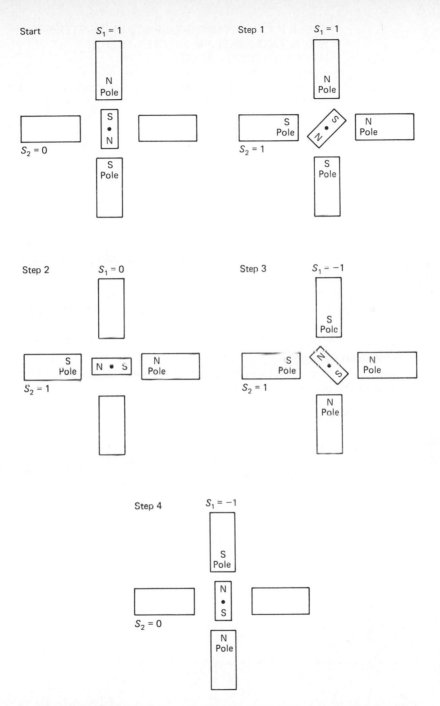

Figure 6.3. Stepping sequence (half-stepping) for a two-phase PM stepper motor for clockwise rotation.

TABLE 6.1 STEPPING SEQUENCE (HALF-STEPPING) FOR A TWO-PHASE PM STEPPER MOTOR WITH TWO ROTOR POLES

Step number	Clockwise rotation		Counterclockwise rotation	
	S_1	S_2	S_1	S_2
1	1	1	1	−1
2	0	1	0	−1
3	−1	1	−1	−1
4	−1	0	−1	0
5	−1	−1	−1	1
6	0	−1	0	1
7	1	−1	1	1
8	1	0	1	0

separate pair of columns is not actually necessary to give the states for the CCW rotation; they are simply given by the CW rotation states but tracked in the opposite direction (bottom to top). Consequently, switching commands may be digitally generated by a simple table lookup procedure with just eight pairs of entries. The clockwise stepping sequence is generated by reading the table in the top-to-bottom direction, and the counterclockwise stepping sequence is generated by reading the same table in the opposite direction.

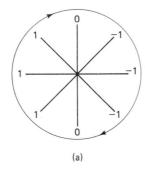

(a)

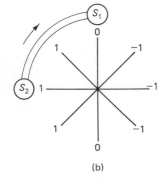

(b)

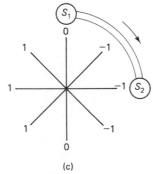

(c)

Figure 6.4. (a) Half-step switching states. (b) Switching logic for clockwise rotation. (c) Switching logic for counterclockwise rotation.

TABLE 6.2 STEPPING SEQUENCE (FULL-STEPPING)
FOR A TWO-PHASE PM STEPPER MOTOR
WITH TWO ROTOR POLES

	State S_1	State S_2	
	1	0	
Clockwise ↓	0	1	
	−1	0	↑ Counterclockwise
	0	−1	

A still more compact representation of switching cycles is also available. Note that in one complete rotation of the rotor, the state of each phase sweeps through one complete cycle of the switching sequence (shown in figure 6.4a) in the clockwise direction. For clockwise rotation of the motor, the state of phase 2 (S_2) *lags* the state of phase 1 (S_1) by two steps (figure 6.4b). For counterclockwise rotation, S_2 *leads* S_1 by two steps (figure 6.4c). Hence, instead of eight pairs of numbers, just eight numbers with a "delay" operation would suffice. Although the logic that generates the switching sequence for a phase winding could be supplied by a microprocessor (a software approach), it is customary to generate it through hardware logic in a device called a *translator*. This approach is more effective because the switching logic is fixed, as noted in the foregoing discussion. We shall say more about the translator in later sections.

The switching sequence given in table 6.1 corresponds to half-stepping, with a step angle of 45°. Full-stepping for the stator-rotor arrangement shown in figure 6.2b corresponds to a step angle of 90°. In this case, only one phase is energized at a time. The corresponding switching sequences, given in table 6.2, are illustrated further in figure 6.5. For half-stepping, both phases have to be energized simultaneously in alternate steps, as is clear from table 6.1.

Now consider the variable-reluctance (VR) stepper motor shown schematically in figure 6.6. The rotor is a nonmagnetized soft-iron bar. If only two phases are used in the stator, there will be ambiguity regarding the direction of rotation. At least three phases would be needed for this two-pole rotor geometry, as shown in figure

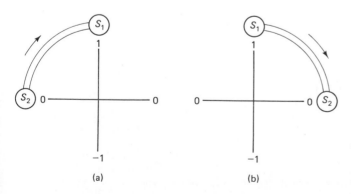

(a)   (b)

Figure 6.5. Full-step switching sequence: (a) for clockwise rotation; (b) for counterclockwise rotation.

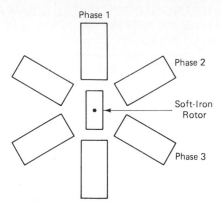

Figure 6.6. Schematic diagram of a three-phase variable-reluctance stepper motor.

6.6. The full-stepping sequence for clockwise rotation is shown in figure 6.7. The step angle is 60°. Only one phase is energized at a time in order to execute full-stepping. With VR stepping motors, the direction of the current is not reversed in the full-stepping sequence; only the states 1 and 0 (on and off) are used for each phase. The half-stepping sequence for clockwise rotation is shown in figure 6.8. In this case, two phases have to be energized simultaneously during some steps. Fur-

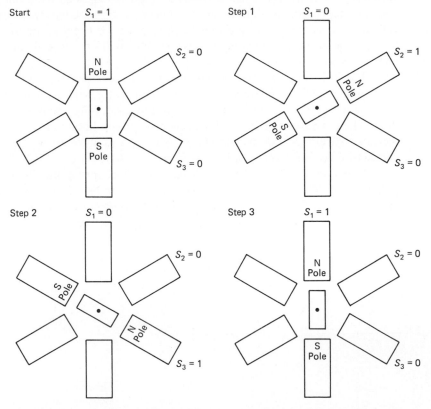

Figure 6.7. Full-stepping sequence for the three-phase VR stepper motor example (step angle = 60°).

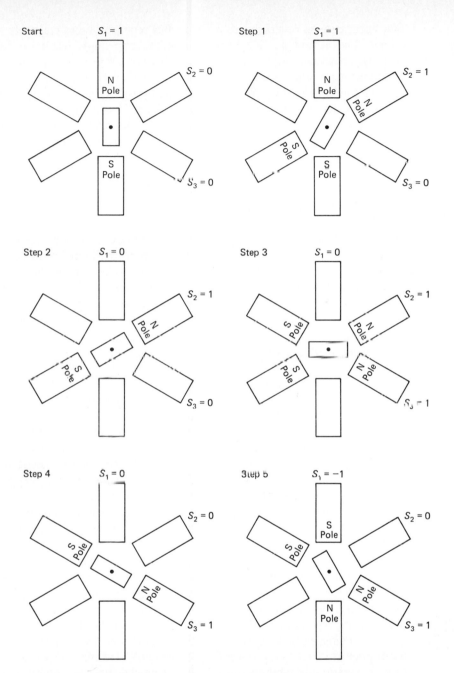

Figure 6.8. Half-stepping sequence for the three-phase VR stepper motor example (Step Angle = 30°).

thermore, current reversals are needed, thus requiring more elaborate switching circuitry. The advantage, however, is that the step angle has been halved to 30°, thereby providing improved motion resolution. When two phases are activated simultaneously, the minimum reluctance position is halfway between the corresponding pole pairs (i.e., 30° from the detent position that is obtained when only one of the two phases is energized), which enables half-stepping. It follows that, depending on the energizing sequence of the phases, either full-stepping (step angle of 60° in this example) or half-stepping (step angle of 30° in the example) would be possible.

6.3 STEPPER MOTOR CLASSIFICATION

Since any actuator that generates stepwise motion can be considered a stepper motor, it is difficult to classify all such devices into a small number of useful categories. For example, devices such as harmonic drives (a class of flexible-gear drives) and pawl-and-ratchet-wheel drives that produce intermittent motions through purely mechanical means are also classified as stepper motors. Of primary interest in today's control application, however, are actuators that generate stepwise motion directly by electromagnetic forces in response to pulse (or digital) inputs. Even for these *electromagnetic incremental actuators,* however, no standardized classification is available.

As mentioned in the preceding section, one widely used classification divides stepping motors into the two groups: permanent-magent (PM) stepping motors and variable-reluctance (VR) stepping motors, depending on whether the rotor is magnetized or not. This nomenclature is somewhat ambiguous and often useless because both types of motors operate by electromagnetic torque generation in a similar manner and undergo changes in reluctance (magnetic resistance) during operation. *Hybrid stepping motors* possess characteristics of both VR and PM stepping motors.

A more practical classification, which is favored in this book, is based on the number of "stacks" of teeth (see figure 6.2a) present on the rotor shaft. Further subclassifications are possible, depending on the tooth pitch (angle between adjacent teeth) of the stator and tooth pitch of the rotor. In a *single-stack stepper motor,* the rotor tooth pitch and the stator tooth pitch generally have to be unequal so that not all teeth in the stator are ever aligned with the rotor teeth at any instant. It is the misaligned teeth that exert the magnetic pull, generating the driving torque. In each motion increment, the rotor turns to the minimum reluctance (stable equilibrium) position corresponding to that particular polarity distribution of the stator.

In *multiple-stack stepper motors,* operation is possible even when the rotor tooth pitch is equal to the stator tooth pitch, provided that at least one stack of rotor teeth is rotationally shifted (misaligned) from the other stacks by a fraction of the rotor tooth pitch. In this design, it is this *interstack misalignment* that generates the drive torque for each motion step. It should be obvious that unequal-pitch multiple-stack steppers are also a practical possibility. In this design, each rotor stack operates as a single-stack stepper motor. The stepper motor classifications described thus far are summarized in figure 6.9.

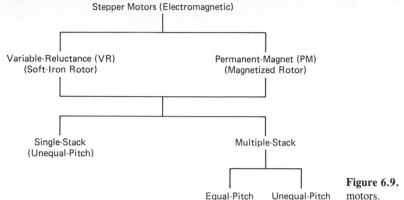

Stepper Motors (Electromagnetic)

Variable-Reluctance (VR)
(Soft-Iron Rotor)

Permanent-Magnet (PM)
(Magnetized Rotor)

Single-Stack
(Unequal-Pitch)

Multiple-Stack

Equal-Pitch   Unequal-Pitch

Figure 6.9. Classifications of stepper motors.

In the next two sections, we shall describe the practical aspects of single-stack and multiple-stack stepper motors. One point to remember is that some form of geometric misalignment of teeth is necessary in both types of motors. A motion step is obtained by simply redistributing (i.e., switching) the polarities of the stator, thereby changing the minimum reluctance detent position of the rotor. Once a stable equilibrium position is reached by the rotor, the stator polarities are switched again to produce a new detent position, and so on. In descriptive examples, it is more convenient to use variable-reluctance stepper motors. However, the principles can be extended in a straightforward manner to cover permanent-magnet stepper motors as well.

6.4 SINGLE-STACK STEPPER MOTORS

Consider the single-stack variable-reluctance stepper motor shown in figure 6.10. The motor has three phases of winding ($p = 3$) in the stator, and there are eight teeth in the soft-iron rotor ($n_r = 8$). The three phases are numbered 1, 2, and 3. Note that each phase represents a group of four stator poles; the total number of stator poles (n_s) is twelve. When Phase 1 is energized, one pair of diametrically opposite poles becomes N (north) poles and the other pair in that phase (located at 90° from the first pair) becomes S (south) poles. Furthermore, a geometrically orthogonal set of four teeth on the rotor align themselves perfectly with these four stator poles. This is a minimum reluctance, stable equilibrium configuration for the rotor (assuming that the other two phases are not activated). Observe, however, that there is a misalignment of 15° between the remaining rotor teeth and stator poles.

If the pitch angle, defined as the angle between adjacent teeth, is denoted by θ (in degrees) and the number of teeth is denoted by n, we have

$$\text{stator pitch } \theta_s = \frac{360°}{n_s}$$

$$\text{rotor pitch } \theta_r = \frac{360°}{n_r}$$

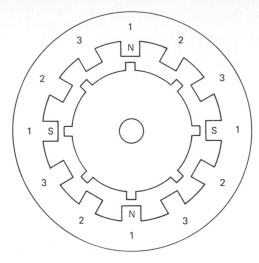

Figure 6.10. Three-phase single-stack VR stepper motor with twelve stator poles (teeth) and eight rotor teeth.

For one-phase-on excitation, the step angle $\Delta\theta$, which should be equal to the smallest misalignment between stator teeth and rotor teeth at any stable equilibrium position, is given by

$$\Delta\theta = \theta_r - \theta_s \qquad (\text{for } \theta_r > \theta_s) \tag{6.1}$$

It is clear that for the arrangement shown in figure 6.10, $\theta_r = 360°/8 = 45°$, $\theta_s = 360°/12 = 30°$, and hence, $\Delta\theta = 45° - 30° = 15°$, as stated earlier.

Now, if Phase 1 is turned off and Phase 2 is turned on, the rotor will turn 15° in the counterclockwise direction to its new minimum reluctance position. If Phase 3 is energized instead of Phase 2, the rotor would turn 15° clockwise. It should be clear that half this step size (7.5°) is also possible with this motor. Suppose, for example, that Phase 1 is on, as before. If Phase 2 is energized while Phase 1 is on, so that two like poles are in adjacent locations, the rotor will turn counterclockwise to the corresponding minimum reluctance position. Since the equivalent field of the two adjoining like poles is halfway between the two poles, two rotor teeth will orient symmetrically about this pair of poles. It is clear that this corresponds to a rotation of 7.5° from the previous detent position, in the CCW direction. For the next half-step (in the CCW direction), Phase 1 is turned off while Phase 2 is on. Thus, in summary, the full-stepping sequence for CCW rotation is 1-2-3-1; for CW rotation, it is 1-3-2-1. The half-stepping sequence for CCW rotation is 1-12-2-23-3-31-1; for CW rotation, it is 1-31-3-23-2-12-1.

Returning to full-stepping, note that since each switching of phases corresponds to a rotation of $\Delta\theta$ and there are p phases, the angle of rotation for a complete switching cycle of p switches is $p.\Delta\theta$. In a switching cycle, the stator polarity distribution returns to the distribution that it had in the beginning. Hence, in one switching cycle (p switches), the rotor should assume a configuration exactly like what it had in the beginning of the cycle. That is, the rotor should turn through a complete pitch angle of θ_r. Hence, the following relationship exists for the one-phase-on case.

$$\theta_r = p.\Delta\theta \tag{6.2}$$

Substituting this in equation 6.1, we have

$$\theta_r = \theta_s + \frac{\theta_r}{p} \qquad \text{(for } \theta_r > \theta_s\text{)} \tag{6.3}$$

where θ_r = rotor tooth pitch angle
θ_s = stator tooth pitch angle
p = number of phases in the stator

Hence, by definition of the pitch angle,

$$\frac{360°}{n_r} = \frac{360°}{n_s} + \frac{360°}{p.n_r}$$

or

$$n_s = n_r + \frac{n_s}{p} \qquad \text{(for } n_s > n_r\text{)} \tag{6.4}$$

where n_r = number of rotor teeth
n_s = number of stator teeth

Finally, the number of steps per revolution is

$$n = \frac{360°}{\Delta\theta} \tag{6.5}$$

Example 6.1

Consider the stepper motor shown in figure 6.10. Given that the number of stator poles $n_s = 12$ and the number of phases $p = 3$, the number of rotor teeth can be calculated from equation 6.4, assuming that $n_r < n_s$:

$$n_r = 12 - \frac{12}{3} = 8$$

This is confirmed by observation. Furthermore, since the rotor pitch $\theta_r = 360°/8 = 45°$, the step angle can be calculated from equation 6.2 as

$$\Delta\theta = \frac{45°}{3} = 15°$$

Note that this is the full-step angle, as observed earlier. Furthermore, the stator pitch is $\theta_s = 360°/12 = 30°$, which confirms that the step angle is $\theta_r - \theta_s = 45° - 30° = 15°$.

Toothed-Pole Construction

The foregoing analysis indicates that the step angle can be reduced by increasing the number of poles in the stator and the number of teeth in the rotor. Obviously, there are practical limitations to the number of poles (windings) that can be incorporated in a stepper motor. A common solution to this problem is to use "toothed" poles in the stator, as shown in figure 6.11a. In this particular case, the stator teeth are

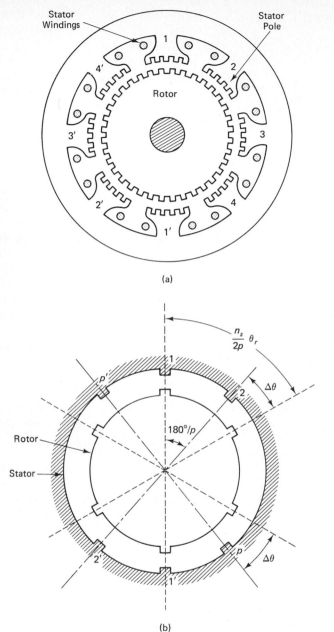

Stator
Windings

Stator
Pole

1

4'

2

Rotor

3'

3

2'

4

1'

(a)

$\frac{n_s}{2p}\theta_r$

$\Delta\theta$

$180°/p$

Rotor

Stator

$\Delta\theta$

(b)

Figure 6.11. A possible toothed-pole construction for a stepper motor: (a) an eight-pole, four-phase motor; (b) schematic diagram for generalizing the step angle equation.

equally spaced but not identical to the rotor teeth. In the toothed-stator construction, n_s represents the number of teeth rather than the number of poles in the stator. The number of rotor teeth has to be increased in proportion. Note that in full-stepping (e.g., one-phase-on) after p number of switchings (steps), where p is the number of phases, the adjacent tooth of the rotor will take the starting position of a particular tooth of the rotor. It follows that the rotor rotates through θ_r (the tooth pitch of the rotor) in p steps. Thus, the relationship $\Delta\theta = \theta_r/p$ still holds. But equation 6.1 has to be modified to accommodate toothed poles. Toothed-stator construction can provide very small angles—$0.72°$, for example, or, more commonly, $1.8°$.

The equations for the step angle given so far in this chapter assume that the number of stator poles is identical to the number of stator teeth. In particular, equations 6.1, 6.3, and 6.4 are obtained using this assumption. These equations have to be modified when there are many teeth on each stator pole. Generalization of the step angle equations for the case of toothed-pole construction can be made by referring to figure 6.11b. This is one possible geometry for a single-stack toothed construction. In this case, the rotor tooth pitch θ_r is not equal to the stator tooth pitch θ_s. Another possibility for a single-stack toothed construction (which is perhaps preferrable from the practical point of view) will be described later. There, θ_r and θ_s are identical, but when all the stator teeth that are wound to one of the phases are aligned with the rotor teeth, all the stator teeth wound to another phase will have a "constant" misalignment with the rotor teeth in the immediate neighborhood. Unlike that case, in the construction used in this case, $\theta_r \neq \theta_s$; hence, only one tooth in a stator pole can be completely aligned with a rotor tooth. Consider the case of $\theta_r > \theta_s$ (i.e., $n_r < n_s$). Note that, typically, we will have $n_s = n_r + 2$, so that when the center tooth of Pole 1 is aligned with a tooth of the rotor, the center tooth of Pole 1' will also have a rotor tooth aligned with it, but the center teeth of all remaining stator poles will not have rotor teeth aligned with them, thereby providing the necessary stator-rotor teeth misalignment to drive the motor. Pole 1 and Pole 1' have opposite polarities and are wound together, representing Phase 1. Since there are $2p$ poles in the stator, p being the number of phases, it is clear from figure 6.11b that the angle between the center tooth of Pole 1 and the center tooth of Pole 2 is $180°/p$. Note that there are $n_s/2p$ stator teeth in this angle, n_s being the total number of teeth in the stator. The angle between the rotor tooth that is aligned with the center tooth of Pole 1 and the rotor tooth nearest to the center tooth of Pole 2 is

$$\frac{n_s}{2p}\,\theta_r$$

where θ_r is the tooth pitch of the rotor. It follows that the stator-rotor tooth misalignment at Pole 2, which has to be equal to the step angle $\Delta\theta$, is given by

$$\Delta\theta = \frac{n_s}{2p}\,\theta_r - \frac{180°}{p}$$

Now, substituting $360°/\theta_s$ for n_s, we get

$$\Delta\theta = \frac{360°}{2p\theta_s}\,\theta_r - \frac{180°}{p} = \frac{360°}{2p\theta_s}(\theta_r - \theta_s)$$

Next, substituting back n_s for $360°/\theta_s$, we have

$$\Delta\theta = \frac{n_s}{2p}(\theta_r - \theta_s) \tag{6.6}$$

Equation 6.6 should be intuitively clear. Specifically, since $(\theta_r - \theta_s)$ is the offset of a rotor tooth with respect to the closest stator tooth within one stator pitch angle, and since there are $n_s/2p$ rotor teeth in the sector made by two adjacent stator poles, we see that the total tooth offset $\Delta\theta$ at the second pole is indeed given by equation 6.6. If there are m poles per phase, the number of stator teeth in the sector is n_s/mp. Hence, in that case, $2p$ in equation 6.6 has to be replaced by mp. We recall that when the stator teeth are interpreted as stator poles, $n_s = 2p$, and then equation 6.6 reduces to equation 6.1, as expected. For the toothed-pole construction, n_s is several times the value of $2p$. In this case, equation 6.6 should be used in place of equation 6.1. For example, equation 6.3 becomes

$$\theta_r = \theta_s + \frac{2\theta_r}{n_s} \qquad (\text{for } \theta_r > \theta_s) \qquad (6.7)$$

and equation 6.4 becomes

$$n_s = n_r + 2 \qquad (6.8)$$

as expected. Equations 6.7 and 6.8 can be directly derived by noting that $\Delta\theta = \theta_r/p$ is true even for the toothed construction and by substituting this in equation 6.6. Note that, in general, p has to be replaced by $n_s/2$ in converting an equation for a nontoothed-pole construction to the corresponding equation for a toothed-pole construction.

Finally, we observe from figure 6.11b that the switching sequence $1\text{-}2\text{-}3\text{-}\cdots\text{-}p$ gives counterclockwise rotations, and the switching sequence $1\text{-}p\text{-}(p-1)\text{-}\cdots\text{-}2$ gives clockwise rotations.

Example 6.2

Consider a simple design example for a single-stack VR stepper. Suppose that the number of steps per revolution, which is a functional requirement, is specified as $n = 200$. This corresponds to a step angle of $\Delta\theta = 360°/200 = 1.8°$. Assume full-stepping. Design restrictions, such as size and the number of poles in the stator, govern ℓ_s, the number of stator teeth per pole. Let us use the typical value of six teeth per pole. Also assume that there are two poles wound to the same stator phase. We are interested in selecting a motor to meet these requirements.

Solution First, we shall derive some useful relationships. Suppose that there are m poles per phase. Hence, there are mp poles in the stator. (*Note:* $n_s = mp\,\ell_s$). Note that in the case of $m = 2$, we obtained equation 6.8. By proceeding in the same way, we obtain the general result

$$n_s = n_r + m \qquad (6.9)$$

This result should be intuitively clear, because when the center tooth of one stator pole is aligned with a rotor tooth, the center teeth of all remaining $(m - 1)$ poles connected to the same phase must also be aligned with the rotor teeth. Dividing equation 6.9 by mp, we get

$$\ell_s = \frac{n_r}{mp} + \frac{1}{p}$$

Now

$$n_r = \frac{360°}{\theta_r} = \frac{360°}{p.\Delta\theta} \qquad \text{(from equation 6.2)}$$

or

$$n_r = \frac{n}{p} \qquad (6.10)$$

Substituting this in the previous result, we get

$$\ell_s = \frac{n}{mp^2} + \frac{1}{p} \qquad (6.11)$$

Now, since $1/p$ is less than 1 for a stepper motor and ℓ_s is greater than 1 for the toothed-pole construction, an approximation for equation 6.11 can be given by

$$\ell_s = \frac{n}{mp^2} \qquad (6.12)$$

where ℓ_s = number of teeth per stator pole
m = number of poles per stator phase
p = number of stator phases
n = number of steps per revolution

In this example, $\ell_s \sim 6$, $m = 2$, and $n = 200$. Hence, from equation 6.16, we have

$$6 \sim \frac{200}{p^2}$$

which gives $p \sim 4$. Note that p has to be an integer. Now, using equation 6.11, we get two possible designs for $p = 4$. First, with the specified values $n = 200$ and $m = 2$, we get $\ell_s = 6.5$, which is slightly larger than the required value of 6. Alternatively, with the required $\ell_s = 6$ and specified $m = 2$, we get $n = 184$, which is slightly smaller than the specified value of 200. Either of these two designs would be acceptable. The second design gives a slightly larger step angle. (Note that $\Delta\theta = 360°/n = 1.96°$ for the second design and $\Delta\theta = 360°/200 = 1.8°$ for the first design.) Summarizing the two designs, we have the following results:

For Design 1:

Number of phases $p = 4$
Number of stator poles = 8
Number of teeth per pole = 6.5
Number of steps per revolution = 200
Step angle = 1.8°
Number of rotor teeth = 50 (from equation 6.10)
Number of stator teeth = 52

For Design 2:

> Number of phases $p = 4$
> Number of stator poles $= 8$
> Number of teeth per pole $= 6$
> Number of steps per revolution $= 184$
> Step angle $= 1.96°$
> Number of rotor teeth $= 46$ (from equation 6.10)
> Number of stator teeth $= 48$

Note: The number of teeth per stator pole (ℓ_s) does not have to be an integer (see Design 1). Since there are interpolar gaps around the stator, it is possible to construct a motor with an integer number of actual stator teeth, even when ℓ_s and n_s are not integers.

Another Toothed Construction

In the foregoing single-stack toothed construction, we have $\theta_r \neq \theta_s$. In an alternative possibility, $\theta_r = \theta_s$, but the stator poles are positioned around the rotor such that when the stator teeth corresponding to one of the phases are fully aligned with the rotor teeth, the stator teeth in another phase will have a constant offset with neighboring rotor teeth. The torque magnitude of this construction is perhaps better because of this uniform tooth offset per phase, but torque ripples would also be stronger (a disadvantage) because of sudden and more prominent changes in magnetic reluctance from pole to pole during phase switching.

To obtain some relations that govern this construction, suppose that figure 6.11a represents a stepper motor of this type. When the stator teeth in Pole 1 (and Pole 1′) are perfectly aligned with the rotor teeth, the stator teeth in Pole 2 (and Pole 2′) will have an offset of $\Delta\theta$ with the neighboring rotor teeth. Suppose that this offset of the stator teeth is in the counterclockwise direction. The pole pitch is given by $360°/pm$. Hence, in this case,

$$\frac{1}{\theta_r}\left[\frac{360°}{pm} + \Delta\theta\right] = r \tag{6.13}$$

where r is the integer number of rotor teeth contained within the angular sector $360°/(pm) + \Delta\theta$. Also

> $\Delta\theta =$ step angle (full-stepping)
> $\theta_r =$ rotor tooth pitch
> $p =$ number of phases
> $m =$ number of stator poles per phase

It should be clear that within two consecutive poles wound to the same phase, there are n_r/m rotor teeth. Since p offsets of magnitude $\Delta\theta$ each will result in a total offset of θ_r, it follows that pr is larger than n_r/m by just one rotor tooth. Accordingly,

$$\frac{n_r}{m} = pr - 1$$

or

$$n_r + m = pmr \qquad (6.14)$$

where n_r is the number of rotor teeth, as given by

$$n_r = \frac{360°}{\theta_r} \qquad (6.15)$$

Note that if we substitute equation 6.14 for r in equation 6.13, we get $\Delta\theta = \theta_r/p$, as expected. Conversely, if we substitute $\Delta\theta = \theta_r/p$ in equation 6.13, we get equation 6.14.

Example 6.3

Consider the full-stepping operation of a stepper motor whose design is governed by equation 6.14. Show that it is not possible to construct a four-phase motor that has fifty rotor teeth (i.e., step angle = 1.8°). Obtain a suitable design for a four-phase motor that uses eight stator poles. Specifically, determine the number of rotor teeth (n_r), the step angle $\Delta\theta$, the number of steps per revolution (n), and the number of teeth per stator pole (ℓ_s).

Solution First, with $n_r = 50$ and $p = 4$, equation 6.14 becomes

$$50 + m = 4mr$$

or

$$m = \frac{50}{(4r - 1)} \qquad (i)$$

Note that m and r should be *natural numbers* (i.e., *positive integers*). Since the smallest such value for r is 1, we see from equation (i) that the largest value for m is 16. None of the sixteen possible values for m (i.e., 1, 2, 3, . . . , 16) in equation (i) will result in a natural number value for r. It follows that $n_r = 50$, $p = 4$ does not provide a realistic design.

Next, consider $p = 4$ and $m = 2$ (i.e., a four-phase motor with eight stator poles). Then equation 6.14 becomes

$$n_r = 8r - 2 \qquad (ii)$$

Hence, one possible design that is close to the previously mentioned case of $n_r = 50$ is realized with $r = 7$. In this case, from equation (ii), we have $n_r = 54$. The corresponding tooth pitch (for both rotor and stator) is

$$\theta_r = \theta_s = \frac{360°}{54} = \frac{20°}{3} \simeq 6.67°$$

The step angle (for full-stepping) is

$$\Delta\theta = \frac{\theta_r}{p} = \frac{20}{3 \times 4} = \frac{5°}{3} \simeq 1.67°$$

The number of steps per revolution is

$$n = \frac{360°}{\Delta\theta} = pn_r = 4 \times 54 = 216$$

$$\text{pole pitch} = \frac{360°}{mp} = \frac{360°}{8} = 45°$$

The maximum number of teeth that could be occupied within a pole pitch is

$$\frac{45°}{\theta_s} = \frac{45° \times 3}{20} = 6.75$$

Hence, the maximum possible number of teeth per pole is

$$\ell_s = 6$$

Practically, however, we might have an interpolar gap of nearly half the pole angle. Hence, a more realistic number for ℓ_s would be the integer value of

$$\frac{1}{\theta_s} \frac{360°}{(8 + 4)}$$

This gives

$$\ell_s = 4$$

Summarizing, we have the following design parameters:

Number of phases $p = 4$
Number of stator poles $mp = 8$
Number of teeth per pole ℓ_s = maximum 6 (typically 4)
Number of steps per revolution (full-stepping) $n = 216$
Step angle $\Delta\theta \simeq 1.67°$
Number of rotor teeth $n_r = 54$
Tooth pitch (both rotor and stator) $\simeq 6.67°$

Microstepping

We have seen how full-stepping or half-stepping can be achieved simply by using an appropriate switching scheme. For example, half-stepping occurs when phase switching alternates between one-phase-on and two-phase-on states. Full-stepping occurs when either one-phase-on switching or two-phase-on switching is used exclusively at every step. Microstepping is achieved by properly changing the phase currents in steps in addition to switching the phases on and off (as in the case of full-stepping and half-stepping). The principle behind this can be understood by considering two identical stator poles (wound with identical windings). When the currents through the windings are identical (in magnitude and direction), the resultant magnetic field will lie symmetrically between the two poles. If the current in one pole is decreased while the other current is kept unchanged, the resultant magnetic field will move closer to the pole with the larger current. Since the detent position (equilibrium position) depends on the position of the resultant magnetic field, it follows that very small step angles can be achieved simply by controlling (varying the relative magnitudes and directions of) the phase currents. Step angles of 1/125 of a full step or smaller could be obtained through microstepping. Motor drive units with the microstepping capability are more costly, but microstepping provides the advantage of accurate motion capabilities, including smoother operation even in the neighborhood of a resonance in the motor-load combination.

6.5 MULTIPLE-STACK STEPPER MOTORS

Both equal-pitch construction ($\theta_r = \theta_s$) and unequal-pitch construction ($\theta_r > \theta_s$ or $\theta_r < \theta_s$) are possible in multiple-stack steppers. In principle, the unequal-pitch construction operates like a cascaded group of single-stack steppers. An advantage of the unequal-pitch construction is that smaller step angles are possible than with an equal-pitch construction of the same size (diameter and number of stacks). But the switching sequence is somewhat more complex for unequal-pitch, multiple-stack stepper motors. First, we shall examine the equal-pitch, multiple-stack construction. Then, the operation of an unequal-pitch, multiple-stack motor should follow directly from the analysis of the single-stack case given in the preceding section.

A longitudinal view of a three-stack stepper motor is shown schematically in figure 6.12. In this example there are three identical stacks of teeth mounted on the same rotor shaft. This is equivalent to connecting three single-stack rotors in cascade. For each rotor stack, there is a toothed stator, with the same pitch angle ($\theta_s = \theta_r$), around it. Each stator segment has several poles. These might appear to be the poles of a standard multiple-phase, single-stack motor, but all the poles in each stator segment are wound to the same phase (in the equal-pitch construction), thus being energized (polarized) or de-energized (depolarized) simultaneously. The misalignment that is necessary to produce the drive torque may be introduced in one of two ways:

1. The teeth in the three stator segments are perfectly aligned, but the teeth in the three rotor stacks are misaligned consecutively by a one-third-pitch angle.
2. The teeth in the three rotor stacks are perfectly aligned, but the teeth in the three stator segments are misaligned consecutively by a one-third-pitch angle.

Suppose that Phase 1 is energized. Then the teeth in Stack 1 will align perfectly with the stator teeth in Phase 1 (segment 1). But the teeth in Stack 2 will be shifted from the stator teeth in Phase 2 (segment 2) by a one-third-pitch angle in one direction, and the teeth in Stack 3 will be shifted from the stator teeth in Phase 3 (segment 3) by a two-thirds-pitch angle in the same direction (or a one-third-pitch angle in the opposite direction). It follows that if Phase 1 is now de-energized and Phase 2 is energized, the rotor will turn through one-third pitch in one direction. If, instead, Phase 3 is turned on after phase 1, the rotor will turn through one-third pitch in the opposite direction. Clearly, the step angle (for full-stepping) is a one-third-pitch angle for the three-stack, three-phase construction. The switching sequence 1-2-3-1

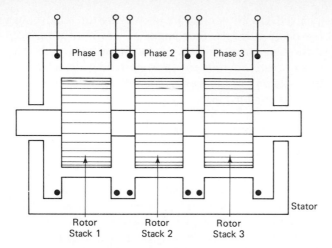

Figure 6.12. Longitudinal view of a three-stack (three-phase) stepper motor.

will turn the rotor in one direction, and the switching sequence 1-3-2-1 will turn the rotor in the opposite direction.

In general, if there are s stacks on the rotor shaft, each rotor stack (or stator segment) is misaligned with the adjacent rotor stack by $(1/s \times \text{pitch angle})$. Hence, the full-stepping step angle is given by

$$\Delta\theta = \frac{\theta}{s} \qquad (6.16)$$

where $\theta = \theta_r = \theta_s$ = tooth pitch angle
s = number of rotor stacks = number of phases

Note that the step angle can be decreased by increasing the number of stacks of rotor teeth. However, this increases the length of the motor and can result in flexural vibration problems (particularly whirling of the shaft), air gap contact problems, and large bearing loads. Half-stepping can be accomplished by energizing two phases at a time. Hence, for one direction, the half-stepping sequence is 1-12-2-23-3-31-1; in the opposite direction, it is 1-13-3-32-2-21-1.

Unequal-pitch, multiple-stack stepper motors are also of practical interest. Very fine angular resolutions (step angles) can be achieved by this design without compromising the length of the motor. Each stator segment has more than one phase (p number of phases) in unequal-pitch stepper motors. Since a step angle of $\theta_r - \theta_s$ is possible (in the case of nontoothed-pole construction) with a single stack, this can be further subdivided into s equal steps using the interstack misalignment. Hence, the overall step angle for an unequal-pitch, multiple-stack stepper motor with non-toothed poles is given by

$$\Delta\theta = \frac{\theta_r - \theta_s}{s} \qquad (\text{for } \theta_r > \theta_s) \qquad (6.17)$$

On the other hand, with a toothed-pole multiple-stack motor, we have

$$\Delta\theta = \frac{n_s(\theta_r - \theta_s)}{2ps} \qquad (6.18)$$

assuming that there are two stator poles per phase. Alternatively, using equation 6.2, we have

$$\Delta\theta = \frac{\theta_r}{p.s} \qquad (6.19)$$

for both toothed-pole and nontoothed-pole motors, where

p = number of phases in each stator segment
s = number of stacks

6.6 OPEN-LOOP CONTROL OF STEPPING MOTORS

In principle, the stepper motor is an open-loop actuator. In its normal operating mode, the stepwise rotation of the motor is synchronized with the command pulse train. This justifies the term *digital synchronous motor*, which is sometimes used to denote the stepper motor. As a result of stepwise (incremental) synchronous operation, positional error in a stepper motor is generally noncumulative; consequently, open-loop control is adequate. An exception is under highly transient conditions near rated torque, when "pulse missing" could be a problem. We shall address this situation in a later section.

The basic components needed for open-loop operation of a stepper motor are identified in figure 6.13a. The *pulse generator* is typically a variable-frequency oscillator. For bidirectional motion, it will generate two pulse trains—the position pulse train and the direction-pulse train, which will be necessary as determined by the required motion trajectory. The position pulses identify the exact times at which angular steps should be initiated. The direction pulses identify the instants at which the direction of rotation should be reversed. Only a position pulse train is needed for unidirectional operation. Generation of the position pulse train for steady-state operation at a constant stepping rate is a relatively simple task. In this case, a single command identifying the stepping rate (pulse rate) would suffice. Then the logic circuitry within the translator will latch onto a constant-frequency oscillator, with the frequency determined by the required speed (stepping rate), and continuously cycle the switching sequence at this frequency. This is a hardware approach to open-loop control of a stepping motor. For steady-state operation, the stepping rate can be set by manually adjusting the knob of a potentiometer connected to the translator. For simple motions (e.g., starting from rest and stopping at a certain angle), the commands that generate the pulse train (commands to the oscillator) can be set manually. Under the more complex transient operating conditions that are present when following intricate motion trajectories, however, microprocessor-based generation of pulse commands, using programmed logic, might be necessary. This is a software approach, which is usually slower than the hardware approach.

The *translator* module has logic circuitry to interpret a pulse train and "translate" it into the corresponding switching sequence for stator field windings (on/off/reverse state for each phase of the stator). The translator also has solid-state switching circuitry (using gates, latches, triggers, etc.) to direct the field currents to

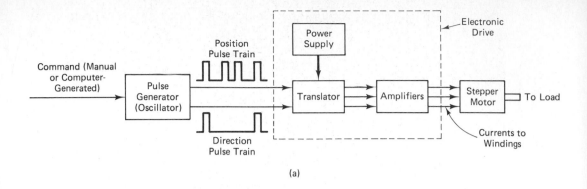

(a)

(b)

Figure 6.13. (a) Open-loop control of a stepper motor. (b) A programmable indexer (courtesy of the Superior Electric Company).

the appropriate phase windings according to the particular switching state. Some translators may possess the capability to generate command pulses at a steady rate. The stepping rate or direction may be changed manually. The translator may not have the capability to keep track of (i.e., count) the number of steps taken by the motor. A device that has all these capabilities, including the standard translator functions, is termed a *preset indexer*. It usually consists of an oscillator, microcircuitry for counting and for various control functions, and a translator in a single package. The required angle of rotation, stepping rate, and direction are set either manually, by turning the corresponding knobs, or by using computer commands in a programmable preset indexer. An external pulse source is not needed in this case. A programmable indexer—consisting of a translator, a pulse source (an oscillator), and microelectronic circuitry for the control of position and speed and for other programmable functions—is shown in figure 6.13b. The indexer can be programmed using a microcomputer or a hand-held programmer (provided with the indexer) through a standard interface (e.g., RS232 serial interface). Control signals within the translator are on the order of 10 mA, whereas phase excitation requires large currents on the order of several amperes. Control signals have to be amplified by using switching amplifiers for phase excitation.

Power to operate the translator (for logic circuitry, switching circuitry, etc.) and to operate phase excitation amplifiers comes from a DC *power supply* (typically 24 V DC). A complete unit that consists of the translator (or preset indexer), amplifiers, and the power supply is termed a *motor-drive system*. Sometimes a pulse source is also included in the drive system. The leads of the output amplifiers on the drive system carry currents to the phase windings on the stator (and to rotor magnetizing coils located on the stator in the case of an electromagnetic rotor) of the stepping motor. The *load* may be connected to the motor shaft directly or through some form of coupling device (e.g., harmonic drive, tooth-timing belt drive, hydraulic amplifier, rack and pinion).

6.7 STEPPER MOTOR RESPONSE

It is useful to examine the response of a stepper motor to a single pulse input before studying the behavior under general stepping conditions. Ideally, when a single pulse is applied, the rotor should instantaneously turn through one step angle ($\Delta\theta$) and stop at that detent position (stable equilibrium position). Unfortunately, the actual single-pulse response is far from this ideal behavior. In particular, the rotor will oscillate for a while about the detent position before settling down. These oscillations result primarily from the interaction of motor load inertia (the combined inertia of rotor, load, etc.) with drive torque. This behavior can be explained using figure 6.14.

Assume single-phase energization (i.e., only one phase is energized at a time). When a pulse is applied to the translator at C, the corresponding stator phase is energized. This generates a torque (due to magnetic attraction), causing the rotor to turn toward the corresponding minimum reluctance detent position (D). The static torque curve (broken line in figure 6.14) represents the torque applied on the rotor from the

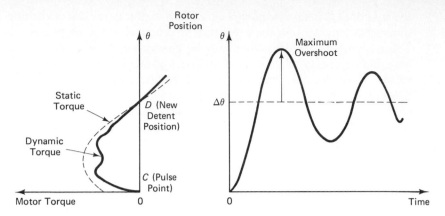

Figure 6.14. Single-pulse response and the corresponding single-phase torque.

energized phase, as a function of the rotor position θ, under ideal conditions when dynamic effects are neglected. Under normal operating conditions, however, there will be induced voltages due to self-induction and mutual induction. Hence, a finite time is needed for the current to build up in the windings once the phases are switched on. Furthermore, there will be eddy currents generated in the rotor. These effects cause the magnetic field to deviate from the static conditions as the rotor moves at a finite speed, thereby making the dynamic torque curve different from the static torque curve, as shown in figure 6.14. The true dynamic torque is somewhat unpredictable because of its dependence on many time-varying factors (rotor speed, rotor position, current level, etc.). The static torque curve is normally adequate to explain many characteristics of a stepper motor, including the oscillations in the single-pulse response.

It is important to note that the static torque is positive at the switching point, but is generally not maximum at that point. To explain this further, consider the three-phase VR stepper motor (with nontoothed poles) shown in figure 6.6. The step angle $\Delta\theta$ for this arrangement is 60°, and the full-step switching sequence for clockwise rotation is 1-2-3-1. Suppose that Phase 1 is energized. The corresponding detent position is denoted by D in figure 6.15a. The static torque curve for this phase is shown in figure 6.15b, with the positive angle measured clockwise from the detent position D. Suppose that we turn the rotor counterclockwise from this stable equilibrium position, using an external rotating mechanism (e.g., by hand). At position C, which is the previous detent position where Phase 1 would have been energized under normal operation, there is a positive torque that tries to turn the rotor to its present detent position D. At position B, the static torque is zero, because the force from the N pole of Phase 1 exactly balances that from the S pole. This point, however, is an *unstable* equilibrium position; a slight push in either direction will move the rotor in that direction. Position A, which is located at a rotor tooth pitch ($\theta_r = 180°$) from position D, is also a "stable" equilibrium position. The maximum static torque occurs at position M, which is located approximately halfway between positions B and D (at an angle $\theta_r/4 = 45°$ from the detent position). The torque at the normal switching position (C) is less than the maximum value.

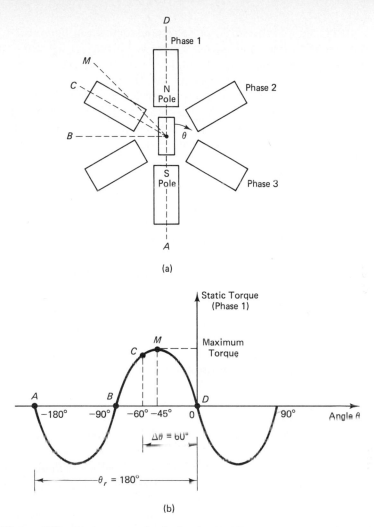

Figure 6.15. Static torque distribution in the VR stepper motor in figure 6.6: (a) schematic diagram; (b) static torque curve for Phase 1.

For simplified analysis, the static torque curve may be considered sinusoidal. In this case, with Phase 1 excited, and with the remaining phases inactive, the static torque distribution T_1 can be expressed as

$$T_1 = -T_{max} \sin n_r \theta \tag{6.20}$$

where θ = angular position in radians, measured from the current detent position (with Phase 1 excited)

n_r = number of teeth on the rotor

T_{max} = maximum static torque

Equation 6.20 can be verified by referring to figure 6.15, where $n_r = 2$. Note that equation 6.20 is valid irrespective of whether the stator poles are toothed or not, even though the example considered in figure 6.15 has nontoothed stator poles.

Returning to the single-pulse response shown in figure 6.14, note that starting from rest at C, the rotor will have positive velocity at the detent position D. Its kinetic energy (or momentum) will take it beyond the detent position. This is the first *overshoot*. Since the same phase is still on, the torque will be negative beyond the detent position; static torque always attracts the rotor to the detent position, which is a stable equilibrium position. The rotor will decelerate because of this negative torque and will attain zero velocity at the point of maximum overshoot. Then the rotor will be accelerated back toward the detent position and carried past this position by the kinetic energy, and so on. This oscillatory motion would continue forever with full amplitude ($\Delta\theta$) if there were no energy dissipation. In reality, however, there are numerous damping mechanisms—such as mechanical dissipation (frictional damping) and electrical dissipation (resistive damping through eddy currents and other induced voltages)—in the stepper motor, that will gradually slow down the rotor, as shown in figure 6.14. Dissipated energy will appear primarily as thermal energy (temperature increase). For some stepper motors, the maximum overshoot could be as much as 80 percent of the step angle. Such high-amplitude oscillations with slow decay rate are clearly undesirable in most practical applications. Adequate damping should be provided by mechanical means (e.g., attaching mechanical dampers), electrical means (e.g., by further eddy current dissipation in the rotor or by using extra turns in the field windings), or by electronic means (electronic switching or multiple-phase energization) in order to suppress these oscillations. The single-pulse response is often modeled using the simple oscillator transfer function.

Now we shall examine the stepper motor response when a sequence of pulses is applied to the motor under normal operating conditions. If the pulses are sufficiently spaced—typically, more than the settling time T_s of the motor (*note*: $T_s \sim$ 4 × motor time constant)—then the rotor will come to rest at the end of each step before starting the next step. This is known as single stepping. In this case, the overall response is equivalent to a cascaded sequence of single-pulse responses; the motor will faithfully follow the command pulses in synchronism. In many practical applications, however, fast responses and reasonably continuous motor speeds (stepping rates) are desired. These objectives can be met, to some extent, by decreasing the motor settling time through increased dissipation (mechanical and electrical damping). This, beyond a certain optimal level of damping, could result in undesirable effects, such as excessive heat generation, reduced output torque, and very sluggish response. Electronic damping, explained in a later section, can eliminate these problems.

Since there are practical limitations to achieving very small settling times, faster operation of a stepper motor would require switching before the rotor settles down in each step. Of particular interest under high-speed operating conditions is *slewing motion*. In this case, the motor operates at steady state in synchronism at a constant pulse rate called the *slew rate*. It is not necessary for the phase switching (i.e., pulse commands) to occur when the rotor is at the detent position of the old phase, but switchings (pulses) should occur in a uniform manner. Since the motor moves in harmony, practically at a constant speed, the torque required for slewing is smaller than that required for transient operation (accelerating and decelerating conditions). A typical displacement time curve under slewing is shown in figure 6.16. The slew rate is given by

$$R_s = \frac{1}{\Delta t} \text{ steps/second} \qquad (6.21)$$

where Δt denotes the time between successive pulses under slewing conditions. Note that Δt could be significantly smaller than the motor settling time T_s. Some periodic oscillation (or hunting) is possible under slewing conditions, as seen in figure 6.16. This is generally unavoidable, but its amplitude can be reduced by increasing damping. The slew rate depends on the external load connected to the motor. Furthermore, motor inertia, damping, and torque rating set an upper limit to the slew rate.

To attain slewing conditions, the stepper motor has to be accelerated from a low speed by *ramping*. This is accomplished by applying a sequence of pulses with a continuously increasing pulse rate $R(t)$. Strictly speaking, ramping represents a linear (straight-line) increase of the pulse rate, as given by

$$R(t) = R_0 + \frac{(R_s - R_0)t}{n.\Delta t} \qquad (6.22)$$

where R_0 = starting pulse rate (typically zero)
R_s = final pulse rate (slew rate)
n = total number of pulses applied

If exponential ramping is used, the pulse rate is given by

$$R(t) = R_s - (R_s - R_0)e^{-t/\tau} \qquad (6.23)$$

If time constant τ of the ramp is equal to $n.\Delta t/4$, a pulse rate of $0.98R_s$ is reached in a total of n pulses. (*Note:* $e^{-4} \approx 0.02$.) In practice, the pulse rate is often increased beyond the slew rate, in a time interval shorter than what is specified for acceleration, and then decelerated to the slew rate by pulse subtraction at the end. In this manner, the slew rate is reached more quickly. In general, during *upramping* (accel-

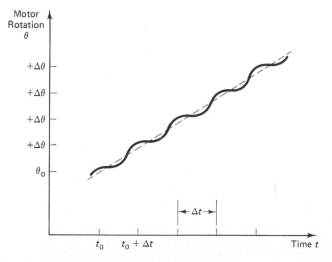

Figure 6.16. Typical slewing response of a stepper motor.

eration), the rotor angle trails the pulse command, and during *downramping* (deceleration), the rotor angle leads the pulse command. These conditions are illustrated in figure 6.17.

In transient operation of stepper motors, nonuniform stepping sequences might be necessary, depending on the complexity of the motion trajectory and the required accuracy. Consider, for example, the three-step drive sequence shown in figure 6.18. The first pulse is applied at *A* when the motor is at rest. The positive torque (curve 1) of the energized phase will accelerate the motor, causing an overshoot beyond the detent position (see broken line). The second pulse is applied at *B*, the point of intersection of the torque curves 1 and 2, which is before the detent position. This switches the torque to curve 2, which is the torque due to the newly energized phase. Fast acceleration is possible in this manner because the torque is kept positive up to the second detent position. The average torque is the maximum when switching occurs at the point of intersection of successive torque curves. This will produce a larger overshoot beyond the second detent position. The third pulse is applied at *C* when the rotor is closest to the required final position. Note that the corresponding torque (curve 3) is relatively small, because the rotor is near its final (third) detent position. As a result, the overshoot from the final detent position is relatively small, as desired. The rotor will then quickly settle down to the final position.

Drive sequences can be designed in this manner to produce virtually any desired motion in stepper motors. The motor controller is programmed to generate the proper pulse train in order to achieve the required phase switchings for a specified motion. Such drive sequences are useful also in compensating for missed pulses (see problem 6.11) and in electronic damping (section 6.9).

Note that to simplify the discussion and illustration, we have used static torque curves in figure 6.18. This assumes instant buildup of current in the energized phase

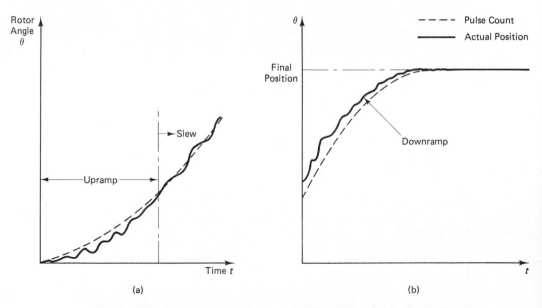

Figure 6.17. Ramping response: (a) accelerating motion; (b) decelerating motion.

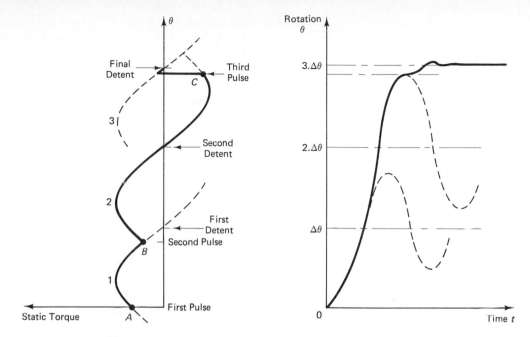

Figure 6.18. Torque response diagram for a three-step drive sequence.

and instant decay of current in the de-energized phase, thus neglecting all induced voltages and eddy currents. In reality, however, the switching torque lines will not be horizontal (instantaneous), with sharp ends, and the entire torque curve will be somewhat irregular. These dynamic torque curves should be used for accurate switching control in practical applications.

The Electrical Time Constant

The electrical time constant of a stepper motor is given by

$$\tau_e = \frac{L}{R} \tag{6.24}$$

where L is the inductance of the energized phase winding and R is the resistance of the energized circuit, including winding resistance. As a result of self-induction, the current in the energized phase does not build up instantaneously when switched on. The larger the electrical time constant, the slower the current buildup. This will result in a lower driving torque in the beginning of each step. Also, because of self-induction, the current does not die out instantaneously when the phase is switched off. The instantaneous voltages caused by self-induction can be high, and they can damage the translator and other circuitry. These harmful effects of induced voltages can be reduced by decreasing the electrical time constant. A convenient way to do this is by increasing resistance R. But we want this increase in R to be effective only during the transient periods (switch-on and switch-off times). During the steady period, we like to have a smaller R, which will give a larger current (and magnetic

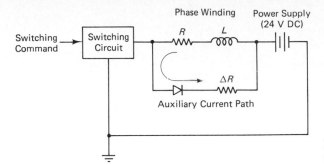

Figure 6.19. A diode circuit for decreasing the electrical time constant.

field), producing a higher torque. This can be accomplished by using a diode and a resistor ΔR, connected in parallel with the phase winding, as shown in figure 6.19. In this case, the current will loop through R and ΔR, as shown, during the switch-on and switch-off periods, thereby decreasing the electrical time constant to

$$\tau_e = \frac{L}{R + \Delta R} \tag{6.25}$$

During steady conditions, however, no current flows through ΔR, as desired.

The electrical time constant is much smaller than the mechanical time constant of a motor. Hence, increasing R is not a very effective way of increasing damping in a stepper motor.

6.8 STATIC POSITION ERROR

If a stepper motor does not support a static load, the equilibrium position under power-on conditions would correspond to the zero-torque (detent) point of the energized phase. If there is a static load T_L, however, the equilibrium position would be shifted to $-\theta_e$, as shown in figure 6.20. The offset angle θ_e is called the static position error.

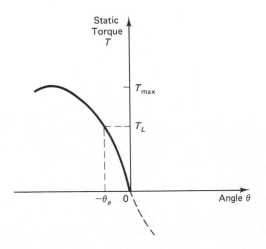

Figure 6.20. Representation of the static position error.

Assuming that the static torque curve is sinusoidal, we can obtain an expression for θ_e. First, note that the static torque curve for each phase is periodic with period $p.\Delta\theta$ (equal to the rotor pitch θ_r), where p is the number of phases and $\Delta\theta$ is the step angle. For example, this relationship is shown for the three-phase case in figure 6.21. Accordingly, the static torque curve may be expressed as

$$T = -T_{max} \sin\left(\frac{2\pi\theta}{p.\Delta\theta}\right) \tag{6.26}$$

where T_{max} denotes the maximum torque. Equation 6.26 can be directly obtained by substituting equation 6.2 in 6.20. Note that under standard switching conditions, equation 6.26 governs for $-\Delta\theta \le \theta \le 0$. With reference to figure 6.20, the static position error is given by,

$$T_L = -T_{max} \sin\left[\frac{2\pi(-\theta_e)}{p.\Delta\theta}\right]$$

or

$$\theta_e = \frac{p.\Delta\theta}{2\pi} \sin^{-1}\left(\frac{T_L}{T_{max}}\right) \tag{6.27}$$

If n denotes the number of steps per revolution, equation 6.27 may be expressed as

$$\theta_e = \frac{p}{n} \sin^{-1}\left(\frac{T_L}{T_{max}}\right) \tag{6.28}$$

It is intuitively clear that the static position error decreases with the number of steps per revolution.

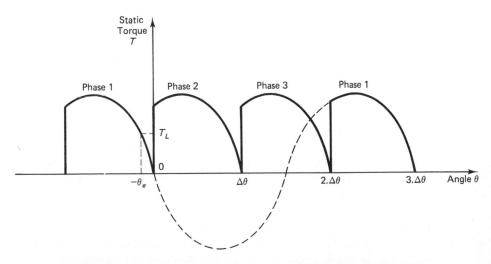

Figure 6.21. Periodicity of the single-phase static torque distribution (a three-phase example).

Example 6.4

Consider a three-phase stepping motor with seventy-two steps per revolution. If the static load torque is 10 percent of the maximum static torque of the motor, determine the static position error.

Solution In this problem,

$$\frac{T_L}{T_{max}} = 0.1, \qquad p = 3, \qquad n = 72$$

Now, using equation 6.25, we have

$$\theta_e = \frac{3}{72} \sin^{-1} 0.1 = 0.0042 \text{ rad} = 0.24°.$$

Note that this is less than 5 percent of the step angle.

6.9 DAMPING OF STEPPER MOTORS

Lightly damped oscillations in stepper motors are undesirable, particularly in applications that require single-step motions or accurate trajectory following under transient conditions. Damping has the advantages of suppressing overshoots and increasing the decay rate of oscillations (i.e., shorter settling time). Unfortunately, heavy damping has drawbacks, such as sluggish response (longer rise time, peak time, or delay), large time constants, and reduction of the net output torque. On the average, however, the advantages of damping outweigh the disadvantages in stepper motor applications. Several techniques have been employed to damp stepper motors. Most straightforward are the conventional techniques of damping that use mechanical and electrical energy dissipation. Mechanical damping is usually provided by a torsional damper attached to the motor shaft. Electrical damping methods include eddy current dissipation in the rotor, the use of magnetic hysteresis and saturation effects, and increased resistive dissipation by adding extra windings to the motor stator. For example, solid-rotor construction has higher hysteresis losses due to magnetic saturation than laminated-rotor construction has. These direct techniques of damping have undesirable side effects, such as excessive heat generation, reduction of the net output torque of the motor, and decreased speed of response. Electronic damping methods have been developed to overcome such shortcomings. These methods are based on employing properly designed switching schemes for phase energization so as to inhibit overshoots in the final step of response. The general drawback of electronic damping is that the associated switching sequences are complex (irregular) and depend on the nature of a particular motion trajectory. The level of damping achieved by this method is highly sensitive to the timing of the switching scheme. Accordingly, a high level of intelligence concerning the actual response of the motor is required to use electronic damping methods effectively. Note, also, that in the design stage, damping in a stepper motor can be improved by judicious choice of values for motor parameters (e.g., resistance of the windings, rotor size, material properties of the rotor, air gap width).

Mechanical Damping

A convenient, practical method of damping stepper motors is to connect an inertia element to the motor shaft through an energy dissipation medium, such as a viscous fluid (e.g., silicone) or a solid friction surface (e.g., brake lining). Two common examples of this type of torsional dampers are the *Houdaille damper* (or *viscous torsional damper*) and the *Lanchester damper*.

The effectiveness of torsional dampers on stepper motors can be examined using a linear dynamic model for the single-step oscillations. From figure 6.22a, it is evident that in the neighborhood of the detent position, the static torque due to the energized phase is approximately linear, and this torque acts as an electromagnetic spring. In this region, the torque can be expressed by

$$T = -K_m \theta \tag{6.29}$$

where θ = angle of rotation measured from the detent position
  K_m = torque constant of the motor

When external dampers are absent, damping forces can come from sources such as bearing friction, resistive dissipation in windings, eddy current dissipation in the rotor, and magnetic hysteresis. If the combined contribution from these internal dissipation mechanisms is represented by a single damping constant C_m, the equation of motion for the rotor near its detent position (equilibrium position) can be written as

$$J_m \frac{d\omega}{dt} = -C_m \omega - K_m \theta \tag{6.30}$$

where J_m = overall inertia of the rotor
  $\omega = d\theta/dt$ = motor speed

Note that for a motor with an external load, the *load inertia* has to be included in J_m.

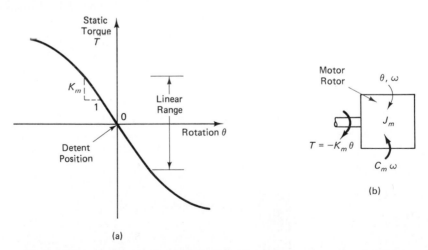

Figure 6.22. Model for single-step oscillations of a stepper motor: (a) linear torque approximation; (b) rotor free-body diagram.

Equation 6.27 is expressed in terms of θ; thus,

$$J_m \ddot{\theta} + C_m \dot{\theta} + K_m \theta = 0 \tag{6.31}$$

The solution of this second-order ordinary differential equation should be obtained using the maximum overshoot point as the initial state:

$$\dot{\theta}(0) = 0 \qquad \text{and} \qquad \theta(0) = \alpha \, \Delta \theta$$

The constant α represents the *fractional overshoot*. Its magnitude can be as high as 0.8. The undamped natural frequency of single-step oscillations is given by

$$\omega_n = \sqrt{\frac{K_m}{J_m}} \tag{6.32}$$

and the damping ratio is given by

$$\zeta = \frac{C_m}{2\sqrt{K_m J_m}} \tag{6.33}$$

With a Houdaille damper attached to the motor (see figure 6.23), the equations of motion are

$$(J_m + J_h)\ddot{\theta} = -C_m \dot{\theta} - K_m \theta - C_d(\dot{\theta} - \dot{\theta}_d) \tag{6.34}$$

$$J_d \ddot{\theta}_d = C_d(\dot{\theta} - \dot{\theta}_d) \tag{6.35}$$

where θ_d = angle of rotation of the damper inertia
J_d = moment of inertia of the damper
J_h = moment of inertia of the damper housing

It is assumed that the damper housing is rigidly attached to the motor shaft. In figure 6.24, a typical response of a mechanically damped stepper motor is compared with the response when the external damper is disconnected. Observe the much faster decay when the external damper is present. One disadvantage of this method of damping, however, is that it always adds inertia to the motor (note the J_h term in equation 6.34). This reduces the natural frequency of the motor (equation 6.32) and, hence, decreases the speed of response (or bandwidth). Other disadvantages include reduction of the effective torque and increased heat generation, which might require a special cooling method.

A stepper motor with a Lanchester damper can be analyzed in a similar manner. The equations of motion are somewhat complex in this case, because the frictional torque is a coulomb type, which has a constant magnitude for a given reaction

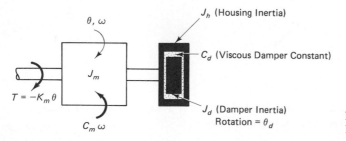

Figure 6.23. A stepper motor with a Houdaille damper.

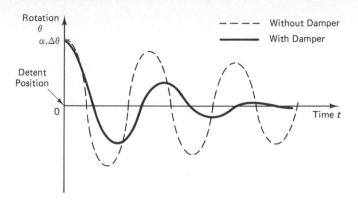

Figure 6.24. Typical single-step response of a stepper motor with a Houdaille damper.

force but acts opposite to the direction of relative motion between the rotor (and damper housing) and the damper inertia element. The reaction force on the friction lining can be adjusted using spring-loaded bolts, thereby changing the frictional torque. There are two limiting states of operation: If the reaction force is very small, the motor is virtually uncoupled (disengaged) from the damper; if the reaction force is very large, the damper inertia will be rigidly attached to the damper housing, thus moving as a single unit. In either case, there is very little dissipation. Maximum energy dissipation takes place under some intermediate condition. For constant-speed operation, by adjusting the reaction force, the damper inertia element can be made to rotate at the same speed as the rotor, thereby eliminating dissipation and torque loss under steady conditions, when damping is usually not needed. This is an advantage of friction dampers.

Electronic Damping

Damping of stepper motor response by electronic switching control is an attractive method of overshoot suppression for several reasons. For instance, it is not an energy dissipation method. In that sense, it is actually an electronic control technique rather than a damping technique. By timing the switching sequence properly, virtually a zero overshoot response could be realized. Another advantage is that the reduction in net output torque is insignificant in this case in comparison to torque losses in direct damping methods. A majority of electronic damping techniques depend on a two-step procedure that is straightforward in principle:

1. Decelerate the last-step response of the motor so as to avoid large overshoots from the final detent position.
2. Energize the final phase (i.e., apply the last pulse) when the motor response is very close to the final detent position (i.e., when the torque is very small).

We could come up with many switching schemes that conform to these two steps. Generally, such schemes differ only in the manner in which response deceleration is brought about (in step 1). Three common methods of response deceleration are

1. **The pulse turn-off method:** Turn off the motor (all phases) for a short time.
2. **The pulse reversal method:** Apply a pulse in the opposite direction (i.e., energize the reverse phase) for a short time.
3. **The pulse delay method:** Maintain the present phase beyond its detent position for a short time.

These three types of switching schemes can be explained using the static torque response curves in figures 6.25 through 6.27. In all three figures, the static torque curve corresponding to the last pulse (i.e., last phase energized) is denoted by 2. The static curve corresponding to the next-to-last pulse is denoted by 1.

   In the pulse turn-off method (figure 6.25), the last pulse is applied at *A*, as usual. This energizes Phase 2, turning off Phase 1. The rotor accelerates toward its final detent position because of the positive torque that is present. At point *B*, which is sufficiently close to the final detent position, Phase 2 is shut off. From *B* to *C*, all phases of the motor are inactive, and the static torque is zero. The motor decelerates during this interval, giving a peak response that is very close to, but below, the final detent position. At point *C*, the last phase (Phase 2) is energized again. Since the corresponding static torque is very small (in comparison to the maximum torque) but positive, the motor will accelerate slowly (assuming a pure inertial load) to the final detent position. By properly choosing the points *B* and *C*, the overshoot can be made sufficiently small. This choice requires a knowledge of the actual response of the motor. The amount of final overshoot can be very sensitive to the timing of the

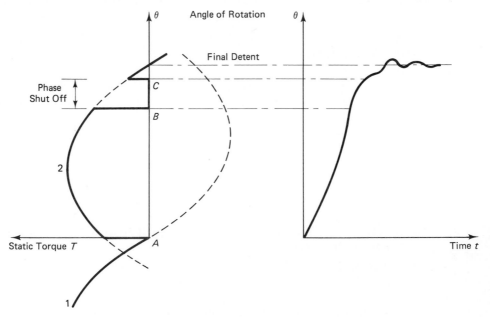

Figure 6.25. The pulse turn-off method of electronic damping.

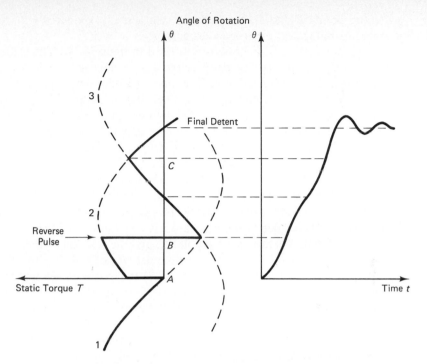

Figure 6.26. The pulse reversal method of electronic damping.

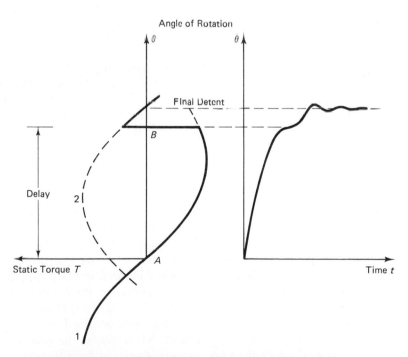

Figure 6.27. The pulse delay method of electronic damping.

switching points B and C. Furthermore, the actual response θ will depend on mechanical damping and other load characteristics as well.

The pulse reversal method is illustrated in figure 6.26. The static torque curve corresponding to the second pulse before last is denoted by 3. As usual, the last phase (Phase 2) is energized at A. The motor will accelerate toward the final detent position. At point B (located at less than half the step angle from A), Phase 2 is shut off and Phase 3 is turned on. (*Note:* The forward pulse sequence is 1-2-3-1, and the reverse pulse sequence is 1-3-2-1.) The corresponding static torque is negative over some duration. (*Note:* For a three-phase stepper motor, this torque is usually negative up to the halfway point of the step angle and positive thereafter.) Consequently, the motor will decelerate first and then accelerate (assuming a pure inertial load); the overall decelerating effect is not as strong as in the previous method. (*Note:* If faster deceleration is desired, Phase 1 should be energized, instead of Phase 3, at B.) At point C, the static torque of Phase 3 becomes equal to that of Phase 2. To avoid large overshoots, Phase 3 is turned off at point C, and the last phase (Phase 2) is energized again. This will drive the motor to its final detent position.

In the pulse delay method (figure 6.27) the last phase is not energized at the detent position of the previous step (point A). Instead, Phase 1 is kept on beyond this point. The resulting negative torque will decelerate the response. If intentional damping is not employed, the overshoot beyond A could be as high as 80 percent. (*Note:* In the absence of any damping, 100 percent overshoot is possible.) When the overshoot peak is reached at B, the last phase is energized. Since the static torque of Phase 2 is relatively small at this point and will reach zero at the final detent position, the acceleration of the motor is slow. Hence, the final overshoot is maintained within a very small value. It is interesting to note that if 100 percent overshoot is obtained with Phase 1 energized, the final overshoot becomes zero in this method, thus producing ideal results.

In all these techniques of electronic damping, the actual response depends on many factors—particularly the dynamic behavior of the load. Hence, the switching points cannot be exactly prespecified unless the true response is known ahead of time (through tests, simulations, etc.). In general, accurate switching might require measuring the actual response and using that information in real time to apply pulses. Note that static torque curves have been used in figures 6.25 through 6.27 to explain electronic damping. In practice, however, currents in the phase windings neither decay nor build up instantaneously following a pulse command (see problem 6.8). Induced voltages, eddy currents, and magnetic hysteresis effects are primarily responsible for this behavior. These factors, in addition to external loads, can complicate the nature of dynamic torque and, hence, the true response of a stepping motor. This can make accurate preplanning of switching points somewhat difficult in electronic damping. In the foregoing discussion, we have assumed that the mechanical damping (including bearing friction) of the motor is negligible and that the load connected to the motor is a pure inertia. In practice, the net torque available to drive the combined rotor load inertia is smaller than the electromagnetic torque generated at the rotor. Hence, in practice, accelerations obtained are not quite as high as figures 6.25 through 6.27 might suggest. Nevertheless, the general characteristics of motor response will be the same as those shown in these figures.

A popular and relatively simple method that may be classified under electronic damping is multiple-phase energization. With this method, two phases are excited simultaneously (e.g., 12-23-31-12). As noted earlier, the step angle remains unchanged. It has been observed, however, that this switching sequence provides a better response (less overshoot) than the single-phase energization method (e.g., 1-2-3-1), particularly for single-stack stepper motors. This can be attributed to the increased magnetic hysteresis and saturation effects of the ferromagnetic materials in the motor as well as higher energy dissipation through eddy currents when two phases are energized simultaneously. Another factor is that multiple-phase excitation results in wider overlaps in magnetic flux between switchings, giving smoother torque transitions. Note, however, that there can be excessive heat generation with this method. This may be reduced, to some extent, by using a lower voltage, typically half the normal voltage, to excite one of the phases (the *damping phase*) while using the full voltage to excite the other phase (the *stepping phase*). The damping phase is the reverse phase, which provides a negative torque to reduce overshoot.

6.10 FEEDBACK CONTROL OF STEPPER MOTORS

Open-loop operation is adequate for many applications of stepper motors, particularly at low speeds and in steady-state operation. The main disadvantage of open-loop control is that the actual response of the motor is not measured; consequently, it is not known whether a significant error is present—because of missed pulses, for example. There are two main reasons for pulse missing:

1. Under variable-speed conditions, if the successive pulses are received at a high frequency (high stepping rate), the phase translator might not respond to a particular pulse, and the corresponding phase would not be energized before the next pulse arrives. This may be due to a malfunction in the drive circuit.
2. Because of a malfunctioning pulse source, a pulse might not actually be generated, even when the motor is operating at well below its rated capacity (low-torque, low-speed conditions). Extra erroneous pulses can also be generated because of faulty drive circuitry.

If a pulse is missed by the motor, the response has to catch up somehow (e.g., by a subsequent overshoot in motion), or else an erratic behavior may result, causing the rotor to oscillate and probably stall eventually. Under very favorable conditions, particularly with small step angles, if a single pulse is missed, the motor will decelerate so that a complete cycle of pulses is missed; then it will lock in with the input pulse sequence again. In this case, the motor will trail the correct trajectory by a rotor tooth pitch angle (θ_r). Accordingly, pulses equal in number to the total phases (p) of the motor (*note:* $\theta_r = p.\Delta\theta$) are missed. It is also possible to lose accuracy by an integer multiple of θ_r because of a single missed pulse in this manner. In general, however, pulse missing can lead to stalling or a highly nonsynchronous response.

As mentioned before, the missing (or dropping) of a pulse can be interpreted in two ways. First, a pulse can be lost between the pulse generator (e.g., a command microcomputer) and the translator. In this case, the logic sequence within the translator that energizes motor phases will remain intact. The next pulse to arrive at the translator will be interpreted as the lost pulse and will energize the phase corresponding to the lost pulse. Hence, a "time delay" is introduced to the command (pulse) sequence. The second interpretation of a missed pulse is that the pulse actually reached the translator, but the corresponding motor phase was not energized because of some hardware problem in the drive circuit. In this case, the next pulse reaching the translator will not energize the phase corresponding to the missed pulse but will energize its own phase. This interpretation is termed *missed phase activation*.

In both interpretations of pulse missing, the motor will decelerate because of the negative torque of the phase that was not switched off. Depending on the timing of subsequent pulses, a negative torque can continue to exist in the motor, thereby subsequently stalling the motor. Motor deceleration due to pulse missing can be explained using the static torque approximation, as shown in figure 6.28. Consider a three-phase motor with one phase on excitation (i.e., only one phase is excited at

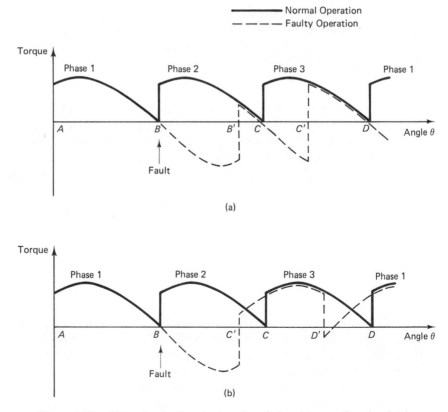

Figure 6.28. Motor deceleration due to pulse missing: (a) case of a missed pulse; (b) case of a missed phase activation.

a given time). Suppose that under normal operating conditions, the motor runs at a constant speed and phase activation is brought about at points A, B, C, D, etc. in figure 6.28, using a pulse sequence sent into the translator. Note that these points are equally spaced (with the horizontal axis being the angle of rotation θ, not time t) because of constant-speed operation. The torque generated by the motor under normal operation, without pulse missing, is shown as a solid line in figure 6.28. Note that Phase 1 is excited at point A, Phase 2 is excited at point B, Phase 3 is excited at point C, and so on. Now let us examine the two cases of pulse missing.

In the first case (figure 6.28a), a pulse is missed at B. Phase 1 continues to be active, providing a negative torque. This slows down the motor. The next pulse is received when the rotor is at position B' (not C) because of rotor deceleration (note that pulses are sent at equal time intervals for constant-speed operation). At point B', Phase 2 (not Phase 3) is excited in this case, because the translator interprets the present pulse as the pulse that was lost. The next pulse is received at C', and so on. The resulting torque is shown by the broken line in figure 6.28a. Since this torque could be much less than the torque in the absence of missed pulses—depending on the locations of points B', C', and so on—the motor might decelerate continuously and finally stall.

In the second case (figure 6.28b), the pulse at B fails to energize Phase 2. This decelerates the motor because of the negative torque generated by the existing Phase 1. The next pulse is received at point C' (not C) because the motor has slowed. This pulse excites Phase 3 (not Phase 2, as in the previous case), because the translator assumes that Phase 2 has been excited by the previous pulse. The subsequent pulse arrives at point D' (not D) because of the slower speed of the motor. The corresponding motor torque is shown by the broken line in figure 6.28b. In this case, also, the net torque can be much smaller than what is required to maintain the normal operating speed, and the motor might stall. To avoid this situation, pulse missing should be detected by response measurement (e.g., using a shaft encoder), and corrective action should be taken by properly modifying the future switching sequence in order to accelerate the motor back into the desired trajectory. In other words, feedback control would be required.

Feedback control is used to compensate for motion errors in stepper motors. A block diagram for a typical closed-loop control system is shown in figure 6.29. This should be compared with figure 6.13a. The noted improvement in the feedback control scheme is that the actual response of the stepper motor is measured and compared with the desired response; if an error is detected, the pulse train to the drive

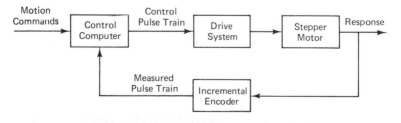

Figure 6.29. Feedback control of a stepper motor.

system is modified appropriately to reduce the error. Typically, an *optical incremental encoder* (chapter 5) is employed as the motion transducer. This device provides two pulse trains that are in phase quadrature, giving both the magnitude and the direction of rotation of the stepper motor. The encoder pitch angle should be made equal to the step angle of the motor for ease of comparison and error detection. Note that when feedback control is employed, the resulting closed-loop system can operate near the rated capacity (torque, speed, acceleration, etc.) of the stepper motor, perhaps exceeding these ratings at times but without introducing excessive error and stability problems (e.g., hunting).

A simple closed-loop device that does not utilize sophisticated control logic is the *feedback encoder–driven stepper motor*. In this case, the drive pulses, except for the very first pulse, are generated by a feedback encoder connected to the motor shaft. This mechanism is particularly useful for steady acceleration and deceleration operations under possible overloading conditions, when there is a likelihood of missing pulses. The principle of operation of a feedback encoder–driven stepper motor may be explained using figure 6.30. The starting pulse is generated externally at the initial detent position O. This will energize Phase 1 and drive the rotor toward the corresponding detent position D_1. The encoder disk is positioned such that the first pulse from the encoder is generated at E_1. This pulse is automatically fed back as the second pulse input to the motor (translator). This pulse will energize Phase 2 and drive the rotor toward the corresponding detent position D_2. During this step, the second pulse from the encoder is generated at E_2, which is automatically fed back as the third pulse input to the motor, energizing Phase 3 and driving the motor toward the detent position D_3, and so on. Note that phase switching occurs (because of an encoder pulse) every time the rotor has turned through a fixed angle $\Delta\theta_s$ from the previous detent position. This angle is termed the *switching angle*. The encoder pulse "leads" the corresponding detent position by an angle $\Delta\theta_L$. This angle is termed the *lead angle*. Note from figure 6.30 that

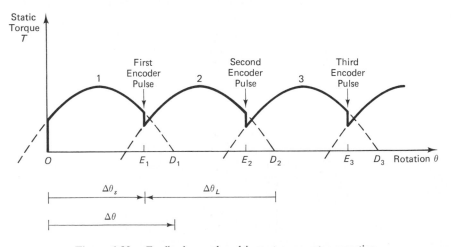

Figure 6.30.   Feedback encoder–driven stepper motor operation.

$$\Delta\theta_s + \Delta\theta_L = 2\Delta\theta \qquad\qquad (6.36)$$

where $\Delta\theta$ denotes the step angle.

For the switching angle position (or lead angle position) shown in figure 6.30, the static torque on the rotor is positive throughout the motion. As a result, the motor will accelerate steadily until damping, other speed-dependent resistive torques, and load torque exactly balance the motor torque. The final steady-state condition corresponds to the maximum speed of operation for a feedback encoder–driven stepper motor. This maximum speed usually decreases as the switching angle is increased beyond the point of intersection of two adjacent torque curves. For example, if $\Delta\theta_s$ is increased beyond $\Delta\theta$, there is a negative static torque from the present phase (before switching) that tends to decelerate the motor somewhat. But the combined effect of the before-switching torque and the after-switching torque is to produce an overall increase in speed until the speed limit is reached. This is generally true, provided that the lead angle $\Delta\theta_L$ is positive (the positive direction, as indicated by the arrowhead in figure 6.30). The lead angle may be adjusted either by physically moving the signal pick-off point on the encoder disk or by introducing a *time delay* into the feedback path of the encoder signal. The former method is less practical, however.

Steady decelerations can be achieved using feedback encoder–driven stepper motors if negative lead angles are employed. In this case, switching to a particular phase occurs when the rotor has actually passed the detent position for that phase. The resulting negative torque will steadily decelerate the rotor, eventually bringing it to a halt. Negative lead angles may be obtained by simply adding a time delay into the feedback path. Alternatively, the same effect (negative torque) can be generated by blanking out (using a blanking gate) the first two pulses generated by the encoder and using the third pulse to energize the phase that would be energized by the first pulse for accelerating operation. This situation is addressed in problem 6.17.

The feedback encoder–driven stepper motor is just a crude form of closed-loop control. Its application is normally limited to steadily accelerating (upramping), steadily decelerating (downramping), and steady-state (constant-speed) operations. More sophisticated feedback control systems require point-by-point comparison of the encoder pulse train with the desired pulse train and injection of extra pulses or extraction (blanking out) of existing pulses at proper instants so as to reduce error. A commercial version of such a feedback controller uses a *count-and-compare* card. More complex applications of closed-loop control include switching control for electronic damping (see figures 6.25 through 6.27), transient drive sequencing (see figure 6.18), and dynamic torque control.

Torque Control through Switching

Under standard operating conditions for a stepper motor, phase switching (of a pulse) occurs at the present detent position. It is easy to see from the static torque diagram in figure 6.31, however, that a higher average torque is possible by advancing the switching time to the point of intersection of the two adjacent torque curves (before

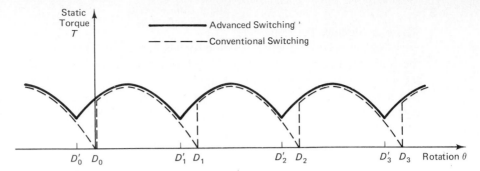

Figure 6.31. The effect of advancing the switching pulses.

and after switching). In the figure, the standard switching points are denoted as D_0, D_1, D_2, and so forth, and the advanced switching points as D_0', D_1', D_2', and so forth. Note that in the case of advanced switching, the static torque always remains greater than the common value at the point of intersection. This confirms what is intuitively clear: Motor torque can be controlled by adjusting the switching point. The actual magnitude of torque, however, will depend on the dynamic conditions that exist. For low speeds, the dynamic torque may be approximated by the static torque curve, making the analysis simpler. As the speed increases, the deviation from the static curve becomes more pronounced, for reasons that were mentioned earlier.

Example 6.5

Suppose that the switching point is advanced beyond the zero-torque point of the switched phase, as shown in figure 6.32. The switching points are denoted by D_0', D_1', D_2', and so forth. Note that although the static torque curve takes negative values in some regions under this advanced switching sequence, the dynamic torque stays positive at all times. The main reason for this is that a finite time is needed for the current in the turned-off phase to decay completely because of induced voltages and eddy current effects.

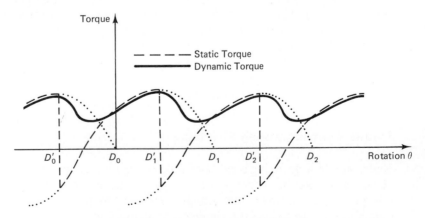

Figure 6.32. Dynamic torque at high speeds.

Stepper Motors   Chap. 6

6.11 STEPPING MOTOR MODELS

In the preceding sections, we have discussed variable-reluctance (VR) stepper motors, which have nonmagnetized soft-iron rotors, and permanent magnet (PM) stepper motors, which have magnetized rotors. *Hybrid stepper motors* are, in fact, a special type of PM stepping motors. Specifically, a hybrid motor has two rotor stacks (see problem 6.21) that are magnetized to have opposite polarities (one rotor stack is the N pole and the other is the S pole). Also, there is a tooth misalignment between the two rotor stacks. As usual, stepping is achieved by switching the phase excitations.

Under steady operation at low speeds, we usually do not need to differentiate between VR motors and PM motors. But under transient conditions, the torque characteristics of the two types of motors can differ substantially. In particular, the torque in a PM motor varies somewhat linearly with the magnitude of the phase current, whereas the torque in a VR motor varies nearly quadratically with the phase current.

A Simplified Model

Under steady-state operation of a stepper motor at low speeds, the motor (magnetic) torque can be approximated by a sinusoidal function, as given by equation 6.20 or equation 6.26. Hence, the simplest model for any type of stepping motor (VR or PM) is a torque source given by

$$T = -T_{\max} \sin n_r \theta \qquad (6.37)$$

or, equivalently,

$$T = -T_{\max} \sin\left(\frac{2\pi\theta}{p.\Delta\theta}\right) \qquad (6.38)$$

where $T_{\max}$ = maximum torque during a step
$\Delta\theta$ = step angle
n_r = number of rotor teeth
p = number of phases

Note that θ is the angular position of the rotor measured from the detent position of the presently excited phase, as indicated in figure 6.33a. Hence, $\theta = -\Delta\theta = -\theta_r/p$ at the previous detent position, where the present phase is switched on, and $\theta = 0$ at the approaching detent position. The θ coordinate frame is then shifted again to a new origin $(+\Delta\theta)$ when the next phase is excited at the approaching detent position in the conventional method of switching. Hence, θ gives the relative position of the rotor during each step. The absolute position is obtained by adding θ to the absolute rotor angle at the approaching detent position.

The motor model is complete with the mechanical dynamic equation for the rotor. With reference to figure 6.33b, Newton's second law gives

$$T - T_L - T_b(\theta, \dot{\theta}) = J\ddot{\theta} \qquad (6.39)$$

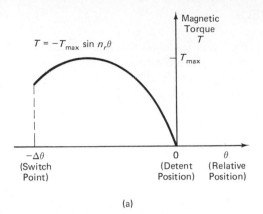

(a)

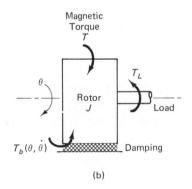

(b)

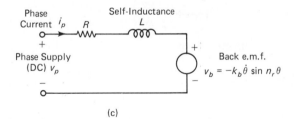

Figure 6.33. Stepper motor models: (a) torque source model; (b) mechanical model; (c) equivalent circuit for an improved model.

where T_L = resisting torque (reaction) on the motor by the driven load
$T_b(\theta, \dot{\theta})$ = dissipative resisting torque (viscous damping torque, frictional torque, etc.) on the motor
J = rotor inertia

Note that T_L will depend on the nature of the external load. Furthermore, $T_b(\theta, \dot{\theta})$ will depend on the nature of damping. If viscous damping is assumed, T_b may be taken as proportional to $\dot{\theta}$. On the other hand, if coulomb friction is assumed, the magnitude of T_b is taken to be constant, and the sign of T_b is the sign of $\dot{\theta}$. In the case of general dissipation (e.g., a combination of viscous, coulomb, and structural damping), T_b is a nonlinear function of both θ and $\dot{\theta}$. Note that the torque source model may be used for both VR and PM types of stepping motors.

An Improved Model

Under high-speed and transient operation of a stepper motor, many of the quantities that were assumed constant in the torque source model will vary with time as well as rotor position. In particular, for a given supply voltage v_p to a phase winding, the associated phase current i_p will not be constant. Furthermore, inductance L in the phase circuit will vary with the rotor position. Also, a voltage v_b (a *back electromotive force,* or e.m.f.) will be induced in the phase circuit because of the magnetic flux changes resulting from the speed of rotation of the rotor (in both VR and PM motors). Hence, an improved dynamic model would be needed to represent the behavior of a stepper motor under high-speed and transient conditions.

Chi (1985) presents sophisticated models based on permeance considerations for both VR and PM stepper motors. A stepper motor model that is a simplified version of those models will be described here. Instead of using rigorous derivations, motor equations are obtained from an equivalent circuit using qualitative considerations.

Since magnetic flux linkage of the phase windings changes as a result of variations in the phase current, a voltage will be induced in the phase windings. Hence, a self-inductance (L) should be included in the circuit. Although a mutual inductance should also be included to account for voltages induced in a phase winding as a result of current variations in the other phase windings, this voltage is usually smaller than the self-induced voltage. Hence, we neglect mutual inductance in this model. Furthermore, flux linkage of the phase windings will change as a result of the motion of the rotor. This induces a voltage v_b (termed a back e.m.f.) in the phase windings. Note that this voltage is present irrespective of whether the rotor is a VR type or a PM type. Also, phase windings will have a finite resistance R. It follows that an approximate equivalent circuit (in particular, neglecting mutual induction) for one phase of a stepper motor may be represented as in figure 6.33c. The phase circuit equation is

$$v_p = Ri_p + L\frac{di_p}{dt} + v_b \qquad (6.40)$$

where v_p = phase supply voltage (DC)
$\quad i_p$ = phase current
$\quad v_b$ = back e.m.f. due to rotor motion
$\quad R$ = resistance in the phase winding
$\quad L$ = self-inductance of the phase winding

The back e.m.f. is proportional to the rotor speed $\dot{\theta}$, and it will also vary with the rotor position θ. The variation with position will be periodic with period θ_r. Hence, using only the fundamental term in a Fourier series expansion, we have

$$v_b = -k_b \dot{\theta} \sin n_r \theta \qquad (6.41)$$

where $\dot{\theta}$ = rotor speed
$\quad \theta$ = rotor position (as defined in figure 6.33a)
$\quad n_r$ = number of rotor teeth
$\quad k_b$ = back e.m.f. constant

Since θ is negative in a conventional step (from $\theta = -\Delta\theta$ to $\theta = 0$), we note that v_b is positive for positive $\dot{\theta}$.

Self-inductance L also varies with the rotor position θ. This variation is periodic with period θ_r. Now, retaining only the constant and the fundamental terms in a Fourier series expansion, we have

$$L = L_0 + L_a \cos n_r \theta \qquad (6.42)$$

where L_0 and L_a are appropriate constants and angle θ is as defined in figure 6.33(a).

Note that equations 6.40 through 6.42 are valid for both types of stepper motors (VR and PM). The torque equation will depend on the type of stepper motor, however.

Torque equation for PM motors. In a permanent-magnet stepper motor, magnetic flux is generated by both the phase current i_p and the magnetized rotor. The flux from the magnetic rotor is constant, but its linkage with the phase windings will be modulated by the rotor position θ. Hence, retaining only the fundamental term in a Fourier series expansion, we have

$$T = -k_m i_p \sin n_r \theta \qquad (6.43)$$

where i_p is the phase current and k_m is the *torque constant* for the PM motor.

Torque equation for VR motors. In a variable-reluctance stepper motor, the rotor is not magnetized; hence, there is no magnetic flux generation from the rotor. The flux generated by the phase current i_p is linked with the phase windings. The flux linkage is modulated by the motion of the VR rotor, however. Hence, retaining only the fundamental term in a Fourier series expansion, the torque equation for a VR stepper motor may be expressed as

$$T = -k_r i_p^2 \sin n_r \theta \qquad (6.44)$$

where k_r is the *torque constant* for the VR motor. Note that torque T depends on the phase current i_p in a quadratic manner in the VR stepper motor.

In summary, to compute the torque T at a given rotor position, we first have to solve the differential equation given by equations 6.40 through 6.42 for known values of the rotor position θ and the rotor speed $\dot{\theta}$ and for a given (constant) phase supply v_p. Initially, as a phase is switched on, the phase current would be zero. The model parameters R, L_0, L_a, and k_b are assumed to be known (experimentally or from the manufacturer's data sheet). Then torque is computed using equation 6.43 for a PM stepper motor or using equation 6.44 for a VR stepper motor. Again, the torque constant (k_m or k_r) is assumed to be known. The simulation of the model then can be completed by using this torque in the mechanical dynamic equation 6.39 to determine the rotor position θ and the rotor speed $\dot{\theta}$.

Model-Based Feedback Control

The foregoing improved motor model is useful in computer simulation of stepper motors—for example, for performance evaluation. Such a model is also useful in model-based feedback control of stepper motors. In this case, the model provides a relationship between motor torque and the motion variables θ and $\dot{\theta}$. Accordingly, we can determine from the model the required phase-switching points in order to generate a desired motor torque (to drive the load). Actual values of θ and $\dot{\theta}$ (e.g., measured using an incremental optical encoder) are used in the model-based computations.

A simple feedback control strategy for a stepper motor will be outlined here. Initially, when the motor is at rest, the phase current $i_p = 0$. Also, $\theta = -\Delta\theta$ and $\dot{\theta} = 0$. As a phase is switched on to drive the motor, the motor equation 6.40 is integrated in real time, using a suitable integration algorithm and an appropriate time step. Simultaneously, the desired position is compared with the actual (measured) position of the load. If the two are sufficiently close, no phase-switching action is taken. But suppose that the actual position lags behind the desired position. Then we compute the present motor torque using model equations 6.41, 6.42, and 6.43 or 6.44 and repeat the computations, assuming (hypothetically) that the excitation is swtiched to one of the two adjoining phases. Since we need to accelerate the motor, we should actually switch to the phase that provides a torque larger than the present torque. If the actual position leads the desired position, however, we need to decelerate the motor. In this case, we switch to the phase that provides a torque smaller than the present torque or we turn off all the phases. Note that the time taken by the phase current to build up to its full value is approximately equal to 4τ, where τ is the electrical time constant for each phase, as approximated by $\tau = L_0/R$. Hence, when a phase is hypothetically switched, numerical integration has to be performed for a time period of 4τ before the torques are compared. It follows that the performance of this control approach will depend on the operating speed of the motor and the computational efficiency of the integration algorithm. At high speeds, less time is available for control computations. Ironically, it is at high speeds that control problems are severe and sophisticated control techniques are needed; hence, hardware implementations of switching are desired. For better control, phase switching has to be based on the motor speed as well as the motor position.

Example 6.6

The desired motion trajectory for a load driven by a PM stepper motor consists of an initial segment of constant acceleration, an intermediate segment of constant speed, and a final segment of constant deceleration to rest. Suppose that the load is an inertia element rigidly connected to the motor. The following parameter values are given:

Equivalent moment of inertia of rotor and load $J = 0.04$ kg.m^2
Equivalent viscous damping constant $b = 0.5$ N.m/rad/s
Number of motor phases $p = 4$
Number of rotor teeth $n_r = 50$
Phase winding parameters: $R = 2.0\ \Omega$, $L_0 = 10.0$ mH, $L_a = 2.0$ mH
Back e.m.f. constant $k_b = 0.05$ V/rad/s
Motor torque constant $k_m = 10$ N.m/A
Phase supply voltage $v_p = 20.0$ V

We have simulated the foregoing control scheme. The desired trajectory and the actual trajectory obtained using feedback control are shown in figure 6.34. We have as-

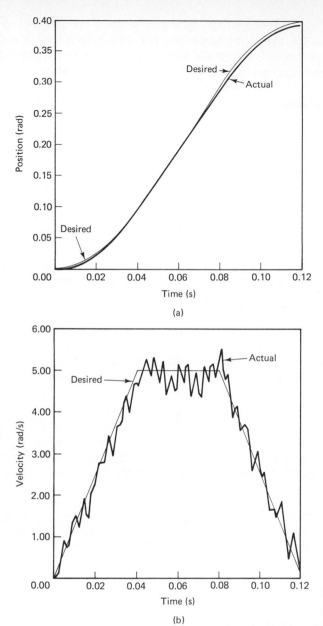

Figure 6.34. Model-based feedback control example: (a) position response; (b) velocity response.

sumed that the measurements of θ and $\dot{\theta}$ are available and that there is no limitation to the switching frequency.

6.12 STEPPER MOTOR SELECTION AND APPLICATIONS

In example 6.2, we discussed a simple design problem that addressed the selection of geometric parameters (number of stator poles, number of teeth per pole, number of rotor teeth, etc.) for a stepper motor. Selection of a stepper motor cannot be made

Stepper Motors Chap. 6

on the basis of geometric parameters alone, however. Torque and speed considerations are often more crucial in the selection process. For example, a faster speed of response is possible if a motor with a larger torque-to-inertia ratio is used. The effort required in selecting a stepper motor for a particular application can be reduced if the selection is done in an orderly manner. The following steps provide some guidelines for the selection process:

- *Step 1:* List the main requirements for the particular application.

This includes operational requirements such as speeds, accelerations, and required accuracy and resolution, and load characteristics, such as size, inertia, fundamental natural frequencies, and resistance torques.

- *Step 2:* Compute the operating torque and stepping rate requirements for the particular application.

Newton's second law is the basic equation employed in this step. Specifically, the required torque rating is given by

$$T = T_R + J_{eq}\frac{\omega_{max}}{\Delta t} \tag{6.45}$$

where T_R = net resistance torque
J_{eq} = equivalent moment of inertia (including rotor, load, gearing, dampers, etc.)
ω_{max} = maximum operating speed
Δt = time taken to accelerate the load to the maximum speed, starting from rest

- *Step 3:* Using the torque versus stepping rate curves for a group of commercially available stepper motors, select a suitable stepper motor.

The torque and speed requirements determined in Step 2 and the accuracy and resolution requirements specified in Step 1 should be used in this step.

- *Step 4:* If a stepper motor that meets the requirements is not available, modify the basic design.

This may be accomplished by changing the speed and torque requirements by adding devices such as gear systems (e.g., harmonic drive) and amplifiers (e.g., hydraulic amplifiers).

- *Step 5:* Select a drive system that is compatible with the motor and that meets the operational requirements in Step 1.

For relatively simple applications, a manually controlled preset indexer or an open-loop system consisting of a pulse source (oscillator) and a translator could be used.

For more complex transient tasks, a microprocessor or a customized hardware controller may be used to generate the desired pulse command in open-loop operation. Further sophistication could be added by using digital processor–based closed-loop control with encoder feedback for tasks that require very high accuracy under transient conditions and for operation near the rated capacity of the motor.

The single most useful piece of information in selecting a stepper motor is the torque versus stepping rate curve. Other parameters that are valuable in the selection process include:

1. The step angle or the number of steps per revolution
2. The static holding torque (starting torque of motor when powered with rated voltage)
3. The maximum slew rate (maximum steady-state stepping rate possible at rated load)
4. The motor torque at the maximum slew rate (pull-out torque)
5. The maximum ramping slope (maximum acceleration and deceleration possible at rated load)
6. The motor time constants (no-load electrical time constant and mechanical time constant)
7. The motor natural frequency (without an external load and near detent position)
8. The motor size (dimensions of poles, stator and rotor teeth, air gap and housing, weight, rotor moment of inertia)
9. The power supply capacity (voltage and power)

There are many parameters that determine the ratings of a stepper motor. For example, the static holding torque decreases with the air gap width and tooth width and increases with the rotor diameter and stack length. Furthermore, the minimum allowable air gap width should exceed the combined maximum lateral (flexural) deflection of the rotor shaft caused by thermal deformations and the flexural loading, such as magnetic pull and static and dynamic mechanical loads. In this respect, the flexural stiffness of the shaft, the bearing characteristics, and the thermal expansion characteristics of the entire assembly become important. Field winding parameters (diameter, length, resistivity, etc.) are chosen by giving due consideration to the required torque, power, electrical time constant, heat generation rate, and motor dimensions. Note that a majority of these are design parameters that cannot be modified in a cost-effective manner during the motor selection stage.

Example 6.7

Common applications of stepper motors include positioning tables (see figure 6.35a and digital x-y plotters. One design of x-y plotters uses a stationary pen with a table driven in two orthogonal (x-y) directions. Note that two stepper motors of nearly equal capacity are required in this application. Suppose that the following parameters are specified:

- A resolution of better than 0.01 in./step
- An operating velocity of greater than 10 in./s to be attained in less than 0.1 s
- Weight of the *x-y* table = 1 lb
- Maximum resistance force (primarily friction) against table motion = 0.5 lb

Select a stepper motor system to satisfy these requirements.

A schematic diagram of the mechanical arrangement for one of the two degrees of freedom is shown in figure 6.35b. A lead screw is used to convert the rotary motion of the motor into rectilinear motion. Free-body diagrams for the rotor and the table are shown in figure 6.36.

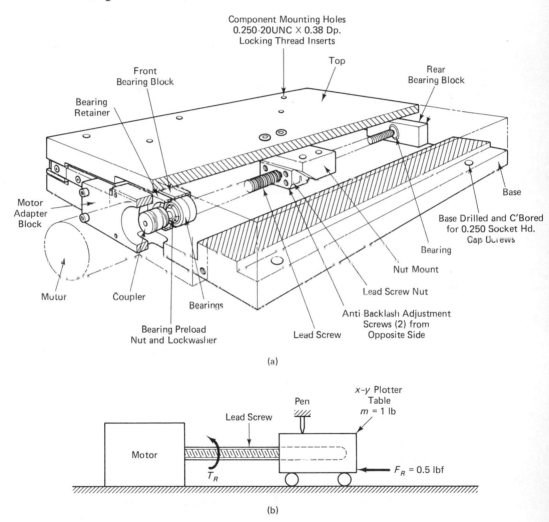

Figure 6.35. (a) A positioning table (courtesy of Daedal Inc.). (b) A stepper motor–driven *x-y* plotter.

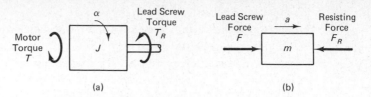

Figure 6.36. Free-body diagrams: (a) rotor; (b) table.

Solution First we shall derive a somewhat generalized relation for this type of application. The equations of motion (from Newton's second law) are

$$\text{For the rotor: } T - T_R = J\alpha \tag{i}$$

$$\text{For the table: } F - F_R = ma \tag{ii}$$

where T = motor torque
  T_R = resistance torque from the lead screw
  J = equivalent moment of inertia of the rotor
  α = angular acceleration of the rotor
  F = driving force from the lead screw
  F_R = external resistance force on the table
  m = equivalent mass of the table
  a = acceleration of the table

Assuming a rigid lead screw without backlash, the compatibility condition is written as

$$a = r\alpha \tag{iii}$$

where r denotes the *gear ratio* (rectilinear motion/angular motion) of the lead screw. The load transmission equation for the lead screw is

$$F = \frac{e}{r} T_R \tag{iv}$$

where e denotes the *fractional efficiency* of the lead screw. Finally, equations (i) through (iv) can be combined to give

$$T = \left(J + \frac{mr^2}{e}\right)\frac{a}{r} + \frac{r}{e} F_R \tag{6.46}$$

In this example, the following values have been specified:

$$m = 1 \text{ lb}, \qquad F_R = 0.5 \text{ lbf}, \qquad a = 10/0.1 = 100 \text{ in./s}^2$$

Let us pick a two-pitch lead screw (i.e., 2 rev/in.) that has 50 percent efficiency. Hence,

$$r = \frac{1}{2 \times 2\pi} = \frac{1}{4\pi} \text{ in./rad} \quad \text{and} \quad e = 0.5$$

For a linear resolution of 0.01 in., the number of steps per revolution is

$$n = \frac{0.5 \text{ in./rev}}{0.01 \text{ in./step}} = 50 \text{ steps/rev}$$

Pick $n = 72$ steps/rev, which corresponds to a commercially available stepper motor. The corresponding rectilinear resolution is

$$\frac{0.5 \text{ in./rev}}{72 \text{ steps/rev}} = 0.008 \text{ in./step}$$

The operating speed of the motor is

$$\omega_{\max} = \frac{10 \text{ in./s} \times 72 \text{ steps/rev}}{0.5 \text{ in./rev}} = 1{,}440 \text{ steps/s}$$

If we pick a motor with rotor inertia $J = 0.5$ lb.in.^{2}, from equation 6.35, the motor torque rating is computed as

$$T = \left(0.5 + \frac{1/(4\pi)^2}{0.5}\right)\frac{100}{1/4\pi \times 386} + \frac{0.5 \times 1}{4\pi \times 0.5} \text{ lbf.in.}$$

(*Note:* $g = 386$ in./s^2.) Hence,

$$T = 28 \text{ oz.in.}$$

Now we should check this value against the speed-torque curve of the selected motor with 72 steps/rev and rotor inertia approximately 0.5 lb.in.^{2}) at the maximum stepping rate of 1,440 steps/s. If the computed torque value lies below the curve, as shown in figure 6.37, the motor is acceptable. Otherwise, we should pick a motor of higher capacity and repeat the computations, or modify the design by using a different lead screw, and so forth.

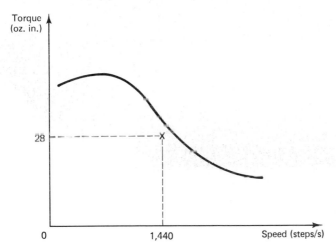

Figure 6.37. Speed-torque curve for a stepper motor.

Stepper Motor Applications

The stepper motor is a low-torque actuator that is particularly suitable for applications that require torques less than 2000 oz.in. For heavy-duty applications, torque amplification is usually necessary. One way to accomplish this is by using a hydraulic actuator in cascade with the motor. The hydraulic valve (typically a rectilinear spool valve as described in chapter 7) that controls the hydraulic actuator (typically a piston-cylinder device) is driven by a stepper motor through suitable gearing for speed reduction as well as for rotary–rectilinear motion conversion. A block diagram for this arrangement is shown in figure 6.38. Torque amplification by an order

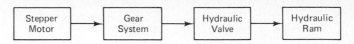

Figure 6.38. Torque amplification of a stepper motor using a hydraulic actuator.

of magnitude is possible with such an arrangement. Of course, the time constant will increase and operating bandwidth will decrease because of the sluggishness of hydraulic components. Also, a certain amount of backlash will be introduced by the gear system. Feedback control will be necessary to reduce the position error that is usually present in open-loop hydraulic actuators.

Stepper motors are incremental actuators. As such, they are ideally suited for digital control applications. High-precision open-loop operation is possible as well, provided that the operating conditions are well within the motor capacity. Early application of stepper motors was limited to low-speed, low-torque drives. With rapid developments in solid-state drives and microprocessor-based pulse generators and controllers, however, high-speed operation under transient conditions at high torques and closed-loop control have become feasible.

There are numerous applications of stepper motors. For example, a stepper motor is particularly suitable in printing applications (including graphic printers, plotters, and electronic typewriters) because the print characters are changed in steps and the printed lines (or paper feed) are also advanced in steps. Stepper motors are used extensively in magnetic tape drives and disk drives of digital computers, particularly to position the read head. In automated manufacturing applications, stepper motors are found as joint actuators and end effector (gripper) actuators of robotic manipulators and as drive units in programmable dies, parts-positioning tables, and tool holders of machine tools (milling machines, lathes, etc.). In automotive applications, pulse windshield wipers, power window drives, power seat mechanisms, automatic carburetor control, process control applications, valve actuators, and parts-handling systems use stepper motors. Other applications of stepper motors include source and object positioning in medical and metallurgical radiography, lens drives in auto-focus cameras, camera movement in computer vision systems, and paper feed mechanisms in photocopying machines.

The advantages of stepper motors include the following:

1. Position error is noncumulative. A high accuracy of motion is possible, even under open-loop control.
2. Large savings in sensor (measuring system) and controller costs are possible when the open-loop mode is used.
3. Because of the incremental nature of command and motion, stepper motors are easily adoptable to digital control applications.
4. No serious stability problems exist, even under open-loop control.
5. Torque capacity and power requirements can be optimized and the response can be controlled by electronic switching.
6. Brushless construction has obvious advantages (see chapter 7).

The disadvantages of stepper motors include the following:

1. They have low torque capacity (typically less than 2,000 oz.in.) compared to DC torque motors.
2. They have limited speed (limited by torque capacity and by pulse-missing problems due to faulty switching systems and drive circuits).
3. They have high vibration levels due to stepwise motion.
4. Large errors and oscillations can result when a pulse is missed under open-loop control.

In most applications, the merits of stepper motors outweigh the drawbacks.

PROBLEMS

6.1. Explain why a two-phase variable-reluctance stepper motor is not a physical reality. A single-stack variable-reluctance stepper motor with nontoothed poles has n_r teeth in the rotor, n_s poles in the stator, and p phases of winding. Show that

$$n_r = \left(1 + \frac{1}{p}\right) n_s \quad (\text{for } n_r > n_s)$$

6.2. For a single-stack stepper motor that has toothed poles, show that, for the case $\theta_r > \theta_s$,

$$\Delta\theta = \frac{n_s}{mp}(\theta_r - \theta_s)$$

$$\theta_r = \theta_s + \frac{m\theta_r}{n_s}$$

$$n_s = n_r + m$$

where $\Delta\theta$ = step angle
θ_r = rotor tooth pitch
θ_s = stator tooth pitch
n_r = number of teeth in the rotor
n_s = number of teeth in the stator
p = number of phases
m = number of stator poles per phase

Assume that the stator teeth are uniformly distributed around the rotor. What are the corresponding equations for the case $\theta_r < \theta_s$?

6.3. For a stepper motor with m stator poles per phase, show that the number of teeth in a stator pole is given by

$$\ell_s = \frac{n}{mp^2} - \frac{1}{p}$$

where n denotes the number of steps per revolution, for the case $n_r > n_s$. (*Hint:* This relation is the counterpart of equation 6.11 for the case $n_r > n_s$.) Pick suitable parameters for a four-phase, eight-pole motor, using this relation, if the step angle is required to be 1.8°. Can the same step be obtained using a three-phase motor?

6.4. Consider the single-stack, three-phase VR stepper motor shown in figure 6.10 ($n_r = 8$ and $n_s = 12$). For this arrangement, compare the following phase-switching sequences:

1-2-3-1

1-12-2-23-3-31-1

12-23-31-12

What is the step angle, and how would you reverse the direction of rotation in each case?

6.5. For a multiple-stack variable-reluctance stepper motor whose rotor tooth pitch angle is not equal to the stator tooth pitch angle (i.e., $\theta_r \neq \theta_s$), show that the step angle may be expressed by

$$\Delta\theta = \frac{\theta_r}{ps}$$

where p = number of phases in each stator segment
s = number of stacks of rotor teeth on the shaft

6.6. Describe the principle of operation of a single-stack VR stepper motor that has toothed poles in the stator. Assume that the stator teeth are uniformly distributed around the rotor. If the motor has five teeth per pole and two pole pairs per phase and provides 500 full steps per revolution, determine the number of phases in the stator. Also determine the number of stator poles, the step angle, and the number of teeth in the rotor.

6.7. Describe the principle of operation of a multiple-stack VR stepper motor that has toothed poles in each stator stack. Show that if $\theta_r > \theta_s$, the step angle of this type of motor is given by

$$\Delta\theta = \frac{n_s}{mps}(\theta_r - \theta_s)$$

where θ_r = rotor tooth pitch
θ_s = stator tooth pitch
n_s = number of teeth in the stator
p = number of phases
m = number of poles per phase
s = number of stacks

Assume that the stator teeth are uniformly distributed around the rotor. What is the corresponding equation if $\theta_r < \theta_s$?

6.8. When a phase winding of a stepper motor is switched on, ideally the current in the winding should instantly reach the full value (hence providing the full magnetic field instantly). Similarly, when a phase is switched off, its current should become zero immediately. It follows that the ideal shape of phase current history is a rectangular pulse sequence, as shown in figure P6.8. In actual practice, however, the pulses deviate from the ideal rectangular shape, primarily because of magnetic induction in the phase windings. Using sketches, indicate how the phase current waveform would deviate from this ideal shape under the following conditions:
(a) Very slow stepping
(b) Very fast stepping at a constant stepping rate
(c) Very fast stepping at a variable (transient) stepping rate
A stepper motor has a phase inductance of 10 mH and a phase resistance of 5 Ω. What is the electrical (L/R) time constant of each phase in a stepper motor? Estimate the stepping rate below which magnetic induction effects can be neglected so that the phase current waveform is almost a rectangular pulse sequence.

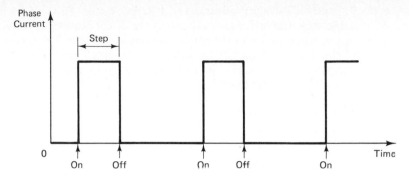

Figure P6.8. Ideal phase current waveform for a stepper motor.

6.9. Consider a stepper motor that has two poles per phase. The pole windings in each phase may be connected either in parallel or in series, as shown in figure P6.9. In each case, determine the required ratings for phase power supply (rated current, rated voltage, rated power) in terms of current i and resistance R, as indicated in figure P6.9a. Note that the power rating should be the same for both cases, as is intuitively clear.

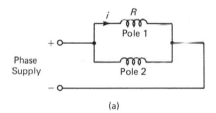

(a)

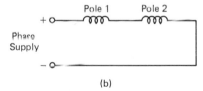

(b)

Figure P6.9. Pole windings in a phase of a stepper motor that has two poles per phase: (a) parallel connection; (b) series connection.

6.10. Define and compare the following pairs of terms in the context of electromagnetic stepper motors:

(a) Pulses and steps
(b) Step angle and resolution
(c) Residual torque and static holding torque
(d) Translator and drive system
(e) PM stepper motor and VR stepper motor
(f) Single-stack stepper and multiple-stack stepper
(g) Stator poles and stator phases
(h) Pulse rate and slew rate

6.11. A stepper motor misses a pulse during slewing (high-speed stepping at a constant rate in steady state). Using a displacement versus time curve, explain how a logic controller could compensate for this error by injecting a special switching sequence.

6.12. A Lanchester damper is attached to a stepper motor. Write equations to describe the single-step response of the motor about the detent position. Assume that the flywheel of the damper is not locked onto its housing at any time. Let T_d denote the magnitude of the frictional torque of the damper. Give appropriate initial conditions. Using a computer simulation, plot the motor response, with and without the damper, for the following parameter values:

Rotor + load inertia J_m = 10.0 lb.in.^{2}
Damper housing inertia J_h = 0.5 lb.in.^{2}
Damper flywheel inertia J_d = 2.0 lb.in.^{2}
Maximum overshoot $\theta(0)$ = 1.0°
Static torque at maximum overshoot = 200 oz.in.
Damping ratio of the motor when the Lanchester damper is disconnected ζ = 0.2
Magnitude of the frictional torque T_d = 50 oz.in.

6.13. Compare and contrast the three electronic damping methods illustrated in figures 6.25 through 6.27. In particular, address the issue of effectiveness in relation to the speed of response and the level of final overshoot.

6.14. In the pulse reversal method of electronic damping, suppose that Phase 1 is energized, instead of Phase 3, at point B in figure 6.26. Sketch the corresponding static torque curve and the motor response. Compare this new method with the pulse reversal method illustrated in figure 6.26.

5.15. A relatively convenient method of electronic damping uses simultaneous multiphase energization, some of the phases being excited with a fraction of the normal operating voltage. A two-phase energization technique has been suggested for a three-phase, single-stack stepper motor. If the standard sequence of switching for forward motion is given by 1-2-3-1, what is the corresponding two-phase energization sequence?

6.16. Briefly discuss the operation of a microprocessor-controlled stepper motor. How would it differ from the standard setup in which a "preset indexer" is employed? Compare and contrast table lookup, programmed stepping, and hardware stepping methods for stepper motor translation.

6.17. Using a static torque diagram, indicate the locations of the first two encoder pulses for a feedback encoder–driven stepper motor for steady deceleration.

6.18. Suppose that the torque produced by a stepping motor when one of the phases is energized can be approximated by a sinusoidal function with amplitude T_{max}. Show that with the advanced switching sequence shown in figure 6.31, the average torque generated is approximately $0.8T_{max}$. What is the average torque generated with conventional switching?

6.19. The torque-speed curve of a stepper motor is approximated by a straight line. The following two parameters are given:

T_0 = torque at zero speed (holding torque)
ω_0 = speed at zero torque (no-load speed)

Suppose that the load resistance is approximated by a rotary viscous damper with damping constant b. Assuming that the motor directly drives the load, without any speed reducers, determine the steady-state speed of the load and the corresponding drive torque of the stepper motor.

6.20. The speed-torque curve of a stepper motor is shown in figure P6.20. Explain the shape of this curve. Suppose that with one phase on, the torque of a stepper motor in the neighborhood of the detent position of the rotor is given by the linear relationship

$$T = -K_m \theta$$

where θ is the rotor displacement measured from the detent position and K_m is the motor torque constant. The motor is directly coupled to an inertial load. The combined moment of inertia of the motor rotor and the inertial load is $J = 0.01$ kg.m^2. If $K_m = 628.3$ N.m/rad, at what stepping rates would you expect dips in the speed-torque curve of the motor-load combination?

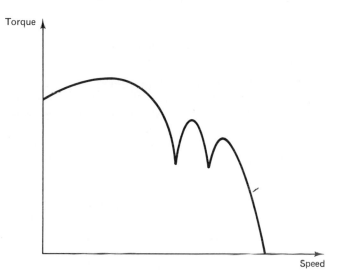

Torque

Speed

Figure P6.20. Typical speed torque curve of a stepper motor.

6.21. Compare the VR stepper motor with the PM stepper motor with respect to the following considerations:
(a) Torque capacity for a given motor size
(b) Holding torque
(c) Complexity of switching circuitry
(d) Step size
(e) Rotor inertia
The hybrid stepper motor possesses characteristics of both the VR and the PM types of stepper motors. Consider a typical construction of a hybrid stepper motor, as shown schematically in figure P6.21. The rotor has two stacks of teeth made of ferromagnetic material, joined together by a permanent magnet, that assign opposite polarities to the two rotor stacks. The tooth pitch is the same for both stacks, but the two stacks have a tooth misalignment of half a tooth pitch $(\theta_r/2)$. The stator may consist of a common tooth stack for both rotor stacks, or it may consist of two tooth stacks that are in complete alignment. The number of teeth in the stator is not equal to the number of teeth in each rotor stack. Note that the stator is made up of several toothed poles that are equally spaced around the rotor. Half the poles are connected to one phase and the other half are connected to the second phase. The current in each phase may be turned on and off or reversed using switching circuitry. The switching sequence for rotation in one direction would be 1^+-2^+-1^--2^-; for rotation in the opposite direction, it would be

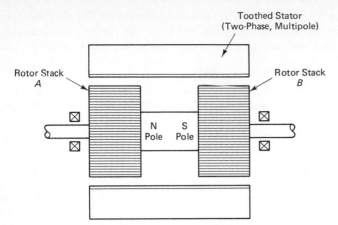

Toothed Stator
(Two-Phase, Multipole)

Rotor Stack
A

Rotor Stack
B

N
Pole

S
Pole

Figure P6.21. Schematic diagram of a hybrid stepper motor.

1^+-2^--1^--2^+, where 1 and 2 denote the two phases and the superscripts $+$ and $-$ denote the direction of current in each phase.

(a) Using a motor that has eighteen teeth in each rotor stack and eight poles in the stator, with two teeth per pole, describe the operation of the motor.

(b) What is the step size of the motor?

6.22. A lectern (or podium) in an auditorium is designed to adjust its height automatically, depending on the height of the speaker. A sonar device measures the height of the speaker and sends a command to the logic hardware controller of a stepper motor that adjusts the lectern vertically through a rack-and-pinion drive. The dead load of the moving parts is supported by a bellows device. A schematic diagram of this arrangement is shown in figure P6.22. The following design requirements have been specified:

Time to adjust a maximum stroke of 1 m = 5 s

Mass of the lectern = 50 kg

Maximum resistance to vertical motion = 5 kg

Displacement resolution = 0.5 cm/step

Select a suitable stepper motor system for this application.

6.23. Using the sinusoidal approximation for static torque in a three-phase variable-reluctance stepper motor, the torques T_1, T_2, and T_3 due to the three phases (1, 2, and 3) activated separately may be expressed as

$$T_1 = -T_{max} \sin n_r \theta$$

$$T_2 = -T_{max} \sin\left(n_r\theta - \frac{2\pi}{3}\right)$$

$$T_3 = -T_{max} \sin\left(n_r\theta - \frac{4\pi}{3}\right)$$

where θ = angular position of the rotor measured from the detent position of Phase 1
n_r = number of rotor teeth

Using the trigometric identity

$$\sin A + \sin B = 2 \sin\left(\frac{A + B}{2}\right) \cos\left(\frac{A - B}{2}\right)$$

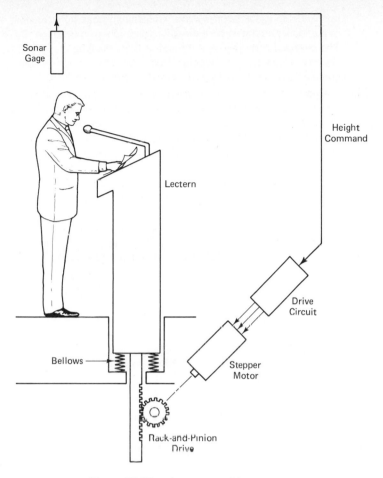

Figure P6.22. An automated lectern.

show that

$$T_1 + T_2 = -T_{max} \sin\left(n_r\theta - \frac{\pi}{3}\right)$$

$$T_2 + T_3 = -T_{max} \sin\left(n_r\theta - \pi\right)$$

$$T_3 + T_1 = -T_{max} \sin\left(n_r\theta - \frac{5\pi}{3}\right)$$

Using these expressions, show that the step angle for the switching sequence 1-2-3 is $\theta_r/3$ and the step angle for the switching sequence 1-12-2-23-3-31 is $\theta_r/6$. Determine the step angle for the two-phase-on switching sequence 12-23-31.

6.24. Some industrial applications of stepper motors call for very high stepping rates under variable load (motor torque) conditions. Since motor torque depends on the current in the phase windings (typically 5 A per phase), one method of obtaining a variable-torque drive is to use an adjustable resistor in the excitation circuit. An alternative method is to use a *chopper drive*. Switchable transistors, diodes, or thyristors are used in a chopper circuit to periodically bypass (chop) the current through a phase winding. The chopped current passes through a free-wheeling diode into the power supply. The chopping interval and chopping frequency are adjustable. Discuss the advantages of chopper drives compared to the resistance drive method.

6.25. A stepper motor with rotor inertia J_m drives a free (no-load) gear train, as shown in figure P6.25. The gear train has two meshed gear wheels. The gear wheel attached to the motor shaft has inertia J_1, and the other gear wheel has inertia J_2. The gear train steps down the motor speed by the ratio $1 : r (r < 1)$. One phase of the motor is energized, and once the steady state is reached, the gear system is turned slightly from the corresponding detent position and released.
(a) Explain why the system will oscillate about the detent position.
(b) What is the natural frequency of oscillation (neglecting mechanical damping) in radians per second?
(c) What is the significance of this frequency in a control system that uses a stepper as the actuator?
(*Hint:* Static torque for the stepper motor may be taken as

$$T = -T_{max} \sin\left(\frac{2\pi\theta}{p.\Delta\theta}\right)$$

with the usual notation.)

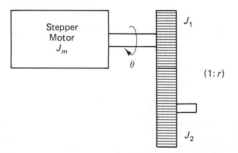

Figure P6.25. A stepper motor–driven gear system.

6.26. Figure P6.26 shows a schematic diagram of a stepper motor. Describe the operation of this motor. What is the step angle of the motor
(a) In full-stepping?
(b) In half-stepping?

6.27. So far, in the problems on toothed single-stack stepper motors, we have assumed that $\theta_r \neq \theta_s$. Now consider the case of $\theta_r = \theta_s$. In a single-stack stepper motor, the necessary stator–rotor tooth misalignment to generate the driving torque is achieved by offsetting the entire group of teeth on a stator pole (not just the central tooth of the pole) by the step angle $\Delta\theta$ with respect to the teeth on the adjacent pole. There are two possibilities. Equations 6.13 and 6.14 govern the case in which the offset is generated by reducing the pole pitch. What are the corresponding equations if the offset of $\Delta\theta$ is realized by increasing the pole pitch? Show that in this case (unlike the case in example 6.3), it is possible to design a four-phase motor that has fifty rotor teeth. Obtain appropriate values for tooth pitch (θ_r and θ_s), full-stepping step angle $\Delta\theta$, number of steps per revolution (n), number of poles per phase (m), and number of stator teeth per pole (ℓ_s) in this design.

6.28. The stepper motor shown in figure 6.11 uses the *balanced pole arrangement*. Specif-

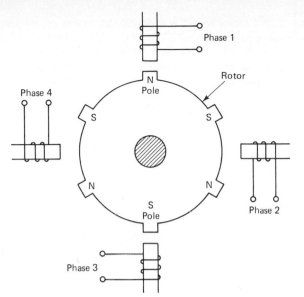

Phase 1

Rotor

N
Pole

Phase 4

S

S

N

N

S
Pole

Phase 2

Phase 3

Figure P6.26. Schematic diagram of a stepper motor.

ically, all the poles wound to the same phase are uniformly distributed around the rotor. In figure 6.11, there are two poles per phase. Hence, the two poles connected to the same phase are placed at diametrically opposite locations. In general, in the case of m poles per phase, the poles connected to the same phase would be located at angular intervals of $360°/m$. What are the advantages of the balanced pole arrangement?

6.29. In connection with phase windings of a stepper motor, explain the following terms:
  (a) Unifilar (or monofilar) winding
  (b) Bifilar winding
  (c) Bipolar winding
  Discuss why the torque characteristics of a bifilar-wound motor are better than those of a unifilar-wound motor at high stepping rates.

6.30. The torque source model may be used to represent both types of stepper motors (VR and PM) at low speeds and under steady operating conditions. What assumptions are made in this model? A stepper motor has an inertial load rigidly connected to its rotor. The equivalent moment of inertia of rotor and load is $J = 5.0 \times 10^{-3}$ kg.m^2. The equivalent viscous damping is $b = 0.5$ N.m/rad/s. The number of phases $p = 4$, and the number of rotor teeth $n_r = 50$. Assume full-stepping (step angle $= 1.8°$). The mechanical model for the motor is

$$T = b\dot{\bar{\theta}} + J\ddot{\bar{\theta}}$$

where $\bar{\theta}$ is the absolute position of the rotor.
  (a) Assuming the torque source model (equation 6.37 or 6.38) with $T_{max} = 100$ N.m, simulate and plot the motor response $\bar{\theta}$ as a function of t for the first ten steps, starting from rest. Assume that in open-loop control, switching is always at the detent position of the present step. You should be careful of the position coordinate, because $\bar{\theta}$ is the absolute position from the starting point and θ is the relative position measured from the approaching detent position of the current step. Plot the response on the phase plane (with speed $\dot{\bar{\theta}}$ as the vertical axis and position $\bar{\theta}$ as the horizontal axis).
  (b) Repeat part (a) for the first 150 steps of motion. Check whether a steady state (speed) is reached or whether there is an unstable response.

(c) Consider the improved PM motor model with torque due to one excited phase being given by equations 6.40 through 6.43, with $R = 2.0 \, \Omega$, $L_0 = 10.0$ mH, $L_a = 2.0$ mH, $k_b = 0.05$ V/rad/s, $v_p = 20.0$ V, and $k_m = 10.0$ N.m/A. Starting from rest and switching at each detent position, simulate the motor response for the first ten steps. Plot $\overline{\theta}$ versus t to the same scale as in part (a). Also, plot the response on the phase plane to the same scale as in part (a). Note that at each switching point, the initial condition of the phase current i_p is zero. Simulation should be done by picking about 100 integration steps for each motor step. In each integration step, first for known θ and $\dot{\theta}$, integrate equation 6.40 along with equations 6.41 and 6.42 to determine i_p. Substitute this in equation 6.43 to compute torque T for the integration step. Then use this torque and integrate the mechanical equation to determine $\overline{\theta}$ and $\dot{\overline{\theta}}$. Repeat this for the subsequent integration steps. After the detent position is reached, repeat the integration steps for the new phase, with zero initial value for current, but using $\overline{\theta}$ and $\dot{\overline{\theta}}$, as computed before, as the initial values for position and speed. Note that $\dot{\theta} = \dot{\overline{\theta}}$.

(d) Repeat part (c) for the first 150 motor steps. Plot the curves to the same scale as in part (b).

(e) Repeat parts (c) and (d), this time assuming a VR motor with torque given by equation 6.44 and $k_r = 1.0$ N.m/A². The rest of the model is the same as for the PM motor.

(f) Suppose that the fifth pulse did not reach the translator. Simulate the open-loop reasponse of the three motor models during the first ten steps of motion. Plot the response of all three motor models (torque soruce, PM, and VR) to the same scale as before. Give both the time history response and the phase plane trajectory for each model.

(g) Suppose that the fifth pulse was generated and translated but the corresponding phase was not activated. Repeat part (f) under these conditions.

(h) If the rotor position is measured, the motor can be accelerated back to the desired response by properly choosing the switching point. Note that the switching point for maximum average torque is the point of intersection of the two adjacent torque curves, not the detent point. Simulate the response under a feedback control scheme of this type to compensate for the missed pulse in parts (f) and (g). Plot the controlled responses to the same scale as for the earlier results. Note that each simulation should be done for all three motor models and the results should be presented as a time history as well as a phase plane trajectory. Also, both pulse losing and phase losing should be simulated in each case. Explain how the motor response would change if the mechanical dissipation were modeled by coulomb friction rather than by viscous damping.

6.31. Piezoelectric stepper motors (Kumada, 1987) are actuators that convert vibrations in a piezoelectric element (say, PZT) generated by an AC voltage (reverse piezoelectric effect—see chapter 3) into rotary motion. Step angles on the order of 0.001° can be obtained by this method. Figure P6.31a shows a piezoelectric stepper motor built at Cambridge University by D. Crawley, G. Amaratunga, and J. Grundy, using Kumada's design. A disassembled view of the motor is shown in figure P6.31b. As the piezoelectric PZT rings vibrate due to an applied AC voltage, vibrations are produced in the conical aluminum disk. An aluminum coupler piece that has two protrusions (feet) on one side and a radially machined beam on the other side is positioned as shown, so that the feet are in contact with the conical disk. Because of this contact of the coupler disk at two ends of a diameter, the disk undergoes radial bending vibrations (contact forces are intermittent), and these vibrations impart twisting (torsional) vibrations onto the beam element. The twisting motion is subsequently converted into a ro-

(a)

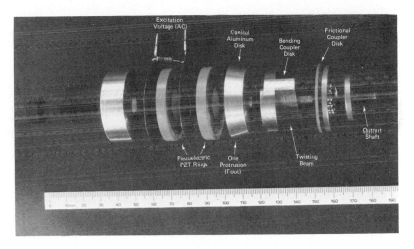

(b)

Figure P6.31. Piezoelectric stepper motor: (a) assembled view; (b) components.

tary motion of the frictional disk, which is frictionally coupled with the top surface of the beam. Essentially, because of the twisting motion, the two top edges of the beam push the frictional disk tangentially in a stepwise manner. This forms the output member of the piezoelectric stepper motor. List several advantages and disadvantages of this motor. Describe an application in which a miniature stepper motor of this type could be used.

REFERENCES

ACARNLEY, P. P. *Stepping Motors: A Guide to Modern Theory and Practice*. Peter Peregrinus, Ltd., Stevenage, England, 1982.

CHI, H. D. "Permeance-Based Step Motor Model Revisited." In *Proceedings of the Fourteenth Annual Symposium on Incremental Motion Control Systems and Devices,* Incremental Motion Control Systems Society, Champaign, Ill., June 1985, pp. 399–410.

DESILVA, C. W. "Heuristic Feedback Control of Variable-Reluctance Stepping Motors." In *Proceedings of the Eighteenth Annual Modeling and Simulation Conference,* Instrument Society of America, 18(4): April 1987, 1389–1393.

KENJO, T. *Stepping Motors and Their Microprocessor Controls*. Clarendon Press, Oxford, England, 1984.

KOZUCHOWSKI, L., and STEVENS, H. G. "Stepping Motors." *Engineering,* May 1979, pp. i–viii.

KUMADA, A. "Ultrasonic Motor Using Bending Longitudinal, and Torsional Vibrations." U.S. Patent No. 4642509, February 1987.

KUO, B. C. (Ed). *Theory and Applications of Step Motors*. West, St. Paul, Minn., 1974.

LEENHOUTS, A. C. "Stepping Motors in Industrial Motion Control." In *Proceedings of the Joint Automatic Control Conference,* WP 10-A, San Francisco, June 1980.

7

Continuous-Drive Actuators

7.1 INTRODUCTION

As discussed in the preceding chapter, stepper motors can be classified as digital actuators, which are pulse-driven devices. Each pulse received at the driver of a digital actuator causes the actuator to move by a predetermined, fixed increment of displacement.

In the early days of analog control, servo actuators were exclusively continuous-drive devices. Since the control signals in this early generation of (analog) control systems generally were not discrete pulses, the use of pulse-driven digital actuators was not feasible in those systems. Direct current (DC) servomotors and servovalve-driven hydraulic and pneumatic actuators were the most widely used types of actuators in industrial control systems, particularly because digital control was not available. Furthermore, the control of alternating current (AC) actuators was a difficult task at that time. Today, AC motors are also widely used as servomotors, employing modern methods of phase voltage control and frequency control through microelectronic drive systems and using field feedback compensation through digital signal processing (DSP) chips. It is interesting to note that actuator control using pulse signals is no longer limited to digital actuators. Pulse width–modulated (PWM) signals are increasingly being used to drive continuous (analog) actuators such as DC servomotors, hydraulic servos, and AC motors. Furthermore, it should be pointed out that electronic-switching commutation in DC motors is quite similar to the method of phase switching used in driving stepper motors.

Although the cost of sensors and transducers is a deciding factor in low-power applications and in situations where precision, accuracy, and resolution are of primary importance, the cost of actuators can become crucial in moderate-to-high-power control applications. It follows that the proper design and selection of actuators can have a significant economical impact in many applications of industrial control.

This chapter will discuss the principles of operation, mathematical modeling, analysis, characteristics, performance evaluation, and methods of control of the more common types of continuous-drive actuators used in control applications. In particular, DC motors, AC induction motors, AC synchronous motors, and hydraulic and pneumatic actuators will be considered.

7.2 DC MOTORS

The DC motor converts direct current (DC) electrical energy into rotational mechanical energy. A major fraction of the torque generated in the rotor (armature) of the motor is available to drive an external load. The DC motor is probably the earliest form of electric motor. Because of features such as high torque, speed controllability over a wide range, portability, well-behaved speed-torque characteristics, and adaptability to various types of control methods, DC motors are still widely used in numerous control applications including robotic manipulators, tape transport mechanisms, disk drives, phonographs, machine tools, and servovalve actuators.

The principle of operation of a DC motor is illustrated in figure 7.1. Consider an electric conductor placed in a steady magnetic field at right angles to the direction of the field. Flux density B is assumed constant. If a DC current is passed through the conductor, the magnetic flux due to the current will loop around the conductor, as shown in the figure. Consider a plane through the conductor, parallel to the direction of flux of the magnet. On one side of this plane, the current flux and the field flux are additive; on the opposite side, the two magnetic fluxes oppose each other. As a result, an imbalance magnetic force F is generated on the conductor, normal to the plane. This force is given by

$$F = Bi\ell \tag{7.1}$$

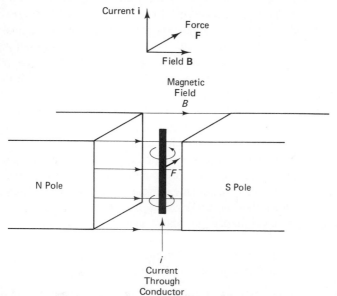

Figure 7.1. Operating principle of a DC motor.

where B = flux density of the original field
i = current through the conductor
ℓ = length of the conductor.

Note that if the field flux is not perpendicular to the length of the conductor, it can be resolved into a perpendicular component that generates the force and to a parallel component that is inactive. The active components of i, B, and F are mutually perpendicular and form a right-hand triad, as shown in figure 7.1. Alternatively, in the vector representation of these three quantities, the vector **F** can be interpreted as the cross product of the vectors **i** and **B**.

If the conductor is free to move, the force will move it at some velocity v in the direction of the force. As a result of this motion in the magnetic field B, a voltage is induced in the conductor. This is known as the back electromotive force, or *back e.m.f.*, and is given by

$$v_b = B\ell v \tag{7.2}$$

By Lenz's law, it follows that the flux due to the back e.m.f. v_b will be opposing the flux due to the original current through the conductor, thereby trying to stop the motion. This is the cause of electrical damping in motors, which we shall discuss later. Equation 7.1 determines the armature torque (motor torque), and equation 7.2 determines the motor speed.

In industrial DC motors, the field flux is usually generated not by a permanent mangnet but electrically in the stator windings—by an electromagnet, as shown schematically in figure 7.2a. Stator poles are constructed from ferromagnetic sheets (i.e., a *laminated construction*). The rotor is typically a laminated cylinder made from a ferromagnetic material. A ferromagnetic core helps concentrate the magnetic flux toward the rotor. The lamination reduces the problem of magnetic hysteresis and limits the generation of eddy currents and associated dissipation within the ferromagnetic material. The rotor has many closely spaced slots on its periphery. These slots carry the rotor windings. The stator windings are powered by the supply voltage v_f, and the rotor windings (armature windings) are powered by the supply voltage v_a.

The rotor in a conventional DC motor is called the *armature* (voltage supply to the armature windings is denoted by v_a). This nomenclature is particularly suitable for electric generators; windings in which the useful voltage is induced (generated) are termed armature windings. According to this nomenclature, armature windings in an AC machine are located in the stator, not in the rotor. Stator windings in a conventional DC motor are termed *field windings*. In an electric generator, the armature moves relative to the magnetic field of the field windings, generating the useful voltage output. In synchronous AC machines, the field windings are the rotor windings. Note that a DC motor may have more than two stator poles and far more conductor slots than what is shown in figure 7.2a. For example, some rotors carry more than 100 conductor slots.

Commutation

In order to maintain the direction of torque in each conductor group (numbered 1, 2, 3, 1′, 2′, and 3′ in figure 7.2a), the direction of current in a conductor has to

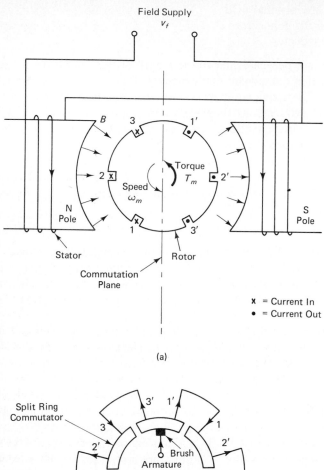

(a)

x = Current In
• = Current Out

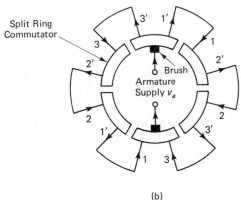

(b)

Figure 7.2. (a) Schematic diagram of a DC motor. (b) Commutator wiring.

change as the conductor crosses the commutation plane. Physically, this may be accomplished by using a split ring and brush commutator, shown schematically in figure 7.2b. The armature voltage is applied to the rotor windings through a pair of carbon brushes that maintain sliding contact with the split ring. The split ring segments, equal in number to the conductor slots in the rotor, are electrically insulated from one another, but the adjacent segments are connected through the armature windings in each opposite pair of rotor slots, as shown. For the rotor position shown in figure 7.2, note that when the split ring rotates in the counterclockwise direction through 30°, the current paths in conductors 1 and 1' reverse but the remaining current paths are unchanged, thus achieving the required commutation. Mechanically, this is possible because the split ring is rigidly mounted on the rotor shaft, as shown in figure 7.3.

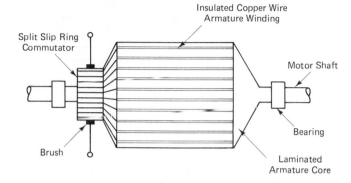

Figure 7.3. Physical configuration of the rotor of a DC motor.

7.3 BRUSHLESS DC MOTORS

In chapter 3, we noted several shortcomings of the slip ring and brush mechanisms used for current transmission through moving members. The main disadvantages include rapid wearout, mechanical loading and heating due to brush friction, contact bounce, excessive noise, and electrical arcing, with the associated hazards in chemical environments, problems of oxidation, and voltage ripples at switching points. Conventional remedies to these problems—such as the use of improved brush designs and modified brush positions to reduce arcing—are inadequate in some applications.

Brushless DC motors are permanent-magnet rotor DC motors that use electronic switching of the current in the stator winding segments to accomplish commutation. In this context, the stator windings of a brushless DC motor can be considered the armature windings. In concept, brushless DC motors are somewhat similar to stepper motors (see chapter 6) and to some types of AC motors. By definition, a DC motor should use a DC supply to power the motor. Furthermore, the torque-speed characteristics of a DC motor are different from those of a stepper motor or AC motor.

We have noted (see figure 7.2) that in conventional DC motors, the magnetic polarity of the stator is fixed and the polarity of the rotor is switched mechanically to obtain the proper direction of motor torque. In brushless DC motors, the polarity of the rotor (permanent magnet) is fixed relative to the rotor itself, and the polarity of the stator is switched by electronic means to achieve the same objective.

Figure 7.4 schematically shows a brushless DC motor and associated commutation circuitry. The rotor is a multiple-pole permanent magnet. Permanent magnets made from rare-earth material are costly but suitable when high torque is required, as in torque motors. Ferrite magnets are more economical but less efficient. The popular two-pole rotor design consists of a diametrically magnetized cylindrical magnet, as shown in figure 7.4. The stator windings are distributed in segments of

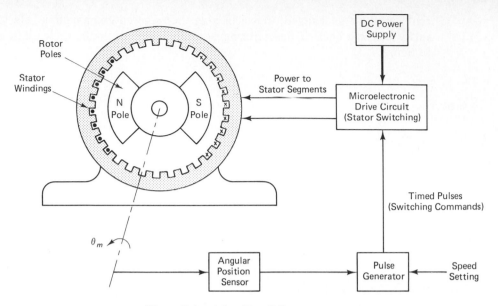

Figure 7.4. A brushless DC motor system.

winding groups around the stator. Each winding segment has a separate supply lead. Figure 7.4 shows a four-segment stator. Two diametrically opposite segments are connected together so that they carry current simultaneously but in opposite directions. Commutation is accomplished by energizing each pair of diametrically opposite segments sequentially, at time instants determined by the rotor position. This commutation could be achieved through mechanical means, using a multiple contact switch driven by the motor itself. Such a mechanism would defeat the purpose, however, because it has most of the drawbacks of regular commutation using split rings and brushes. Modern brushless motors almost exclusively use microelectronic switching for commutation. For constant-speed operation, open-loop switching may be used. In this case, speed setting is provided as the input to a timing pulse generator. It generates a pulse sequence starting at zero pulse rate and increasing (ramping) to the final rate that corresponds to the speed setting. Each pulse causes the driver circuit, which has proper switching circuitry, to energize a pair of stator segments. In this way, the input pulse signal activates the stator segments sequentially, thereby generating a stator field that rotates at a speed that is determined by the pulse rate. This rotating magnetic field would accelerate the rotor to its final speed. A separate command (or a separate pulse) is needed to reverse the direction of rotation, which is accomplished by reversing the switching sequence.

   Under transient motions, it is necessary to know the actual position of the rotor for accurate switching. An angular position sensor (e.g., a shaft encoder or a Hall effect sensor) may be used for this purpose, as shown in figure 7.4. By switching the stator segments at the proper instants, it is possible to maximize the motor torque. To understand this, consider a brushless DC motor that has two rotor poles and four stator winding segments. Let us number the stator segments as in figure 7.5a and define the rotor angle θ_m. The typical shape of the static torque curve of the motor

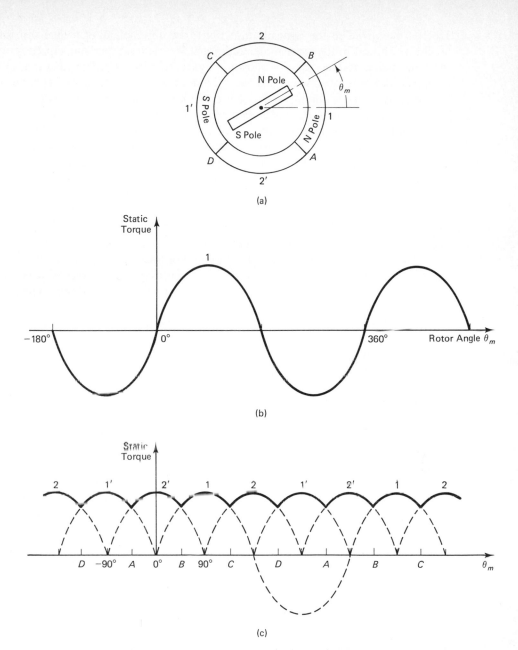

Figure 7.5. (a) A brushless DC motor. (b) Static torque curve with no switching (one stator segment energized). (c) Switching sequence for maximum average torque.

when segment 1 is energized (segment 1′ being automatically energized in the opposite direction) is shown in figure 7.5b as a function of θ_m. The torque distribution when segment 2 is energized would be identical, but it would be shifted to the right through 90°. Similarly, if segment 1′ is energized in the positive direction (segment

1 being energized in the opposite direction), the corresponding torque distribution would be shifted to the right by an additional 90°, and so on. The superposition of these individual torque curves is shown in figure 7.5c. It should be clear that to maximize the motor torque, switching has to be done at the points of intersection of the torque curves corresponding to the adjacent stator segments (as for stepper motors). These switching points are indicated as A, B, C, and D in figure 7.5c. Under transient motions, position measurement (e.g., using a shaft encoder) would be required to determine these switching positions accurately. An effective solution is to locate Hall effect sensors at switching points (fixed) around the stator and use the voltage pulses generated in these sensors when a rotor pole passes each sensor to activate switching of the field windings. Note from figure 7.5c that positive average torque is possible even if the switching positions are shifted from these ideal locations by less than 90° to either side. It follows that the motor torque can be controlled by adjusting the switching locations with respect to the actual position of the rotor.

The smoothness and magnitude of the motor torque, the accuracy of operation, and motor controllability can be improved by increasing the number of winding segments in the stator. This increases the number of power lines and the complexity of the commutation circuitry, however.

Brushless DC motors are used in general-purpose applications as well as in servo systems. Motors in the range 0.1–1 hp, operating at speeds up to 7,200 rpm, are used in computer peripherals and as positive displacement pump drives.

7.4 DC MOTOR EQUATIONS

Equivalent circuits for the stator and the rotor of a conventional DC motor are shown in figure 7.6a. Since the field flux is proportional to field current i_f, we can express the magnetic torque of the motor as

$$T_m = k i_f i_a \tag{7.3}$$

which directly follows from equation 7.1. Next, in view of equation 7.2, the back e.m.f. in the armature of the motor is given by

$$v_b = k' i_f \omega_m \tag{7.4}$$

The following notation has been used:

i_f = field current

i_a = armature current

ω_m = angular speed of the motor

and k and k' are motor constants that depend on factors such as the rotor dimensions, the number of turns in the armature winding, and the permeability (inverse of reluctance) of the magnetic medium. Note that in the case of ideal electrical-to-mechanical energy conversion at the rotor, we have $T_m \omega_m = v_b i_a$ with consistent units (e.g., torque in Newton-meters, speed in radians per second, voltage in volts, and current in amperes). Then we note that the constant k and k' are identical.

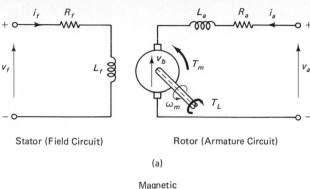

Stator (Field Circuit)   Rotor (Armature Circuit)

(a)

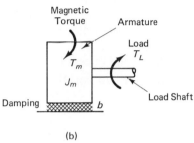

(b)

Figure 7.6. (a) The equivalent circuit of a conventional DC motor (separately excited). (b) Armature mechanical loading diagram.

The field circuit equation is obtained by assuming that the stator magnetic field is not affected by the rotor magnetic field (i.e., the stator inductance is not affected by the rotor) and that there are no eddy current effects in the stator; thus,

$$v_f = R_f i_f + L_f \frac{di_f}{dt}$$ (7.5)

where v_f = supply voltage to the stator
R_f = resistance of the field winding
L_f = inductance of the field winding

The equation for the armature rotor circuit is

$$v_a = R_a i_a + L_a \frac{di_a}{dt} + v_b$$ (7.6)

where v_a = supply voltage to the armature
R_a = resistance of the armature winding
L_a = leakage inductance in the armature winding

Note that the primary inductance or *mutual inductance* in the armature winding is represented in the back e.m.f. term. The leakage inductance, which is usually neglected, represents the fraction of the armature flux that is not linked with the stator and is not used in the generation of useful torque. This includes self-inductance in the armature.

The mechanical equation of the motor is obtained by applying Newton's second law to the rotor. Assuming that the motor drives a load that requires a load torque T_L to operate, and that the frictional resistance in the armature can be modeled as viscous, we have (see figure 7.6b)

$$J_m \frac{d\omega_m}{dt} = T_m - T_L - b\omega_m \qquad (7.7)$$

where J_m = moment of inertia of the rotor
b = equivalent damping constant for the rotor

Note that the load torque may be due, in part, to the inertia of the external load that is coupled to the motor shaft. If the coupling flexibility is neglected, the load inertia may be directly added to (lumped with) the rotor inertia after accounting for any speed reducers that might be present. In general, a separate set of equations is necessary to represent the dynamics of the external load.

Equations 7.3 through 7.7 form the dynamic model for a DC motor. In obtaining this model, we have made several assumptions and approximations. In particular, we have either approximated or neglected the following factors:

1. Coulomb friction and associated dead band effects
2. Magnetic hysteresis (particularly in the stator core)
3. Magnetic saturation
4. Eddy current effects
5. Nonlinear constitutive relations for magnetic induction
6. Brush contact resistance, finite-width contact of brushes, and other types of noise and nonlinearities in split ring commutators
7. The effect of the rotor magnetic flux (armature flux) on the stator magnetic flux (field flux)

Steady-State Characteristics

In selecting a motor for a given application, its steady-state characteristics are a major determining factor. In particular, steady-state torque-speed curves are employed. In the separately excited case shown in figure 7.6a, where the armature circuit and field circuit are excited by separate independent voltage sources, it can be shown that the steady-state torque-speed curve is a straight line. To verify this, we set the time derivatives in equations 7.5 and 7.6 to zero; this corresponds to steady-state conditions. It follows that i_f is constant for a fixed voltage v_f. By substituting equations 7.3 and 7.4 in 7.6, we get

$$v_a = \frac{R_a}{ki_f} T_m + k' i_f \omega_m$$

Under steady-state conditions in the field circuit, we have, from equation 7.5,

$$i_f = \frac{v_f}{R_f}$$

It follows that the steady-state torque-speed characteristics of a separately excited DC motor may be expressed as

$$\omega_m + \frac{R_a R_f^2}{kk' v_f^2} T_m = \frac{R_f v_a}{k' v_f} \qquad (7.8)$$

Now, since v_a and v_f are constant at steady state, on defining constants T_s and ω_0, the foregoing relation can be expressed as

$$\frac{\omega_m}{\omega_0} + \frac{T_m}{T_s} = 1$$

where ω_0 = no-load speed (at steady state, assuming zero damping)
T_s = stalling torque of motor

It should be noted from the preceding equation that if there is no damping ($b = 0$), the steady-state magnetic torque of the motor is equal to the load torque. In practice, however, there is mechanical damping on the rotor, and the load torque is smaller than the motor torque. In particular, the motor will stall at a load torque smaller than T_s. The idealized characteristic curve given by equation 7.8 is shown in figure 7.7.

Example 7.1

A load is driven at constant power under steady-state operating conditions, using a separately wound DC motor with constant supply voltages to the field and armature windings. Show that, in theory, two operating points are possible. Also show that one of the operating points is stable and the other one is unstable.

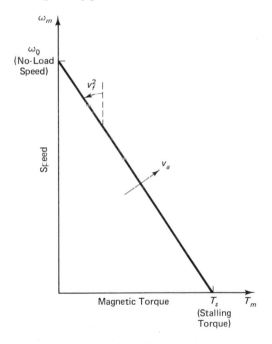

Figure 7.7. Steady-state speed-torque characteristics of a separately-wound DC motor.

Solution As shown in figure 7.7, the steady-state characteristic curve of a DC motor that has windings that are separately excited by constant voltage supplies is a straight line. The constant-power curve for the load is a hyperbola, because $T\omega_m$ = constant in this case. The two curves shown in figure 7.8 intersect at points P and Q. At point P, if there is a slight decrease in the speed of operation, the motor (magnetic) torque will increase to T_{PM} and the load torque demand will increase to T_{PL}. But since $T_{PM} > T_{PL}$, the

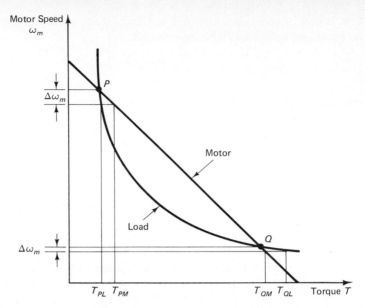

Figure 7.8. Operating points for a constant-power load driven by a DC motor.

system will accelerate back to point P. It follows that point P is a stable operating point. Alternatively, at point Q, if the speed drops slightly, the magnetic torque of the motor will increase to T_{QM} and the load torque demand will increase to T_{QL}. However, in this case, $T_{QM} < T_{QL}$. As a result, the system will decelerate further, subsequently stalling the system. Therefore, it can be concluded that point Q is an unstable operating point.

The shape of the steady-state speed-torque curve will be modified if a common voltage supply is used to excite both the field winding and the armature winding. Here, the two windings have to be connected together. Three such arrangements are

1. A shunt-wound motor
2. A series-wound motor
3. A compound-wound motor

In the shunt-wound motor, the armature winding and the field winding are connected in parallel. In the series-wound motor, they are connected in series. In the compound-wound motor, part of the field winding is connected with the armature winding in series and the other part is connected in parallel. These three winding arrangements for a DC motor are shown in figure 7.9. Note that in the shunt-wound motor at steady state, the back e.m.f. v_b depends directly on the supply voltage. Since the back e.m.f. is proportional to the speed, it follows that speed controllability is good with the shunt-wound configuration. Since the relation between v_b and the supply voltage is coupled through both the armature winding and the field winding in the series-wound motor, speed controllability is relatively poor. But in this case, a relatively large current flows through both windings at low speeds of the mo-

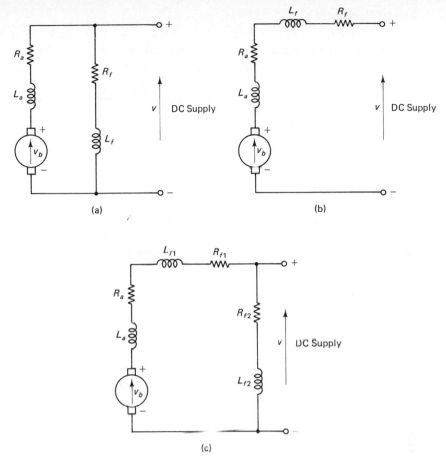

Figure 7.9. (a) A shunt wound motor. (b) A series-wound motor. (c) A compound-wound motor.

tor, giving a higher starting torque. Also, the operation is approximately constant-power in this case. These properties are summarized in table 7.1. Since both speed controllability and higher starting torque are desirable characteristics, compound-wound motors are used to obtain a performance between the two extremes.

TABLE 7.1 THE EFFECTS OF WINDING CONFIGURATION ON THE STEADY-STATE CHARACTERISTICS OF A DC MOTOR

DC motor type	Field coil resistance	Speed controllability	Starting torque
Shunt-wound	High	Good	Average
Series-wound	Low	Poor	High
Compound-wound	Parallel high, series low	Average	Average

Speed Regulation

Variation in the operating speed of a motor due to changes in the external load is measured by the percentage speed regulation. Specifically,

$$\text{percentage speed regulation} = \frac{(\omega_0 - \omega_f)}{\omega_f} \times 100\% \qquad (7.9)$$

where ω_0 = no-load speed
ω_f = full-load speed

This is a measure of the speed stability of a motor; the smaller the percentage speed regulation, the more stable the operating speed under varying load conditions (particularly with load disturbances). In the shunt-wound configuration, the back e.m.f., and hence the rotating speed, depends directly on the supply voltage. Consequently, the armature current and the related motor torque have virtually no effect on the speed. For this reason, the percentage speed regulation is relatively small for shunt-wound motors, resulting in improved speed stability.

Example 7.2

An automated guideway transit (AGT) vehicle uses a series-wound DC actuator in its magnetic suspension system. If the desired control bandwidth of the active suspension (in terms of the suspension force) is 40 Hz, what is the required minimum bandwidth for the input voltage signal?

Solution The actuating force is

$$F = ki_a i_f = ki^2$$

where i denotes the common current through both windings of the actuator. Consider a harmonic component

$$v(\omega) = v_0 \sin \omega t$$

of the input voltage to the windings, where ω denotes the frequency of the chosen frequency component. The field current is given by

$$i(\omega) = i_0 \sin(\omega t + \phi)$$

at this frequency, where ϕ denotes the phase shift. The corresponding actuating force is

$$F = ki_0^2 \sin^2(\omega t + \phi) = ki_0^2[1 - \sin(2\omega t + 2\phi)]/2$$

It follows that there is an inherent frequency doubling in the suspension system. As a result, the required minimum bandwidth for the input voltage signal is 20 Hz.

Example 7.3

Consider the three types of winding connections for DC motors, shown in figure 7.9. Derive equations for the steady-state torque-speed characteristics in the three cases. Sketch the corresponding characteristic curves. Using these curves, discuss the behavior of the motor in each case.

Solution

Shunt-Wound Motor

Note that at steady state, the inductances are not present in the motor equivalent circuit. For the shunt-wound DC motor (figure 7.9a), the field current is

$$i_f = v/R_f = \text{constant} \tag{i}$$

The armature current is

$$i_a = [v - v_b]/R_a \tag{ii}$$

The back e.m.f. for a motor speed of ω_m is given by

$$v_b = k' i_f \omega_m \tag{iii}$$

Substituting equations (i), (ii), and (iii) in the motor magnetic torque equation

$$T_m = k i_f i_a \tag{iv}$$

we get

$$\omega_m + \left(\frac{R_a R_f^2}{k k' v^2}\right) T_m = \frac{R_f}{k'} \tag{7.10}$$

Note that equation 7.10 represents a straight line with a negative slope of magnitude

$$\left[\frac{R_a R_f^2}{k k' v^2}\right]$$

Since this magnitude is typically small, it follows that good speed regulation (constant-speed operation and relatively less sensitivity of the speed to torque changes) can be obtained using shunt-wound motors. The characteristic curve for the shunt-wound DC motor is shown in figure 7.10a. The starting T_s is obtained by setting $\omega_m = 0$ in equation 7.10. The no-load speed ω_0 is obtained by setting $T_m = 0$ in the same equation. The corresponding expressions are tabulated in table 7.2. Note that if the input voltage v is increased, the starting torque will increase but the no-load speed will remain unchanged, as sketched in figure 7.10a.

Series-Wound Motor

At steady state, for the series-wound DC motor shown in figure 7.9b, the field current is equal to the armature current; thus,

$$i_a = i_f = \frac{v - v_b}{R_a + R_f} \tag{v}$$

The back e.m.f. is given by equation (iii), as before. The motor magnetic torque is given by

$$T_m = k i_f^2 \tag{vi}$$

From these relations, we get the following equation for the steady-state speed-torque relation of a series-wound motor:

$$\omega_m = \frac{v}{k'} \sqrt{\frac{k}{T_m}} - \frac{R_a + R_f}{k'} \tag{7.11}$$

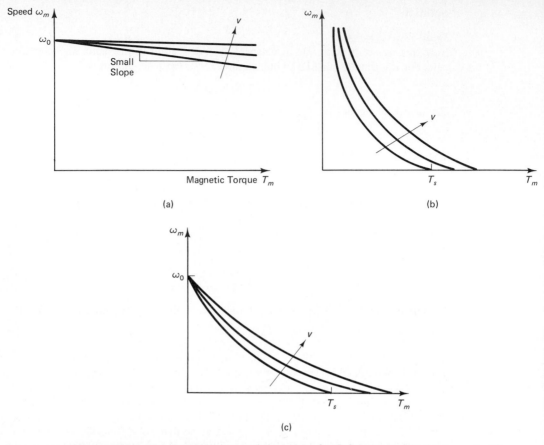

Figure 7.10. Torque-speed characteristic curves for DC motors: (a) shunt-wound; (b) series-wound; (c) compound-wound.

This equation is sketched in figure 7.10b. Note that the starting torque, as given in table 7.2, increases with the input voltage v. In the present case, the no-load speed is infinite. For this reason, the motor will coast at low loads. It follows that speed regulation in series-wound motors is poor. Starting torque and low-speed operation are satisfactory, however.

Compound-Wound Motor
Figure 7.9c gives the equivalent circuit for a compound-would DC motor. Note that part of the field winding is connected in series with the rotor winding and the other part is connected in parallel with the rotor circuit. Under steady-state conditions, the currents in the two parallel branches of the circuit are given by

$$i_a = i_{f1} = \frac{v - v_b}{R_a + R_{f1}} \tag{vii}$$

$$i_{f2} = \frac{v}{R_{f2}} \tag{viii}$$

Note that the total field current that generates the stator field is

TABLE 7.2 COMPARISON OF DC MOTOR WINDING TYPES

Winding type	No-load speed ω_0	Starting torque T_s
Shunt-wound	$\dfrac{R_f}{k'}$	$\dfrac{kv^2}{R_a R_f}$
Series-wound	∞	$\dfrac{kv^2}{(R_a + R_f)^2}$
Compound-wound	$\dfrac{R_{f2}}{k'}$	$\dfrac{kv^2}{R_a + R_{f1}}\left[\dfrac{1}{R_a + R_{f1}} + \dfrac{1}{R_{f2}}\right]$

$$i_f = i_{f1} + i_{f2} \tag{ix}$$

which, in view of equations (vii), (viii), and (iii), becomes

$$i_f = v\left[\frac{1}{R_a + R_{f1}} + \frac{1}{R_{f2}}\right] - \frac{k' i_f \omega_m}{R_a + R_{f1}}$$

Consequently,

$$i_f = v\left[\frac{1}{R_a + R_{f1}} + \frac{1}{R_{f2}}\right] \bigg/ \left[1 + \frac{k' \omega_m}{R_a + R_{f1}}\right] \tag{x}$$

The motor magnetic torque is given by

$$T_m = k i_f i_a = k i_f \frac{v - v_b}{R_a + R_{f1}} = k i_f \frac{v - k' i_f \omega_m}{R_a + R_{f1}} \tag{xi}$$

Finally, by substituting equation (x) in (xi), we get the steady-state torque-speed relationship; thus,

$$T_m = \frac{kv^2\left(\dfrac{1}{R_a + R_{f1}} + \dfrac{1}{R_{f2}}\right)\left[1 - k'\omega_m\left(\dfrac{1}{R_a + R_{f1}} + \dfrac{1}{R_{f2}}\right) \bigg/ \left(1 + \dfrac{k'\omega_m}{R_a + R_{f1}}\right)\right]}{(R_a + R_{f1})\left(1 + \dfrac{k'\omega_m}{R_a + R_{f1}}\right)} \tag{7.12}$$

This equation is sketched in figure 7.10c. The expressions for the starting torque and the no-load speed are given in table 7.2.

By comparing the foregoing results, we can conclude that good speed regulation and high starting torques are available from a shunt-wound motor, and nearly constant-power operation is possible with a series-wound motor. The compound-wound motor provides a trade-off.

The Electrical Damping Constant

Newton's second law governs the dynamic response of a motor. In equation 7.7, for example, b denotes the mechanical (viscous) damping constant and represents mechanical dissipation of energy. As is intuitively clear, mechanical damping torque opposes motion—hence the negative sign in the $b\omega_m$ term in equation 7.7. Note, further, that the magnetic torque T_m of the motor is also dependent on speed ω_m. In particular, the back e.m.f. which depends on ω_m, produces a magnetic field that tends to oppose the motion of the motor rotor. This acts as a damper, and the corresponding damping constant is given by

$$b_e = -\frac{\partial T_m}{\partial \omega_m} \tag{7.13}$$

This parameter is termed the electrical damping constant. Caution should be exercised when experimentally measuring b_e. Note that in constant-speed tests, the inertia torque of the rotor will be zero; there is no torque loss due to inertia. Torque measured at the motor shaft includes torque reduction due to mechanical dissipation (mechanical damping) within the rotor, however. Hence the magnitude of the slope of the speed-torque curve obtained by steady-state tests is equal to $b_e + b_m$, where b_m is the equivalent viscous damping constant representing mechanical dissipation at the rotor.

Example 7.4

Split-field series-wound DC motors are sometimes used as servo actuators. A motor circuit for this arrangement, under steady-state conditions, is shown in figure 7.11. The field winding is divided into two identical parts and supplied by a differential amplifier (such as a push-pull amplifier) such that the magnetic fields in the two winding segments oppose each other. In this manner, the difference in the two input voltage signals (i.e., an error signal) is employed in driving the motor. Split-field DC motors are used in low-power applications. Determine the electrical damping constant of the motor shown in figure 7.11.

Solution Suppose that $v_1 = \bar{v} + (\Delta v/2)$ and $v_2 = \bar{v} - (\Delta v/2)$, where $\bar{v}$ is a constant representing the average supply voltage. Hence,

$$v_1 - v_2 = \Delta v \tag{i}$$

$$v_1 + v_2 = 2\bar{v} \tag{ii}$$

The motor is controlled using the differential voltage Δv. In a servo actuator, this corresponds to a feedback error signal. Using the notation shown in figure 7.11, the field current and armature current are given by

$$i_f = i_{f1} - i_{f2} \tag{iii}$$

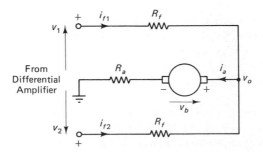

Figure 7.11. A split-field series-wound DC motor.

Continuous-Drive Actuators   Chap. 7

$$i_a = i_{f1} + i_{f2} \tag{iv}$$

Hence, the motor magnetic torque can be expressed as

$$T_m = ki_a i_f = k(i_{f1} + i_{f2})(i_{f1} - i_{f2}) \tag{v}$$

Also, the node voltage is

$$v_o = v_1 - i_{f1}R_f = v_2 - i_{f2}R_f$$

Hence,

$$i_{f1} - i_{f2} = \frac{v_1 - v_2}{R_f} = \frac{\Delta v}{R_f} \tag{vi}$$

and

$$2v_o = v_1 + v_2 - R_f(i_{f1} + i_{f2}) \tag{vii}$$

But

$$v_o = v_b + i_a R_a = k'i_f \omega_m + i_a R_a$$

where v_b denotes the back e.m.f. in the rotor. Hence, in view of equations (iii) and (iv), we have

$$v_o = k'(i_{f1} - i_{f2})\omega_m + (i_{f1} + i_{f2})R_a \tag{viii}$$

Substitute equation (viii) in (vii); thus,

$$\frac{v_1 + v_2}{2} = \frac{R_f}{2}(i_{f1} + i_{f2}) + k'(i_{f1} - i_{f2})\omega_m + R_a(i_{f1} + i_{f2}) \tag{ix}$$

Substitute (ii) and (vi) in (ix); thus,

$$\bar{v} = \frac{k'}{R_f}\Delta v\,\omega_m + \left(R_a + \frac{R_f}{2}\right)(i_{f1} + i_{f2}) \tag{x}$$

Substitute (vi) in (v); thus,

$$T_m = \frac{k}{R_f}(i_{f1} + i_{f2})\,\Delta v \tag{xi}$$

Substitute (xi) in (x); thus,

$$\bar{v} = \frac{k'}{R_f}\Delta v\,\omega_m + \left(R_a + \frac{R_f}{2}\right)\frac{R_f}{k\,\Delta v}T_m$$

or

$$T_m + \frac{kk'\,\Delta v^2}{R_f^2(R_a + R_f/2)}\omega_m = \frac{k\bar{v}\,\Delta v}{R_f(R_a + R_f/2)} \tag{7.14}$$

This is a linear relationship between T_m and ω_m. According to equation 7.13, the electrical damping constant for a split-field series-wound DC motor is given by

$$b_e = \frac{kk'\,\Delta v^2}{R_f^2(R_a + R_f/2)} \tag{7.15}$$

Note that the damping is zero under balanced conditions ($\Delta v = 0$). But damping increases quadratically with the differential voltage Δv.

7.5 CONTROL OF DC MOTORS

We have noted that by using proper winding arrangements, DC motors can be operated over a wide range of speeds and torques. Because of this adaptability, DC motors are particularly suitable as variable-drive actuators. In the past, AC motors were employed almost exclusively in constant-speed applications, because variable-speed control of AC motors, by conventional means, is much more difficult than variable-speed control of DC motors. For this reason, DC motors have dominated in industrial control applications for many decades.

The function of a conventional servo system that uses a DC motor as the actuator is almost exclusively motion control (position and speed control). There are applications, however, that require torque control, directly or indirectly, but they usually require more sophisticated control techniques. Control of a DC motor is accomplished by controlling either the stator field flux or the armature flux. If the armature and field windings are connected through the same circuit (see figure 7.9), both techniques apply simultaneously. Specifically, the two methods of control are,

1. Armature control
2. Field control

In armature control, the field current in the stator circuit is kept constant and the input voltage v_a to the rotor circuit is varied in order to achieve a desired performance (i.e., to reach specified values of position, speed, torque, etc.). Since v_a directly determines the motor back e.m.f., after allowance is made for the impedance drop due to resistance and leakage inductance of the armature circuit, it follows that armature control is particularly suitable for speed manipulation over a wide range of speeds (typically, 10 dB or more). The motor torque can be kept constant simply by keeping the armature current constant, because the field current is virtually a constant in the case of armature control (see equation 7.3).

In field control, the armature voltage (and current) is kept constant and the input voltage v_f to the field circuit is varied. From equation 7.3, it can be seen that since i_a is kept more or less constant, the torque will vary in proportion to the field current i_f. Also, since the armature voltage is kept constant, the back e.m.f. will remain virtually unchanged. Hence, it follows from equation 7.4 that the speed will be inversely proportional to i_f. This means that by increasing the field voltage, the motor torque can be increased while the motor speed is decreased, so that the output power will remain more or less constant in field control. Field control is particularly suitable for constant-power drives under varying torque-speed conditions, such as those present in tape transport mechanisms.

DC Servomotors

If the system characteristics and loading conditions are very accurately known, it is possible, in theory, to schedule the input signal to a motor (e.g., the armature voltage or field voltage) so as to obtain a desired response (e.g., motion trajectory or torque) from it. In this case of *open-loop control* or *computed-input control* (sometimes inappropriately referred to as feedforward control), parameter variations, model uncertainties, and external disturbances can produce errors that will build up (integrate) rapidly and will display unstable behavior. This is not acceptable in control system implementations. Feedback control is used to reduce these errors and to improve the control system performance, particularly with regard to stability, robustness, accuracy, and speed of response. In feedback control systems, response variables are sensed and fed back to the driver end of the system so as to reduce the response error. In DC servomotor systems, both angular position and speed might be measured (using shaft encoders, tachometers, resolvers, RVDTs, potentiometers, etc.) and compared with the desired position and speed. The error signals (desired response − actual response) are conditioned and compensated using analog circuitry or are processed by a digital control hardware processor or control computer and are supplied to drive the servomotor toward the desired response. For accurate position control, both position feedback and velocity feedback are usually needed. For speed control, velocity feedback alone might be adequate, but position error can build up. On the other hand, if only position feedback is used, a large error in velocity is possible, even when the position error is small. Under certain conditions (e.g., high gains, large time delays), the control system could even become marginally stable or unstable with position feedback alone. For this reason, DC servo systems normally employ tachometer feedback (velocity feedback) in addition to other types of feedback. In some commercial servomotors, the motor and the tachometer are usually available as a single package, with a common housing. Some servomotors have a built-in encoder for position and speed measurement in digital control systems.

Motion control (position and speed control) requires indirect control of motor torque. In applications where torque itself is a primary output (e.g., metal forming operations, micromanipulation, and tactile operations, as mentioned in chapter 4) and in situations where small motion errors could produce large unwanted forces (e.g., in parts assembly), direct control of motor torque would be necessary. In some torque control applications, this is accomplished using armature current feedback or field current feedback, because the armature current and the field current determine the motor torque. We have noted in chapter 4, however, that the motor torque is not equal to the load torque, or the torque transmitted through the output shaft of the motor. Hence, for precise torque control, direct measurement of torque (e.g., using strain gage sensors or inductive sensors) would be required.

A schematic representation of an analog DC servo system is given in figure 7.12. The actuator is a DC motor. The sensors might include a tachometer to measure angular speed, a potentiometer to measure angular position, and a strain gage torque sensor, which is optional. The process (the system that is driven) is represented by the load block in the figure. Signal-conditioning (filters, amplifiers, etc.) and compensating (lead, lag, etc.) circuitry are represented by a single block. The power supply to the servo amplifier (and to the motor) is not shown in the figure. The motor and tachometer are usually available as an integral unit with a common shaft. The position sensor (RVDT, potentiometer, resolver, etc.) is normally attached to the load.

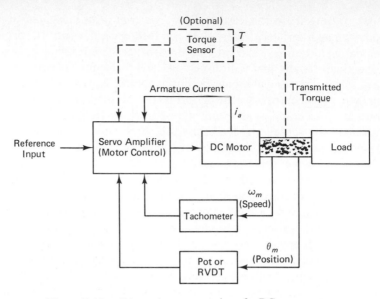

Figure 7.12. Schematic representation of a DC servo system.

Armature Control

In armature-controlled DC motors, the armature voltage is used as the control input, while keeping the conditions in the field circuit constant. In particular, the field current i_f is assumed constant. Consequently, equations 7.3 and 7.4 can be written as

$$T_m = k_m i_a \tag{7.16}$$

$$v_b = k'_m \omega_m \tag{7.17}$$

The parameters k_m and k'_m are termed the *torque constant* and the *back e.m.f. constant*, respectively. Note that with consistent units, $k_m = k'_m$ in the case of ideal electrical-to-mechanical energy conversion at the rotor. In the Laplace domain, equation 7.6 becomes

$$v_a - v_b = (L_a s + R_a)i_a \tag{7.18}$$

Note that, for convenience, time domain variables (functions of t) are used to denote their Laplace transforms (functions of s). It is understood, however, that the time functions are not identical to the Laplace functions. Also, in the Laplace domain, the mechanical equation 7.7 becomes

$$T_m - T_L = (J_m s + b_m)\omega_m \tag{7.19}$$

where J_m and b_m denote the moment of inertia and the rotary viscous damping constant, respectively, of the motor rotor. Equations 7.16 through 7.19 can be represented in block diagram form, as in figure 7.13. Note that the speed ω_m is taken as the motor output. If the motor position θ_m is considered the output, it is obtained by passing ω_m through an integration block $1/s$. Note, further, that the load torque T_L,

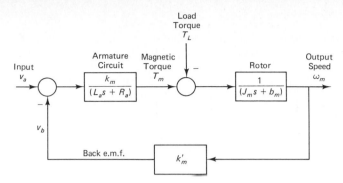

Figure 7.13. Open-loop block diagram for an armature-controlled DC motor.

which is the useful (effective) torque transmitted to the load that is being driven, is an (unknown) input to the system. Usually, T_L increases with ω_m because a larger torque is necessary to drive a load at a higher speed. If a linear (and dynamic) relationship exists between T_L and ω_m at the load, a feedback path can be completed from the output speed to the input load torque through a proper load transfer function (load block). The system shown in figure 7.13 is not a feedback control system. The feedback path representing the back e.m.f. is a natural feedback that is characteristic of the process (DC motor); it is not an external control feedback loop.

The overall transfer relation for the system is given by

$$\omega_m = \frac{k_m}{\Delta(s)} v_a - \frac{(L_a s + R_a)}{\Delta(s)} T_L \qquad (7.19)$$

where $\Delta(s)$ is the characteristic polynomial of the system, given by

$$\Delta(s) = (L_a s + R_a)(J_m s + b_m) + k_m k_m' \qquad (7.20)$$

This is a second-order polynomial in s. The *electrical time constant* of the armature is

$$\tau_a = \frac{L_a}{R_a} \qquad (7.21)$$

The mechanical response of the rotor is governed by the *mechanical time constant*,

$$\tau_m = \frac{J_m}{b_m} \qquad (7.22)$$

Usually, τ_m is at least one order of magnitude larger than τ_a. Hence, τ_a can be neglected in comparison to τ_m for most practical purposes. In that case, the transfer functions in equation 7.19 become first order.

Note that the characteristic polynomial is the same for both transfer functions in equation 7.19, regardless of the input (v_a or T_L). This should be the case, because $\Delta(s)$ determines the natural response of the system and does not depend on the system input. True time constants are obtained by solving $\Delta(s) = 0$ and taking the reciprocal of the magnitudes of the real parts of the two roots (eigenvalues). These are not the same as τ_a and τ_m because of the presence of the $k_m k_m'$ term (coupling term) in $\Delta(s)$.

Example 7.5

Determine an expression for the dominant time constant of an armature-controlled DC motor. What is the speed response of the motor to a unit step input in armature voltage in the absence of a load?

Solution   By neglecting the electrical time constant in equation 7.20, we have the characteristic ploynomial

$$\Delta(s) = R_a(J_m s + b_m) + k_m k'_m$$

This is expressed as

$$\Delta(s) = k'(\tau s + 1)$$

where τ is the overall dominant time constant. It follows that the dominant time constant is given by

$$\tau = \frac{R_a J_m}{(R_a b_m + k_m k'_m)} \tag{7.23}$$

With $T_L = 0$, the motor transfer relation is

$$\omega_m = \frac{k}{(\tau s + 1)} v_a \tag{7.24}$$

where the gain is

$$k = \frac{k_m}{(R_a b_m + k_m k'_m)} \tag{7.25}$$

The speed response to a unit step change in v_a, with zero initial conditions, is

$$\omega_m(t) = k(1 - e^{-t/\tau}) \tag{7.26}$$

This is a nonoscillatory response. In practical situations, some oscillations will be present in the free response because, invariably, a load inertia is coupled to the motor through a shaft that has some flexibility (is not rigid).

Since open-loop operation can lead to excessive error and instability, velocity feedback is usually employed in speed control. In velocity feedback, motor speed is sensed using a device such as a tachometer and is fed back to the controller, which compares it with the desired speed, and the error is used to correct the deviation. Usually, additional dynamic compensation has to be provided through analog circuits or by digital processing. The error signal is passed through the compensator in order to improve the performance of the control system.

In position control, the motor angle θ_m is the output. In this case, the open-loop system has a free integrator; the characteristic polynomial is $s(\tau s + 1)$. This is a marginally stable system. In particular, if a slight disturbance or model error is present, it will be integrated out, which can lead to a diverging error in the motor angle. This unstable behavior cannot be corrected using velocity feedback alone. Position feedback is needed to remedy the problem. Then position and velocity feedback gains can be chosen so as to obtain the desired response (speed of response, overshoot limit, accuracy, etc.). Control system design involves selection of proper parameter values for sensors and other components in the control system.

Example 7.6

Often, analytical modeling is not feasible, and modeling using experimental data might be the only available recourse. In particular, experimentally determined steady-state torque-speed characteristics may be used to determine (approximately) a dynamic model for a motor. To illustrate this approach, consider an armature-controlled DC motor. Sketch steady-state speed-load torque curves using input voltage (armature voltage) v_a as a parameter that is constant for each curve but varies from curve to curve. Obtain an equation to represent these curves. Now consider an armature-controlled DC motor driving a load of inertia J_L, connected directly to the rotor through a shaft that has torsional stiffness k_L. The viscous damping constant at the load is b_L. Obtain the transfer function for the load position θ_L.

Solution From equation 7.8, the steady-state speed-torque curves for a separately excited DC motor are given by

$$T_m + \left[\frac{kk'v_f^2}{R_a R_f^2}\right]\omega_m = \left[\frac{kv_f}{R_a R_f}\right]v_a \tag{7.27}$$

Since v_f is constant for armature-controlled motors, we can define constants k_m and b_e as

$$k_m = \frac{kv_f}{R_f} \tag{7.28}$$

and,

$$b_e = \frac{kk'v_f^2}{R_a R_f^2} = \frac{k_m k_m'}{R_a} \tag{7.29}$$

k_m' being defined similarly. Hence, equation 7.27 becomes

$$T_m + b_e\omega_m = \frac{k_m}{R_a}v_a \tag{7.30}$$

Note that k_m is the torque constant defined by equation 7.16, k_m' is the back e.m.f. constant defined by equation 7.17, and b_e is the electrical damping constant defined by equation 7.13. Note, however, that the torque T_{Ls} supplied to the load at steady-state (constant-speed) conditions is less than the motor magnetic torque T_m because of the presence of mechanical dissipation. Specifically,

$$T_{Ls} = T_m - b_m\omega_m \tag{7.31}$$

It should be noted that if the motor speed is not constant, the output torque of the motor will be further affected because some torque is used up in accelerating (or decelerating) the rotor intertia. Obviously, this does not enter into constant-speed tests. Now, by substituting equation 7.31 in 7.30, we have

$$T_{Ls} + (b_m + b_e)\omega_m = \frac{k_m}{R_a}v_a \tag{7.32}$$

Note that we measure T_{Ls}, not T_m, in (constant-speed) motor tests. It follows from equation 7.32 that the steady-state speed-torque curves for an armature-controlled DC motor are parallel straight lines with a negative slope of magnitude $b_m + b_e$. These curves are sketched in figure 7.14. Note that once a characteristic curve is obtained experi-

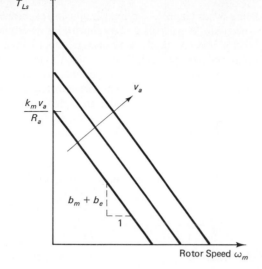

Load Torque T_{Ls}

$\dfrac{k_m v_a}{R_a}$

v_a

$b_m + b_e$

1

Rotor Speed ω_m

Figure 7.14. Steady-state speed-torque curves for an armature-controleed DC motor.

mentally, the parameters $b_m + b_e$ and k_m/R_a can be directly extracted. Once this is accomplished, equation 7.32 is completely known and can be used for modeling the control system.

The system given in this example is shown in figure 7.15. Suppose that θ_m denotes the motor angle. Newton's second law gives the rotor equation:

$$T_m - k_L(\theta_m - \theta_L) - b_m \dot{\theta}_m = J_m \ddot{\theta}_m \tag{i}$$

and the load equation:

$$-k_L(\theta_L - \theta_m) - b_L \dot{\theta}_L = J_L \ddot{\theta}_L \tag{ii}$$

Substituting equation 7.30 in (i) and (ii), and taking Laplace transforms, we get

$$\frac{k_m}{R_a} v_a + k_L \theta_L = [J_m s^2 + (b_m + b_e)s + k_L]\theta_m \tag{iii}$$

$$k_L \theta_m = (J_L s^2 + b_L s + k_L)\theta_L \tag{iv}$$

As usual, we use the same symbols to denote Laplace transforms and their time functions. By substituting equation (iv) in (iii) and after straightforward algebraic manipulation, we obtain the system transfer function

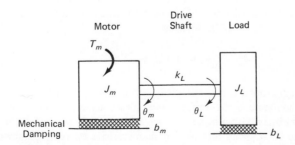

Motor

Drive Shaft

Load

T_m

k_L

J_m

J_L

θ_m

θ_L

Mechanical Damping

b_m

b_L

Figure 7.15. A motor driving an inertial load (example 7.6).

$$\frac{\theta_L}{v_a} =$$

$$\frac{k_L k_m / R_a}{s[J_m J_L s^3 + \{J_L(b_m + b_e) + J_m b_L\}s^2 + \{k_L(J_L + J_m) + b_L(b_m + b_e)\}s + k_L b_L + k_L(b_m + b_e)]}$$

(7.33)

Note that k_m / R_a and $b_m + b_e$ are the experimentally determined parameters. The mechanical parameters k_L, b_L, and J_L, are assumed to be known. Notice the free integrator present in the transfer function given by equation 7.33. This gives a pole (eigenvalue) at the origin of the s-plane ($s = 0$). It represents the rigid-body mode of the system, implying that the load is not externally restrained by a spring.

We have seen that for some winding configurations, the speed-torque curve is not linear. Thus, the slope of a characteristic curve is not constant. Hence, an experimentally determined model would be valid only for an operating region in the neighborhood of the point where the slope was determined.

Example 7.7

A DC motor uses 2 hp under no-load conditions to maintain a constant speed of 600 rpm. The motor torque constant $k_m = 1$ V.s, the rotor moment of inertia $J_m = 0.1$ kg.m^2, and the armature circuit parameters are $R_u = 10\ \Omega$ and $L_a = 0.01$ H. Determine the electrical damping constant, the mechanical damping constant, the electrical time constant of the armature circuit, the mechanical time constant of the rotor, and the true time constants of the motor.

Solution With the same units, $k_m' = k_m$. Hence, from equation 7.29, the electrical damping constant is

$$b_e = \frac{k_m^2}{R_a} - \frac{1}{10} - 0.1 \text{ N.m/rad/s}$$

The power absorbed by the motor at no-load conditions is

$$2 \text{ hp} = 2 \times 746 \text{ W} = 1,492 \text{ W}$$

and the corresponding speed is

$$\omega_m = \frac{600}{60} \times 2\pi \text{ rad/s} = 20\pi \text{ rad/s}$$

This power is used in electrical and mechanical damping at constant speed ω_m. Hence,

$$(b_m + b_e)\omega_m^2 = 1,492$$

$$b_m + b_e = \frac{1,492}{(20\pi)^2} = 0.38 \text{ N.m/rad/s}$$

It follows that, the mechanical damping constant is

$$b_m = 0.38 - 0.1 = 0.28 \text{ N.m/rad/s}$$

From equations 7.21 and 7.22,

$$\tau_a = \frac{0.01}{10} = 0.001 \text{ s}$$

$$\tau_m = \frac{0.1}{0.28} = 0.36 \text{ s}$$

Note that τ_m is several orders larger than τ_a. In view of equation 7.20, the characteristic polynomial of the motor transfer function can be written as

$$\Delta(s) = R_a[b_m(\tau_a s + 1)(\tau_m s + 1) + b_e] \qquad (7.34)$$

The poles (eigenvalues) are given by $\Delta(s) = 0$. Hence,

$$0.28(0.001s + 1)(0.36s + 1) + 0.1 = 0$$

or

$$s^2 + 1010s + 3,800 = 0$$

Solving for the motor eigenvalues, we get

$$\lambda_1 = -3.8 \quad \text{and} \quad \lambda_2 = -1,006$$

Note that the two poles are real and negative. This means that any disturbance in the motor speed will die out exponentially without oscillations.

The time constants are given by the reciprocals of the magnitudes of the real parts of the eigenvalues. Hence, the true time constants are

$$\tau_1 = 1/3.8 = 0.26 \text{ s}$$
$$\tau_2 = 1/1006 = 0.001 \text{ s}$$

The small time constant τ_2, which derives primarily from the electrical time constant of the armature circuit, can be neglected for all practical purposes. The larger time constant τ_1 comes not only from the mechanical time constant τ_m (rotor inertia/mechanical damping constant) but also from the electrical damping constant (back e.m.f. effect) b_e. Hence, τ_1 is not equal to τ_m, even though the two are of the same order of magnitude.

Field Control

In field-controlled DC motors, armature current is kept constant and field voltage is used as the control input. Since i_a is constant, equation 7.3 can be written as

$$T_m = k_a i_f \qquad (7.35)$$

where k_a is the electromechanical torque constant for the motor. The back e.m.f. relation and the armature circuit equation are not used in this case. Equations 7.5 and 7.7 are written in the Laplace form as

$$v_f = (L_f s + R_f)i_f \qquad (7.36)$$
$$T_m - T_L = (J_m s + b_m)\omega_m \qquad (7.37)$$

Equations 7.35 through 7.37 can be represented by the open-loop block diagram given in figure 7.16. The transfer relationship is

$$\omega_m = \frac{k_a}{(L_f s + R_f)(J_m s + b_m)}v_f - \frac{1}{(J_m s + b_m)}T_L \qquad (7.38)$$

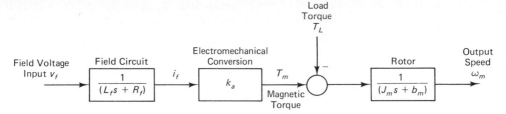

Figure 7.16. Open-loop block diagram for a field-controlled DC motor.

In this case, the electrical time constant originates from the field circuit and is given by

$$\tau_f = \frac{L_f}{R_f} \tag{7.39}$$

The mechanical time constant τ_m can be defined as in equation 7.22. The characteristic polynomial of the open-loop field-controlled motor is

$$\Delta(s) = (L_f s + R_f)(J_m s + b_m) \tag{7.40}$$

It follows that τ_f and τ_m are the true time constants of the system, unlike in armature control. As in the armature-controlled DC motor, the electrical time constant is very small and can be neglected in comparison to the mechanical time constant. Again, the speed and angular position of the motor have to be measured and fed back for accurate motion control.

Phase-Locked Control

Phase-locked control is a modern approach to controlling DC motors. A block diagram of a phase-locked servo system is shown in figure 7.17. The position command is generated according to the desired motion of the motor, using a controlled signal generator (e.g., a voltage-controlled oscillator) or using digital means (e.g., a microprocessor). This reference signal is in the form of a pulse train that is quite analogous to the output signal of an incremental encoder (chapter 5). The rotation of the motor (or load) is measured using an incremental encoder. This pulse train forms the feedback signal. The reference signal and the feedback signal are supplied to a *phase*

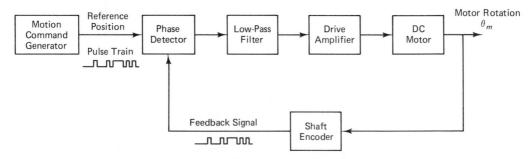

Figure 7.17. Schematic diagram of a phase-locked servo.

detector, which generates a signal representing the phase difference between the two signals and possibly some unwanted high-frequency components. The unwanted components are removed using a low-pass filter, and the resulting (error) signal is supplied to the drive amplifier of the motor. The error signal drives the motor so as to correct any deviations from the desired motion.

This is a *phase control* method. The objective here is to maintain a fixed phase difference (ideally, a zero phase difference) between the reference signal and the position signal. Under these conditions, the two signals are phase-locked together. Any deviation from the locked conditions will generate an error signal that will bring the motor motion back in phase with the reference command. In this manner, deviations due to external disturbances, such as load changes on the motor, are also corrected.

One method of determining the phase difference of two pulse signals is by detecting the edge transitions, as noted in chapter 5. An alternative method is to take the product of the two signals and then low-pass filter the result. To illustrate this second method, suppose that the primary (harmonic) components of the reference pulse signal and the response pulse signal are ($u_0 \sin \theta_u$) and ($y_0 \sin \theta_y$), respectively, where

$$\theta_u = \omega t + \phi_u$$

$$\theta_y = \omega t + \phi_y$$

Note that ω is the frequency of the two pulse signals (assumed to be the same) and ϕ denotes the phase angle. The product signal is

$$p = u_0 y_0 \sin \theta_u \sin \theta_y$$

$$= \tfrac{1}{2} u_0 y_0 [\cos(\theta_u - \theta_y) - \cos(\theta_u + \theta_y)]$$

Consequently,

$$p = \tfrac{1}{2} u_0 y_0 \cos(\phi_u - \phi_y) - \tfrac{1}{2} u_0 y_0 \cos(2\omega t + \phi_u + \phi_y) \qquad (7.41)$$

Low-pass filtering will remove the high-frequency component of frequency 2ω, leaving the signal

$$e = \tfrac{1}{2} u_0 y_0 \cos(\phi_u - \phi_y) \qquad (7.42)$$

This is a nonlinear function of the phase difference ($\phi_u - \phi_y$). Note that by applying a $\pi/2$ phase shift to the original two signals, we also could have determined $\tfrac{1}{2} u_0 y_0 \sin (\phi_u - \phi_y)$. In this manner, the magnitude and sign of ($\phi_u - \phi_y$) are determined. Very accurate position control can be obtained by driving this phase difference to zero. This is the objective of phase-locked control; the phase angle of the output is locked to the phase angle of the command signal. In more sophisticated phase-locked servos, the frequency differences are also detected—for example, using pulse counting—and compensated. This is analogous to the classic proportional plus derivative (PD) control. It is clear that phase-locked servos are velocity-control devices as well, because velocity is proportional to the pulse frequency. When the two pulse signals are synchronized, the velocity error also approaches zero, subject to the available resolution of the control system components. Typically, speed error levels of ±0.002 percent or less are possible using phase-locked servos. Also, the overall cost of a phase-locked servo system is usually less than that of a conventional analog servo system, because less expensive solid-state devices replace bulky analog control circuitry.

Thyristor Control

We have noted that the control of DC motors is usually accomplished by controlling the supply voltage to either the armature circuit or the field circuit. A dissipative method of achieving this involves using a variable resistor in series with the supply source to the circuit. This method, besides being wasteful, has other disadvantages. Notably, the heat generated at the control resistor has to be removed promptly to avoid malfunction and damage due to high temperatures. A much more desirable way to control the supply voltage is by using solid-state devices. Specifically, by varying the on time and off time of a solid-state switch, the supply voltage to the motor circuit can be "chopped" so that the average supply voltage is controlled. A solid-state switch frequently used in motor control circuits is the thyristor.

The thyristor is also known as a *silicon-controlled rectifier, a solid-state controlled rectifier, a semiconductor-controlled rectifier,* or simply an SCR. It is a pellet made of four layers (pnpn) of semiconductor material (e.g., silicon with a trace of dope material). It has three terminals—the anode, the cathode, and the gate—as shown in figure 7.18a. The anode and the cathode are connected to the circuit that carries the load current i. When the gate potential v_g is less than or equal to zero with respect to the cathode, the thyristor cannot conduct in either direction ($i = 0$). When v_g is made positive, the thyristor will conduct from anode to cathode but not

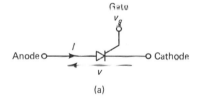

(a)

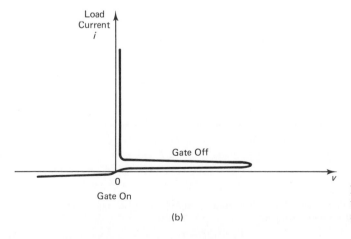

(b)

Figure 7.18. (a) Symbol for a thyristor. (b) Ideal characteristic curve of a thyristor.

in the opposite direction (i.e., it acts like a basic diode). In other words, a positive firing signal (i.e., a positive trigger voltage) v_g will close (turn on) the switch. To open (turn off) the switch again, we not only have to make v_g zero (or slightly negative) with respect to the cathode, but also the load current from the anode to the cathode has to be zero (or slightly negative). This is the natural mode of operation of a thyristor. When the supply voltage is DC, it does not drop to zero; hence, the thyristor would be unable to turn itself off. In this case, a *commutating circuit* that can make the voltage across the thyristor slightly negative has to be employed. This is called forced commutation (as opposed to natural commutation) of a thyristor. Note that when a thyristor is conducting, it offers virtually no resistance, and the voltage drop across the thyristor can be neglected for practical purposes. An idealized voltage-current characteristic curve of a thyristor is shown in figure 7.18b. Solid-state switching devices are lossless (or nondissipative) in nature.

A basic thyristor circuit, using a DC supply, that may be used in DC motor

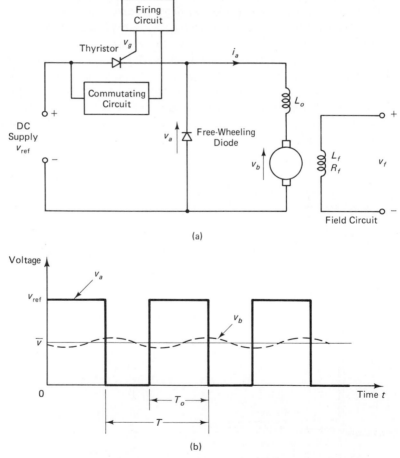

(a)

(b)

Figure 7.19. (a) A thyristor (SCR) control circuit (chopper) for a DC motor with a DC supply. (b) Circuit voltage signals.

control is shown in figure 7.19a. The DC supply voltage is v_{ref}, the voltage pulse signal supplied to the armature circuit is v_a, and the back e.m.f. in the motor is v_b. The nature of these voltages is shown in figure 7.19b. Since the supply voltage to the armature circuit is chopped, the circuit in figure 7.19a is usually known as a *chopper circuit*. The field circuit is separately excited, as shown. The armature resistance R_a is neglected to provide a qualitative explanation of the voltages appearing in various parts in the circuit; R_a should be included in a more accurate model. The inductance L_o includes the usual armature leakage inductance, self-inductance, and so forth (denoted by L_a), and an external inductance that is needed to avoid large fluctuations in armature current i_a, since v_a is pulsating. Alternatively, a series-wound motor in which the field inductance L_f is connected in series with the armature may be used to increase L_o. A free-wheeling diode provides a path for i_a during the off period of v_a so as to avoid large voltage buildup in the armature.

Initially, the voltage v_g applied to the gate terminal will close (turn on) the SCR, allowing v_{ref} to be applied to the armature circuit. Since v_{ref} is DC, however, the SCR will not open (turn off) by itself. Hence, a commutating circuit that is capable of applying a slightly negative voltage to the anode of the SCR is needed. The commutating circuit usually consists of a capacitor that is charged to provide the required voltage, a diode, and a thyristor.

The average voltage $\overline{v}$ supplied to the armature circuit is the average of v_a. This is given by

$$\overline{v} = \frac{T_o}{T} v_{ref} \qquad (7.43)$$

where T_o = on time of the supplied voltage pulse
T = pulse period

Note that $\overline{v}$ can be varied by changing either T_o (pulse width modulation or PWM) or T (pulse frequency modulation, or PFM). This method of pulsing control is employed in chopper drive circuits of DC motors.

Note that if v_a were a constant, there would not be a potential drop across the inductor L_o, and v_b would be a constant equal to v_a. It follows that the average value (i.e., the DC component) of v_b is equal to the average value of v_a, which is denoted by $\overline{v}$ in equation 7.43. When the motor speed is properly regulated (i.e., speed fluctuations are small), and assuming that the conditions in the field circuit are steady, the back e.m.f. v_b is nearly a constant. Hence,

$$v_b \sim \overline{v} \qquad (7.44)$$

Note that the voltage across L_o is $v_a - v_b$. It follows that

$$L_o \frac{di_a}{dt} = v_a - v_b \qquad (7.45)$$

Now, in view of equation 7.44, we can write

$$L_o \frac{di_a}{dt} = v_a - \overline{v} \qquad (7.46)$$

or

$$i_a = \frac{1}{L_o} \int (v_a - \overline{v}) \, dt \qquad (7.47)$$

Equation 7.47 tells us that the change in the armature current is proportional to the area between the v_a curve and the $\overline{v}$ line shown in figure 7.19b. In particular, starting from a steady-state value of i_a, the armature current will rise by

$$\Delta i_a = \frac{1}{L_o}(v_{\text{ref}} - \overline{v})T_o$$

over the time period T_o in which the thyristor is on. Substituting equation 7.43, we get

$$\Delta i_a = \frac{v_{\text{ref}}}{L_o T}(T - T_o)T_o \qquad (7.48)$$

Then, over the time period $(T - T_o)$ during which the thyristor is off, the armature current drops by

$$\frac{1}{L_o}\overline{v}(T - T_o)$$

which is equal to

$$\frac{v_{\text{ref}}}{L_o T}(T - T_o)T_o$$

So the armature current will go back to i_a. This cycle will repeat over the subsequent pulse cycles of duration T. It follows that the fluctuation in the armature current is given by equation 7.48. For a given T, this amplitude is maximum when $T_o = T/2$. Hence,

$$(\Delta i_a)_{\text{max}} = \frac{T v_{\text{ref}}}{4 L_o} \qquad (7.49)$$

Note that the current fluctuations can be reduced by increasing L_o and decreasing T for a given supply voltage v_{ref}. Note, further, that since $v_{\text{ref}} - \overline{v}$ is constant, it follows from equation 7.47 that the armature current increases or decreases linearly with time.

Example 7.8

Consider the chopper circuit given in figure 7.19a. The chopper frequency is 200 Hz, the series inductance in the armature circuit is 50 mH, and the supply DC voltage to the chopper is 100 V. Determine the amplitude of the maximum (worst case) fluctuation in the armature current.

Solution   Since we are interested in the worst case of current fluctuations, we use equation 7.49. We have

$$T = \frac{1}{200} \, \text{s}, \qquad L_o = 0.05 \, \text{mH}, \qquad v_{\text{ref}} = 100 \, \text{V}$$

Substituting values, we get

$$(\Delta i_a)_{max} = \frac{100}{200 \times 4 \times 0.05} = 2.5 \text{ A}$$

High-power DC motors are usually driven by rectified AC supplies (single-phase or three-phase). In this case, also, motor control can be accomplished by using thyristor circuits. Full-wave circuits are those that use both the positive and negative parts of an AC supply voltage. A full-wave single-phase control circuit for a DC motor is shown in figure 7.20. It uses two thyristors. For convenience, the armature resistance R_a is not shown, although it is important in the analysis of the circuit. The inductance L_o contains the usual armature component L_a as well as an external inductor that is connected in series with the armature to reduce surges in the armature current. Furthermore, a free-wheeling diode is used to provide a current path when the two thyristors are turned off. Two additional diodes are provided to complete the current path for each half of the supply wave period.

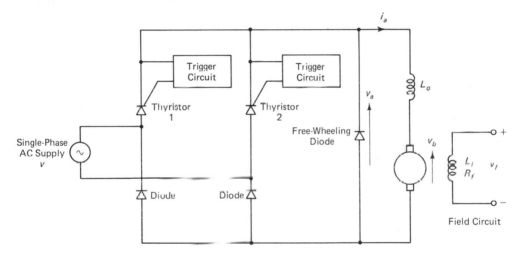

Figure 7.20. Full-wave single-phase control circuit for a DC motor.

Various signals through the circuit are sketched in figure 7.21. The supply voltage v is shown in figure 7.21a. The broken line in this figure is the voltage v_a that would result if the two thyristors were replaced by diodes. Figure 7.21b shows the DC voltage v_a supplied to the armature circuit and the back e.m.f. v_b across the armature; T_o is the firing time of each thyristor since the time when the voltage supplied to a thyristor begins to build from zero. Specifically, during the positive half of the supply voltage v, thyristor 1 will be triggered after time T_o, and during the negative half of the supply voltage, thyristor 2 will be triggered after time T_o. Note that in this full-wave circuit, the negative half of v also appears as positive in v_a (across the free-wheeling diode). The back e.m.f. v_b is reasonably constant, and so is the armature current i_a. Note that the voltage across L_o is $v_a - v_b$. Hence, the armature current variation is given by the area between the v_a and v_b curves, as is clear from equation 7.45. As in the DC supply case, the motor can be controlled either by vary-

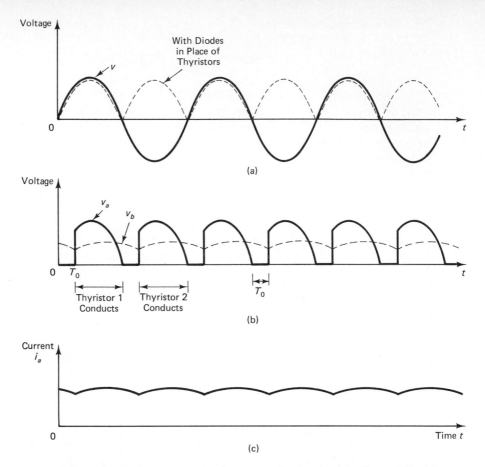

Figure 7.21. Voltages and armature current for the circuit in figure 7.20: (a) supply voltage; (b) voltage to the armature circuit and back e.m.f.; (c) armature current.

ing the thyristor firing time T_o for a given supply frequency or by varying the supply frequency for a given firing time.

The two diodes in the supply part of the circuit in figure 7.20 may be replaced by two thyristors. In that case, the two thyristors in each current path (i.e., for positive v and negative v) have to be triggered simultaneously. The operation of a three-phase control circuit for a DC motor can be analyzed by direct extension of the concepts presented here.

Servo Amplifiers

Servo amplifiers are used to drive motors (mostly DC motors) in servo control applications. Modern servo amplifiers use pulse-width modulation (PWM) to drive servo-motors efficiently under variable-speed conditions, without incurring excessive power losses. Integrated microelectronic design makes them compact, accurate, and inexpensive. A typical servo amplifier card consists of the following components, connected in series:

1. A differential amplifier
2. A servo preamplifier
3. A PWM amplifier

Leads for an external power supply (15 V DC is typical) are provided. The reference input signal and the feedback signal (from a tachometer or position sensor) are connected to the input leads of the differential amplifier. The resulting difference (error signal) is conditioned (and amplified) by the servo preamplifier and is used as the modulating signal to the PWM amplifier. The reference switching frequency of the PWM amplifier is high (on the order of 25 kHz), and pulse-width modulation may be accomplished through switching control, as explained in the foregoing section. The PWM signal from the amplifier (at a typical voltage of 10 V) is used to drive the servomotor—for example, using the PWM signal to energize the field windings of a brushless DC motor. Current feedback may be used to improve the servo amplifier performance.

7.6 TORQUE MOTORS

Conventionally torque motors are high-torque DC motors with permanent-magnet stators. These actuators characteristically possess a linear (straight-line) torque-speed relationship, primarily because of their high-strength permanent-magnet stators. The magnet should have high flux density per unit volume of the magnet material, yielding a high torque/mass ratio for a torque motor. Furthermore, *coercivity* (resistance to demagnetization) should be high and the cost has to be moderate. Rare-earth materials (e.g., samarium cobalt, $SmCo_5$) possess most of these desirable characteristics, although their cost could be high. Conventional ferrite magnets and alnico (aluminum-nickel-cobalt) or ceramic magnets provide a relatively low torque/mass ratio. Hence, rare-earth magnets are widely used in torque motor and servomotor applications. A comparison of typical rare-earth motors and alnico motors that are commercially available is given in table 7.3. When operating at high torques (e.g., several thousand lbf.ft), the motor speeds have to be quite low for a given level of power. For this reason, torque motors are particularly suitable for direct-drive applications (e.g., direct-drive robot arms) without needing additional speed reducers and gears. Since gear drives introduce undesirable effects—such as backlash, additional inertia loading, and higher friction—torque motors are suitable for high-precision applications that require high-torque drives. Torque motors are usually more expensive than the conventional types of DC motors. This is not a major drawback, however, because torque motors are often custom-made and are supplied as units that can be directly integrated with the process (load) within a common housing. For example, the stator might be integrated with one link of a robot arm and the rotor with

TABLE 7.3 COMPARISON OF COMMERCIAL RARE-EARTH MOTORS AND ALNICO MOTORS

Motor size classification	Magnet type	Motor dimensions (with standard hubs and flanges)			Mass (kg)	Peak torque (N.m)	Torque/mass ratio (N.m/kg)
		Outer diameter (mm)	Inner diameter (mm)	Length (mm)			
Small	Rare earth	81	29	60	1.52	6.8	4.5
	Alnico	72	23	64	1.31	1.7	1.3
Medium	Rare earth	183	100	32	2.70	15.0	5.5
	Alnico	183	100	34	3.05	8.2	2.7
Large	Rare earth	228	136	42	4.44	27.2	6.1
	Alnico	228	136	41	4.34	14.9	3.4
Extra large	Rare earth	646	523	152	100.10	952.0	9.5
	Alnico	734	415	165	100.10	585.0	5.8

Source: Adapted from H. Asada and T. Kanade, "Design of Direct-Drive Mechanical Arms," Robotics Institute Report CMU-RI-TR-81-1, Carnegie Mellon University, April 1981.

the next link, thus forming a common joint in a direct-drive robot. Torque motors are widely used as valve actuators in hydraulic servovalves where large torques and very small displacements are required.

Brushless torque motors have permanent-magnet rotors and wound stators. Consider a brushless DC motor with electronic commutation. The output torque can be increased by increasing the number of magnetic poles. Since direct increase of the magnetic poles has serious physical limitations, a toothed construction, as in stepping motors, could be employed for this purpose. Torque motors of this type have toothed ferromagnetic stators with field windings on them. Their rotors are similar in construction to those of variable-reluctance (VR) stepping motors.

Harmonic Drives

A straightforward way to increase the output torque of a motor is to employ a gear system with high gear reduction. This has several disadvantages, however. For example, backlash in gears would be unacceptable in high-precision applications. Frictional loss of torque, wear problems, and the need for lubrication must also be considered. Furthermore, the mass of the gear system reduces the overall torque/mass ratio and the useful bandwidth of the actuator.

A harmonic drive is a special type of gear reducer that provides very large speed reductions (e.g., 200:1) without backlash problems. The harmonic drive is often integrated with conventional motors to provide very high torques, particularly in direct-drive and servo applications. The principle of operation of a harmonic drive is shown in figure 7.22. The fixed and rigid housing of the drive has internal teeth. An annular spline has external teeth that can mesh with the internal teeth of the housing in a limited region when pressed with a radial force. The external radius of the annular spline is slightly smaller than the internal radius of the fixed spline. The annular spline usually undergoes some elastic deformation during the meshing pro-

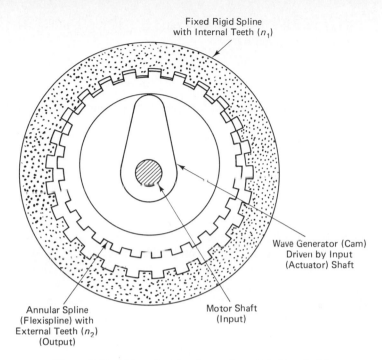

Figure 7.22. The principle of operation of a harmonic drive.

cess. Hence, it is called a flexispline. The rotation of the flexispline is the output of the drive; hence, it is linked with the driven load. The input shaft (motor shaft) drives the wave generator (represented by a cam in figure 7.22). The wave generator motion brings about controlled meshing between the rigid spline and the flexispline through a slight amount of elastic deformation. Hence, backlash is not present. Suppose that

$$n_1 = \text{number of teeth (internal) in the rigid spline}$$

$$n_2 = \text{number of teeth (external) in the flexispline}$$

It follows that

$$\text{tooth pitch of the rigid spline} = \frac{2\pi}{n_1} \text{ radians}$$

$$\text{tooth pitch of the flexispline} = \frac{2\pi}{n_2} \text{ radians}$$

Further, suppose that n_1 is slightly smaller than n_2. Hence, during one tooth engagement, the flexispline rotates through $(2\pi/n_1 - 2\pi/n_2)$ radians in the direction of rotation of the wave generator. During one full rotation of the wave generator, there will be a total of n_1 tooth engagements in the rigid spline. Hence, the rotation of the flexispline during one rotation of the wave generator is

$$n_1 \left(\frac{2\pi}{n_1} - \frac{2\pi}{n_2} \right) = \frac{2\pi}{n_2}(n_2 - n_1)$$

It follows that the gear reduction ratio ($r : 1$) representing the ratio (input speed/output speed) is given by

$$r = \frac{n_2}{n_2 - n_1} \qquad (7.50)$$

We can see that by making n_1 very close to n_2, very high gear reductions can be obtained.

An inherent shortcoming of a harmonic drive is that the motion of the output device (flexispline) is eccentric (or epicyclic). This problem is not serious when the eccentricity is small and is further reduced because of the flexibility of the flexispline. For improved performance, however, this epicyclic rotation has to be reconverted into a concentric rotation. This may be done using various means, including flexible coupling and pin-slot transmissions. The output device of a pin-slot transmission is a flange that has pins arranged on the circumference of a circle centered at the axis of the output shaft. The input to the pin-slot transmission is the flexispline motion, which is transmitted through a set of holes on the flexispline. The pin diameter is smaller than the hole diameter, the associated clearance being adequate to take up the eccentricity in the flexispline motion. This principle is shown schematically in figure 7.23. Alternatively, pins could be attached to the flexispline and the slots on the output flange. The eccentricity problem can be eliminated altogether by using a double-ended cam in place of the single-ended cam wave generator shown in figure 7.22. With this new arrangement, meshing takes place at two diametrical ends simultaneously, and the flexispline is deformed elliptically in doing this. But the center of rotation of the flexispline now coincides with the center of the input shaft. This double-mesh design is more robust and is widely used in industrial harmonic drives.

Traction drives (or friction drives) employ frictional coupling to eliminate backlash and overloading problems. These are not harmonic drives. In a traction drive, the drive member (input roller) is frictionally engaged with the driven mem-

A = Flexispline Center
B = Output (Load) Shaft Center

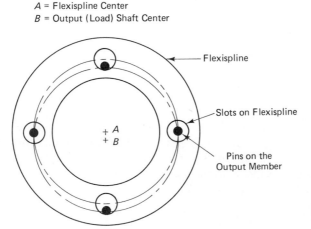

Figure 7.23. The principle of a pin-slot transmission.

Continuous-Drive Actuators   Chap. 7

ber (output roller). The disadvantages of traction drives include indeterminance of the speed ratio under slipping (overload) conditions and large size and weight for a specified speed ratio.

The harmonic drive principle can be integrated with an electric motor in a different manner. Suppose that the flexispline is made of an electromagnetic material, as the rotor of a VR stepper motor. Instead of the mechanical wave generator (cam) in figure 7.22, suppose that a rotating magnetic field with some polarity is generated around the fixed spline; the magnetic attraction will cause the tooth engagement between the flexispline and the fixed spline. In this case, the flexispline is the motor rotor and the fixed spline is the motor stator. The motor speed ω_m is a fraction of the speed of the rotating magnetic field ω_f and is given by

$$\omega_m = \frac{n_2 - n_1}{n_2} \omega_f \qquad (7.51)$$

Very low speeds and high torques can be obtained in this manner. Note that this type of harmonic drive motor can be operated either as a stepper motor or as an AC synchronous motor. Motor speed can be controlled by varying the frequency of the AC voltage supply.

7.7 MOTOR SELECTION CONSIDERATIONS

Torque and speed are the two primary considerations in choosing a motor for a particular application. Motor manufacturers' data that are usually available to users include the following specifications:

1. Mechanical specifications

   - Mechanical time constant
   - No-load speed
   - Speed at rated load
   - No-load acceleration
   - Rated torque
   - Rated output power
   - Frictional torque
   - Damping constant
   - Dimensions and weight
   - Armature moment of inertia

2. Electrical specifications

   - Electrical time constant
   - Input power
   - Armature resistance and inductance

- Field resistance and inductance
- Compatible drive circuit specifications (voltage, current, etc.)

3. General specifications

- Brush life and motor life
- Efficiency
- Operating temperature and other environmental conditions
- Heat transfer characteristics
- Mounting configuration
- Coupling methods

When a specific application calls for large speed variations (e.g., speed tracking over a range of 10 dB or more), armature control is preferred. Note, however, that at low speeds (typically, half the rated speed), poor ventilation and associated temperature buildup can cause problems. At very high speeds, mechanical limitations and heating due to frictional dissipation become determining factors. For constant-speed applications, shunt-wound motors are preferred. Finer speed regulation may be achieved using a servo system with encoder or tachometer feedback or with phase-locked operation. For constant-power applications, the series-wound or compound-wound motors are preferable to shunt-wound units. If the shortcomings of mechani-

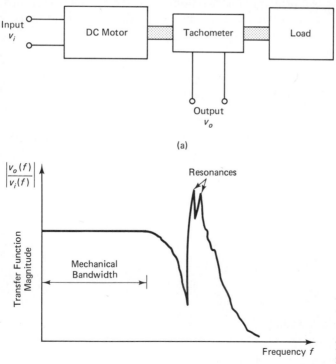

Figure 7.24. Determination of the mechanical bandwidth of a DC motor: (a) test setup; (b) test result.

cal commutators and limited brush life are crucial, brushless DC motors should be used.

For high-speed and transient operations of a DC motor, its mechanical time constant (or mechanical bandwidth) is an important consideration. This is limited by armature and load inertia, shaft flexibility, and the dynamics of the mounted instrumentation, such as tachometers and encoders. The mechanical bandwidth of a DC motor can be determined by simply measuring the velocity transducer signal v_o for a transient drive signal v_i and computing the ratio of their Fourier spectra. This is illustrated in figure 7.24. (A better way of computing this transfer function is by the cross-spectral density method.) The flat region of the resulting frequency transfer function (magnitude) plot determines the mechanical bandwidth of the motor.

A simple way to determine the operating conditions of a motor is by using its torque-speed curve, as illustrated in figure 7.25. The minimum torque T_{min} is limited mainly by loading considerations. The minimum speed ω_{min} is determined primarily by operating temperature. These boundaries select the useful region of the operating curves. The optimal operating points are those that fall within this segment on the torque-speed curve. Note that an upper limit on speed may also be imposed to take transmission limitations into account.

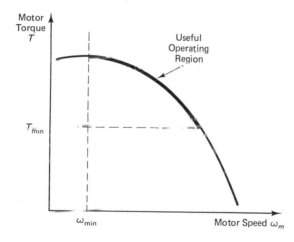

Figure 7.25. Representation of the useful operating region for a DC motor.

7.8 INDUCTION MOTORS

With the widespread availability of alternating current (AC) as an economical form of power supply for operating industrial machinery, much attention has been given to the development of AC motors. Because of the rapid progress made in this area, AC motors have managed to replace DC motors in many industrial applications until the recent revival of the DC motor, particularly in control system applications. However, AC motors are still generally more attractive than conventional DC motors, especially in heavy-duty (high-power) applications. Some advantages of AC motors are

1. Cost-effectiveness

2. Convenient power source (standard AC supply)
3. No commutator and brush mechanisms needed in some types
4. Lower power dissipation, lower rotor inertia, and light weight in some designs
5. Virtually no electric arcing (less hazardous in chemical environments)
6. Constant-speed operation without servo control (in synchronous machines)
7. No drift problems in AC amplifiers in supply circuits (unlike DC amplifiers)
8. High reliability

The primary disadvantages include

1. Lower starting torque
2. Auxiliary starting device needed for some motors
3. Difficulty of variable-speed control (except when modern thyristor-control devices and field feedback compensation techniques are used)

We shall discuss two basic types of AC motors: (1) induction motors (asynchronous motors) and (2) synchronous motors.

Rotating Field

The operation of an AC motor can be explained using the concept of a rotating magnetic field. A rotating field is generated by a set of windings uniformly distributed on a circle and excited by AC supplies with uniform phase differences. To illustrate this, consider a standard three-phase supply. The voltage in each phase is 120° out of phase with the voltage in the next phase. The phase voltages can be represented by

$$v_1 = a \cos \omega_p t$$

$$v_2 = a \cos\left(\omega_p t - \frac{2\pi}{3}\right) \qquad (7.52)$$

$$v_3 = a \cos\left(\omega_p t - \frac{4\pi}{3}\right)$$

where ω_p is the frequency of each phase of the AC supply (line frequency). Note that v_1 leads v_2 by $2\pi/3$ radians and v_2 leads v_3 by the same angle. Furthermore, since v_1 leads v_3 by $4\pi/3$ radians, it is correct to say that v_1 lags v_3 by $2\pi/3$ radians. Hence, v_1 leads $-v_3$ by $(\pi - 2\pi/3) = \pi/3$. Now consider a group of three windings, each of which has two segments (a positive segment and a negative segment) uniformly arranged around a circle, as shown in figure 7.26, in the order v_1, $-v_3$, v_2, $-v_1$, v_3, $-v_2$. Note that each winding segment has a phase difference of $\pi/3$ (or 60°) from the adjacent segment. The physical (geometric) spacing of adjacent winding segments is also 60°. Now, consider the time interval $\Delta t = \pi/(3\omega_p)$. At the end of a time interval Δt, the status of $-v_3$ becomes identical to that of v_1 in the beginning of the time interval. Similarly, the status of v_2 after Δt becomes that of $-v_3$ in the beginning, and so on. In other words, the voltage status (and hence the magnetic field status) switches from one segment to the adjacent segment in time intervals of Δt.

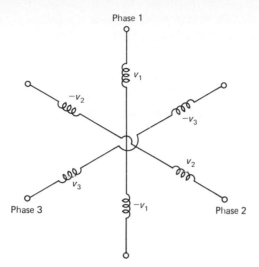

Phase 1

v_1

$-v_2$

$-v_3$

$-v_3$

v_2

v_3

Phase 3

$-v_1$

Phase 2

Figure 7.26. The generation of a rotating magnetic field using a single three-phase winding set.

This means that the field appears to rotate physically around the circle at angular velocity ω_p. It is not necessary for the three sets of three-phase windings to be distributed over the entire 360° angle of the circle. Suppose, instead, that these three sets (six segments) of windings are distributed within the first 180° of the circle, at 30° apart, and a second three sets are distributed similarly within the remaining 180°. Then, the field would appear to rotate at half the speed ($\omega_p/2$), because in this case, Δt is the time taken for the field to rotate through 30°, not 60°. Hence, the general formula for the angular speed ω_f of a rotating field is

$$\omega_f = \frac{\omega_p}{n} \qquad (7.53)$$

where ω_p = frequency of the AC supply in each phase (line frequency)
n = number of three-phase winding sets used (number of pole pairs per phase)

Note that when $n = 1$, there are two coils ($+ve$ and $-ve$) for each phase (i.e., there are two poles per phase). Similarly, when $n = 2$, there are four coils for each phase. Hence, n denotes the number of "pole pairs" per phase in a stator. In this manner, the speed of rotating field can be reduced to a fraction of the line frequency simply by adding more sets of windings. These windings occupy the stator of an AC motor. It is this rotating field that generates the driving torque by interacting with the rotor windings. The nature of this interaction determines whether a particular motor is an induction motor or a synchronous motor.

Example 7.9

Another way to interpret the concept of a rotating magnetic field is to consider the resultant field due to the individual magnetic fields in the stator windings. Consider a single set of three-phase windings arranged geometrically as in figure 7.26. Suppose that the magnetic field due to phase 1 is denoted by $a \sin \omega_p t$. Show that the resultant magnetic field has an amplitude of $3a/2$ and that the field rotates at speed ω_p.

Solution The magnetic field vectors in the three sets of windings are shown in figure 7.27a. These can be resolved into two orthogonal components, as shown in figure 7.27b. The component in the vertical direction is

$$a \sin \omega_p t - a \sin\left(\omega_p t - \frac{2\pi}{3}\right) \cos \frac{\pi}{3} - a \sin\left(\omega_p t - \frac{4\pi}{3}\right) \cos \frac{\pi}{3}$$

$$= a \sin \omega_p t - \frac{a}{2}\left[\sin\left(\omega_p t - \frac{2\pi}{3}\right) + \sin\left(\omega_p t - \frac{4\pi}{3}\right)\right]$$

$$= a \sin \omega_p t - a \sin(\omega_p t - \pi) \cos \frac{\pi}{3}$$

$$= a \sin \omega_p t + \frac{a}{2}[\sin \omega_p t]$$

$$= \frac{3a}{2} \sin \omega_p t$$

Note that we used the trigonometric identities

$$\sin A + \sin B = 2 \sin\left(\frac{A + B}{2}\right) \cos\left(\frac{A - B}{2}\right)$$

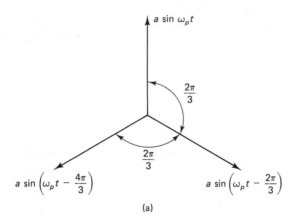

(a)

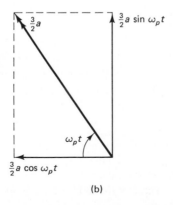

(b)

Figure 7.27. An alternative interpretation of a rotating magnetic field: (a) magnetic fields of the windings; (b) resultant magnetic field.

and

$$\sin(A - \pi) = -\sin A$$

in deriving this result. The horizontal component (to the left) of the magnetic fields is

$$a \sin\left(\omega_p t - \frac{4\pi}{3}\right) \sin \frac{\pi}{3} - a \sin\left(\omega_p t - \frac{2\pi}{3}\right) \sin \frac{\pi}{3}$$

$$= \frac{\sqrt{3}}{2} a \left[\sin\left(\omega_p t - \frac{4\pi}{3}\right) - \sin\left(\omega_p t - \frac{2\pi}{3}\right) \right]$$

$$- \sqrt{3} \, a \cos(\omega_p t - \pi) \sin\left(-\frac{\pi}{3}\right)$$

$$= \frac{3a}{2} \cos \omega_p t$$

Here we have used the trigonometric identities

$$\sin A - \sin B = 2 \cos \frac{A + B}{2} \sin \frac{A - B}{2}$$

$$\cos(A - \pi) = -\cos A$$

$$\sin -A = -\sin A$$

The resultant of the two orthogonal components is a vector of magnitude $3a/2$, making an angle $\omega_p t$ with the horizontal component, as shown in figure 7.27b. It follows that the resultant magnetic field has a magnitude of $3a/2$ and rotates in the clockwise direction at speed ω_p radians per second.

Induction Motor Characteristics

The stator windings of an induction motor generate a rotating magnetic field, as explained in the previous section. The rotor windings are purely secondary windings that are not energized by an external voltage. For this reason, no commutator-brush devices are needed in induction motors (see figure 7.28). The core of the rotor is made of ferromagnetic laminations in order to concentrate the magnetic flux and to minimize dissipation. The rotor windings are embedded in the axial direction on the surface of the rotor and are interconnected in groups. The rotor windings may consist of uninsulated copper (or any other conductor) bars (*a cage rotor*) or wire with one or more turns in each slot (*a wound rotor*). First, consider a stationary rotor. The rotating field in the stator intercepts the rotor windings, thereby generating an induced current due to mutual induction or transformer action (hence the name *induction motor*). The resulting secondary flux interacts with the primary, rotating flux, producing a torque in the direction of rotation of the stator field. This torque drives the rotor. As the rotor speed increases, initially the motor torque also increases (moderately) because of secondary interactions between the stator circuit and the rotor circuit, even though the relative magnitude between the rotating field speed and the rotor speed decreases, thus reducing the direct transformer action. (*Note:* The relative speed is termed the *slip rate*.) Soon, maximum torque will be reached.

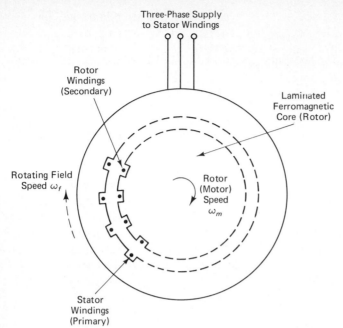

Three-Phase Supply
to Stator Windings

Rotor
Windings
(Secondary)

Laminated
Ferromagnetic
Core (Rotor)

Rotating Field
Speed ω_f

Rotor
(Motor)
Speed
ω_m

Stator
Windings
(Primary)

Figure 7.28. Schematic diagram of an induction motor.

A further increase in rotor speed (a decrease in slip rate) will decrease the motor torque sharply, until at synchronous speed (zero slip rate), the motor torque becomes zero. This behavior of an induction motor is illustrated by the typical characteristic curve given in figure 7.29. From starting torque T_s to maximum torque (breakdown torque) T_{max}, the motor behavior is unstable because, for example, an incremental increase in speed will cause an increase in torque that will further increase the speed,

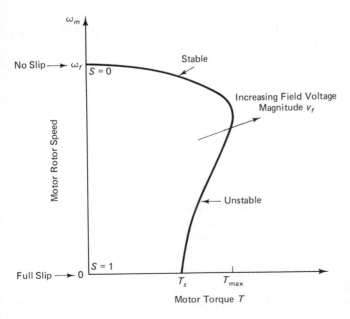

ω_m

No Slip → ω_f $S = 0$

Stable

Increasing Field Voltage
Magnitude v_f

Motor Rotor Speed

Unstable

Full Slip → 0 $S = 1$

T_s T_{max}

Motor Torque T

Figure 7.29. Torque-speed characteristic curve of an induction motor.

and an incremental reduction in speed will bring about a reduction in torque that will further reduce the speed. The portion of the curve from T_{max} to a zero-torque (no-load or synchronous) condition represents stable operation. Under normal operating conditions, an induction motor should operate in this region.

The fractional slip S for an induction motor is given by

$$S = \frac{\omega_f - \omega_m}{\omega_f} \tag{7.54}$$

Even when there is no external load, the synchronous operating condition ($S = 0$) is not achieved at steady state because of the presence of frictional torque that opposes the rotor motion. When an external torque (load torque) T_L is present, under normal operating conditions, the slip rate will further increase so as to increase the motor torque to support this load torque. As is clear from figure 7.29, in the stable region of the characteristic curve, the induction motor is quite insensitive to torque changes; a small change in speed would require a very large change in torque (in comparison with an equivalent DC motor). For this reason, an induction motor is relatively insensitive to load variations and can be regarded as a constant-speed machine. Note that if the rotor speed is increased beyond the synchronous speed (i.e., $S < 0$), the motor becomes a generator.

The Torque-Speed Relationship

It is instructive to determine the torque-speed relationship for an induction motor. This relationship provides insight into possible control methods for induction mo-

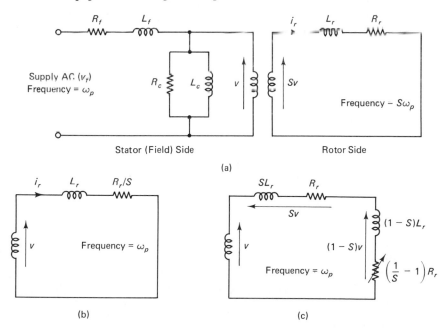

Figure 7.30. (a) Stator and rotor circuits for an induction motor. (b) Rotor circuit referred to the stator side. (c) Representation of available mechanical power using the rotor circuit.

tors. Stator and rotor equivalent circuits for one phase of an induction motor are shown in figure 7.30a. The circuit parameters are

R_f = stator coil resistance
L_f = stator leakage inductance
R_c = stator core iron loss resistance (eddy current effects, etc.)
L_c = stator core (magnetizing) inductance
L_r = rotor leakage inductance
R_r = rotor coil resistance

The magnitude of the supply AC voltage for each phase of the stator windings is v_f at the line frequency ω_p. The rotor current due to the induced e.m.f. is i_r. After allowing for the voltage drop due to stator resistance and stator leakage inductance, the voltage available for mutual induction is denoted by v. This is also the induced voltage in the secondary (rotor) winding at standstill, assuming the same number of turns. This induced voltage changes linearly with slip S, because the induced voltage is proportional to the relative velocity of the rotating field with respect to the rotor $(\omega_f - \omega_m)$, as is evident from equation 7.2. Hence, the induced voltage in the rotor winding (secondary winding) is Sv. Note, further, that at standstill ($S = 1$), the frequency of the induced voltage in the rotor is ω_p. At synchronous speed of rotation ($S = 0$), this frequency is zero because the magnetic field is fixed and constant relative to the rotor in this case. Now, assuming a linear variation of frequency of the induced voltage between these two extremes, we note that the frequency of the induced voltage in the rotor circuit is $S\omega_p$. These observations are indicated in figure 7.30a.

Using the frequency domain (complex) representation for the out-of-phase currents and voltages, the rotor current (complex) i_r is given by

$$i_r = \frac{Sv}{(R_r + jS\omega_p L_r)} = \frac{v}{(R_r/S + j\omega_p L_r)} \tag{7.55}$$

From equation 7.55, it is clear that the rotor circuit can be represented by a resistance R_r/S and an inductance L_r, excited by voltage v at frequency ω_p in series. This is, in fact, the rotor circuit referred to the stator side, as shown in figure 7.30b. This circuit can be grouped into two parts, as shown in figure 7.30c. The inductance SL_r and resistance R_r with a voltage drop Sv are identical to the rotor circuit in figure 7.30a. Note that SL_r has to be used for circuit equivalence here instead of L_r because of the frequency difference in the two circuits. The second voltage drop $(1 - S)v$ in figure 7.30c represents the back e.m.f. due to rotor–stator field interaction; it generates the capacity to drive an external load (mechanical power). It follows that the available mechanical power, per phase, of an induction motor is given by $i_r^2(1/S - 1)R_r$. Hence,

$$T_m \omega_m = p i_r^2 \left(\frac{1}{S} - 1\right) R_r \tag{7.56}$$

where T_m = motor torque generated in the rotor
  ω_m = motor rotor speed
  p = number of supply phases
  i_r = magnitude of the current in the rotor

The magnitude of the current in the rotor circuit is obtained from equation 7.55; thus,

$$i_r = \frac{v}{\sqrt{R_r^2/S^2 + \omega_p^2 L_r^2}} \qquad (7.57)$$

By substituting equation 7.57 in 7.56, we get

$$T_m = pv^2 \frac{S(1-S)}{\omega_m} \frac{R_r}{(R_r^2 + S^2\omega_p^2 L_r^2)} \qquad (7.58)$$

Now, from equations 7.53 and 7.54, we can express the number of pole pairs per phase of stator winding as

$$n = \frac{\omega_p}{\omega_m}(1-S) \qquad (7.59)$$

Equation 7.59 is substituted in 7.58; thus,

$$T_m = \frac{pnv^2 S R_r}{\omega_p(R_r^2 + S^2\omega_p^2 L_r^2)} \qquad (7.60)$$

If the resistance and leakage inductance in the stator are neglected, v is approximately equal to the stator excitation voltage v_f. This gives the torque-slip relationship:

$$T_m = \frac{pnv_f^2 S R_r}{\omega_p(R_r^2 + S^2\omega_p^2 L_r^2)} \qquad (7.61)$$

Note that by using equation 7.59, it is possible to express S in equation 7.61 in terms of rotor speed ω_m. This results in a torque-speed relationship that gives the characteristic curve shown in figure 7.29. Specifically, we employ the fact that the motor speed ω_m is related to slip through

$$S = \frac{\omega_p - n\omega_m}{\omega_p} \qquad (7.62)$$

Note, further, from equation 7.61 that the motor torque is proportional to the square of the supply voltage v_f.

Example 7.10

In the derivation of equation 7.61, we assumed that the number of effective turns per phase in the rotor is equal to that in the stator. This assumption is generally not valid, however. Determine how the equation should be modified in the general case. Suppose that

$$r = \frac{\text{number of effective turns per phase in the rotor}}{\text{number of effective turns per phase in the stator}}$$

Solution At standstill ($S = 1$), the induced voltage in the rotor is rv and the induced current is i_r/r. Hence, the impedance in the rotor circuit is given by

$$Z_r = \frac{rv}{i_r/r} = r^2\frac{v}{i_r} \tag{7.63}$$

or

$$Z_r = r^2 Z_{\text{req}}$$

It follows that the true rotor impedance (or resistance and inductance) simply has to be divided by r^2 to obtain the equivalent impedance. In this general case of $r \neq 1$, the resistance R_r and inductance L_r should be replaced by $R_{\text{req}} = R_r/r^2$ and $L_{\text{req}} = L_r/r^2$ in equation 7.61.

Example 7.11

Consider a three-phase induction motor that has one pole pair per phase. The equivalent resistance and leakage inductance in the rotor circuit are 8 Ω and 0.06 H, respectively. The motor supply is 115 V in each phase at 60 Hz. Compute the torque-speed curve for the motor.

Solution In this example, $R_r = 8$ Ω, $L_r = 0.06$ H, $v_f = 115$ V, $n = 1$, $p = 3$, and $\omega_p = 60 \times 2\pi$ rad/s. Now, using equation 7.61 along with 7.62, we can compute the torque-speed curve. The result is shown in figure 7.31.

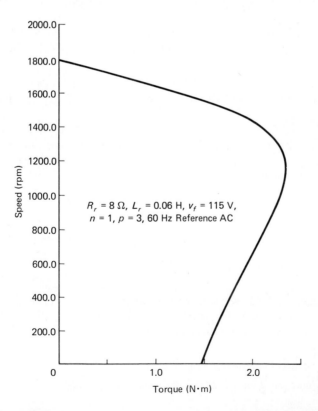

$R_r = 8$ Ω, $L_r = 0.06$ H, $v_f = 115$ V, $n = 1$, $p = 3$, 60 Hz Reference AC

Speed (rpm)

Torque (N·m)

Figure 7.31. Torque-speed curve for an induction motor example.

7.9 INDUCTION MOTOR CONTROL

DC motors are widely used in servo control applications because of their simplicity and flexible speed-torque capabilities; DC motors are easy to control, and they operate efficiently over a wide range of speeds. The initial cost and the maintenance cost of a DC motor are generally higher than those for a comparable AC motor. AC motors are most common in medium- to high-power drive applications. Of late, much effort has been invested in developing improved control methods for AC motors, and significant progress is seen in this area. Today's AC motors with thyristor control and field feedback compensation can provide speed control comparable to the capabilities of DC servomotors (e.g., $1:20$ range of speed variation, or 26 dB).

Since fractional slip S determines motor speed ω_m, equation 7.61 suggests several possibilities for controlling an induction motor. Four methods for induction motor control are

1. Excitation frequency control (ω_p)
2. Supply voltage control (v_f)
3. Rotor resistance control (R_r)
4. Pole changing (n)

Excitation Frequency Control

Excitation frequency control can be accomplished using a *thyristor circuit*. As discussed earlier, a thyristor (or SCR) is a semiconductor device that possesses very effective and nondissipative switching characteristics. Thyristors can handle high voltages and power levels. The principle of operation of the thyristor was discussed in the section on DC motor control (section 7.5).

A variable-frequency AC output can be generated from a DC supply using an *inverter circuit*. A single-phase inverter circuit is shown in figure 7.32. Thyristors 1

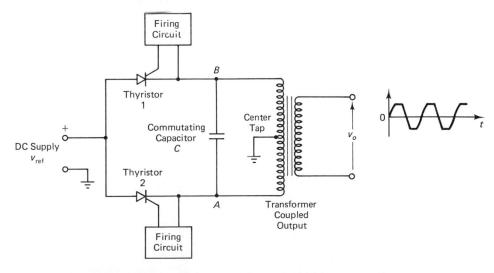

Figure 7.32. A single-phase inverter circuit for frequency control.

and 2 are gated by their firing circuits according to the required frequency of the output voltage v_o. The primary winding of the output transformer is center-tapped. A DC supply v_{ref} is applied to the circuit as shown. If both thyristors are not conducting the voltage across the capacitor C is zero. Now, if thyristor 1 is gated (fired), the current in the upper half of the primary winding will build to its maximum and the voltage across that half will reach v_{ref} (since the voltage drop across thyristor 1 is very small). As a result of the corresponding change in the magnetic flux, a voltage v_{ref} (approximately) will be induced in the lower half of the primary winding, complementing the voltage in the upper half. Accordingly, the voltage across the primary winding (or across the capacitor) is approximately $2v_{ref}$. Now, if thyristor 2 is fired, the voltage at point A becomes v_{ref}. Since the capacitor is already charged to $2v_{ref}$, the voltage at point B becomes $3v_{ref}$. This means that a voltage of $2v_{ref}$ is applied across thyristor 1 in the nonconducting direction. As a result, thyristor 1 will be turned off. Then, as before, a voltage $2v_{ref}$ is generated in the primary winding, but in the opposite direction, because it is thyristor 2 that is conducting now. In this manner, approximately rectangular AC voltage v_o is generated at the circuit output. The frequency of the voltage is equal to the inverse of the firing interval between the two thyristors. A three-phase inverter can be formed by triplicating the single-phase inverter and by phasing the firing times appropriately.

The block diagram in figure 7.33a shows a frequency control system for an induction motor. A standard three-phase supply is rectified to provide the DC supply to the three-phase inverter circuit. The firing of the thyristors is controlled by a microprocessor or hardware controller. If the control requirements are simple, a variable-frequency oscillator may be used instead. The controller may use external commands and feedback signals to generate the firing logic according to some program or hardware logic. Feedback signals may include shaft encoder readings (motor angle) for speed control and stator current (particularly for motor torque control). A typical control strategy is shown in figure 7.33b. In this case, the control processor provides a two-mode control. In the initial mode, the torque is kept constant while accelerating the motor. In the next mode, the power is kept constant while further increasing the speed. Both modes of operation can be achieved by frequency control. Modern frequency controllers for induction motors use pulse-width modulation (PWM), as for DC motors; and a single integrated circuit (IC) chip containing about 30,000 circuit elements can function as a hardware controller for an induction motor.

AC motors with frequency control are employed in many applications, including variable-flow control of pumps and fans, industrial manipulators (robots, hoists, etc.), process plants, and flexible operation of production machinery for flexible (variable-output) production. In particular, they can function as servomotors.

Voltage Control

From equation 7.61, it is seen that the torque of an induction motor is proportional to the square of the supply voltage level. It follows that an induction motor can be controlled by varying the supply voltage. This may be done in several ways. For example, amplitude modulation of the supply AC, using a ramp generator, will accom-

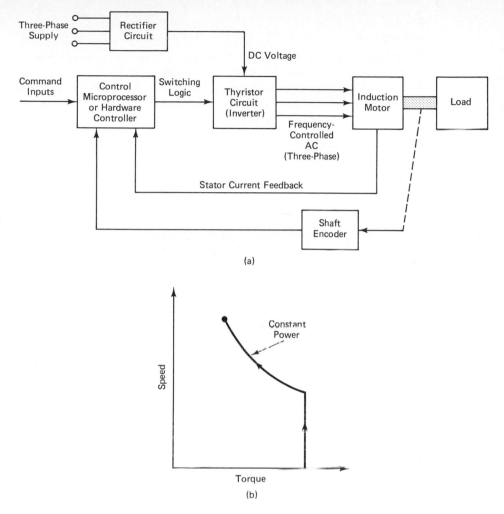

Figure 7.33. (a) Variable-frequency control of an induction motor. (b) A typical control strategy.

plish this objective directly by varying the supply amplitude. Alternatively, periodically (at high frequency) introducing zero-voltage regions (blanking out) in the AC supply using a thyristor circuit with firing delays will, as in pulse width modulation, accomplish voltage control by varying the root-mean-square (rms) value of the supply voltage. Voltage control methods are appropriate for small induction motors, but they provide poor efficiency when control over a wide speed range is required. Frequency control methods are recommended in low-power applications. An advantage of voltage control methods over frequency control methods is lower stator copper loss.

Example 7.12

Show that the fractional slip versus motor torque characteristics of an induction motor, at steady state, may be expressed by

$$T_m = \frac{aSv_f^2}{[1 + (S/S_b)^2]} \qquad (7.64)$$

Identify the parameters a and S_b. Show that S_b is the slip corresponding to the break-down torque (maximum torque) T_{max}. Obtain an expression for T_{max}.

An induction motor with parameter values $a = 4 \times 10^{-3}$ N.m/V^2 and $S_b = 0.2$ is driven by an AC supply that has line frequency 60 Hz. Stator windings have two pole pairs per phase. Initially, the line voltage is 500 V. The motor drives a mechanical load that can be represented by an equivalent viscous damper with damping constant $b = 0.265$ N.m/rad/s. Determine the operating point (torque and speed) for the system. Suppose that the supply voltage is dropped by 50 percent (to 250 V) using a voltage control scheme. What is the new operating point? Is this a stable operating point? In view of your answer, comment on the use of voltage control in induction motors.

Solution First, we note that equation 7.61 can be expressed as equation 7.64, with

$$a = \frac{pn}{\omega_p R_r} \qquad (7.65)$$

$$S_b = \frac{R_r}{\omega_p L_r} \qquad (7.66)$$

The breakdown torque is the peak torque and is obtained using

$$\frac{\partial T_m}{\partial \omega_m} = 0$$

But

$$\frac{\partial T_m}{\partial \omega_m} = \frac{\partial T_m}{\partial S} \frac{\partial S}{\partial \omega_m} = -\frac{1}{\omega_f} \frac{\partial T_m}{\partial S}$$

The last step used equation 7.54. It follows that the breakdown torque is given by

$$\frac{\partial T_m}{\partial S} = 0$$

Now, differentiating equation 7.64 with respect to S and equating to zero, we get

$$\left[1 + \left(\frac{S}{S_b} \right)^2 \right] - S \left[2 \frac{S}{S_b^2} \right] = 0$$

or

$$1 - \left(\frac{S}{S_b} \right)^2 = 0$$

Hence, $S = S_b$ corresponds to the breakdown torque. Substituting in equation 7.64, we have

$$T_{max} = \tfrac{1}{2} a S_b v_f^2 \qquad (7.67)$$

Next, using the given parameter values, the speed-torque curve is plotted as shown in figure 7.34 for the two cases $v_f = 500$ V and $v_f = 250$ V. Note that with $S_b = 0.2$, we have, from equation 7.67, $(T_{max})_1 = 100$ N.m and $(T_{max})_2 = 25$ N.m. These values are confirmed from the curves in figure 7.34. The load curve is

$$T_m = b\omega_m$$

or

$$T_m = b\omega_f \frac{\omega_m}{\omega_f}$$

Now, from equation 7.53, the synchronous speed is given by

$$\omega_f = \frac{60 \times 2\pi}{2} \text{ rad/s} = 188.5 \text{ rad/s}$$

Hence,

$$b\omega_f = 0.265 \times 188.5 = 50 \text{ N.m}$$

This is the slope of the load line shown in figure 7.34. The points of intersection of the load line and the motor characteristic curve are the steady-state operating points. They are, for case 1 ($v_f = 500$ V),

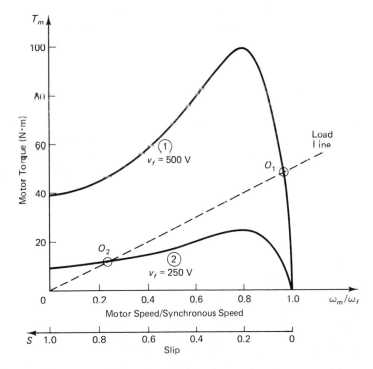

Figure 7.34. Speed-torque curves for induction motor voltage control (example 7.12).

Operating torque = 48 N.m
Operating slip = 4%
Operating speed = 1,728 rpm

and for case 2 (v_f = 250 V),

Operating torque = 12 N.m
Operating slip = 77%
Operating speed = 414 rpm

Note that when the supply voltage is halved, the torque drops by a factor of four and the speed drops by about 76 percent. But what is worse is that the new operating point (O_2) is in the unstable region (from $S = S_b$ to $S = 1$) of the motor characteristic curve. It follows that large drops in supply voltage are not feasible, and the efficiency of the motor can drop significantly with voltage control.

Rotor Resistance Control

It can be seen, from equation 7.61, that an induction motor can be controlled by varying R_r. Since this is a dissipative technique, it is also a wasteful method. It was a commonly used method for induction motor control prior to the development of more efficient thyristor circuits, digital signal processing (DSP) chips, and related control techniques. The rotor of an induction motor has a closed circuit (resistive-inductive) that is not connected to an external power supply, unlike a DC motor. In the wound-rotor design, windings are usually arranged and connected as polyphase groups (e.g., a delta-configuration (Δ) or star configuration (Y) in three-phase motors), just like the stator windings, but without a supply voltage. The current in the rotor circuit is generated purely by magnetic induction, but it determines the torque-speed characteristics of the motor. The motor response is controlled by changing the rotor resistance. This can be accomplished by connecting a variable resistance to each phase externally through a slip ring and brush arrangement, as shown schematically in figure 7.35 for the three-phase star (Y) connection. Rotor resistance control

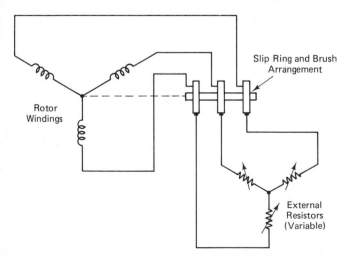

Figure 7.35. Rotor resistance control of an induction motor (three-phase).

     Continuous-Drive Actuators   Chap. 7

has all the disadvantages of the voltage control method. In particular, the motor efficiency drops considerably when motor operation over a wide range of speeds is needed.

Pole-Changing Control

The number of pole pairs per phase in the stator windings (n) is a parameter in the speed-torque equation 7.61. It follows that an induction motor can be controlled by changing n. This can be accomplished by switching the supply connections in the stator windings in some manner. The principle is illustrated in figure 7.36. Consider the windings in one phase of the stator. With the coil currents as in figure 7.36a, the magnetic fields in the alternate pairs of adjacent coils cancel out. When the coil currents are as in figure 7.36b, all adjacent pairs of coils have complementary magnetic fields. As a result, the number of pole pairs per phase is doubled when the stator is switched from the configuration in figure 7.36a to that in 7.36b. Note that when the stator windings are switched into a certain configuration of poles, the same switching should be done simultaneously to the rotor windings. This results in an additional complexity in the case of a wound rotor. In a squirrel-cage rotor, a separate

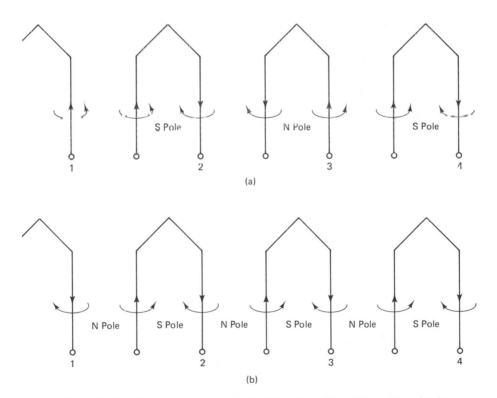

Figure 7.36. Pole-changing control of an induction motor. The number of pole pairs per phase is doubled when the currents in stator windings 1 and 3 are reversed, changing from (a) to (b).

switching mechanism is not necessary for the rotor because it automatically reacts to configure itself according to the winding configuration in the stator. For this reason, cage rotor induction motors are better suited for pole-changing control.

Field Feedback Control

A novel method for controlling AC motors is through field feedback compensation. This approach can be explained using the equivalent circuit shown in figure 7.30c. Note that this circuit separates the rotor-equivalent impedance into two parts—a nonproductive part and a torque-producing part, as discussed previously. In the present method of control, the magnetic field component associated with the first part is sensed and compensated for in the stator current. As a result, only the second part, corresponding to the back e.m.f., will remain, and the AC motor will behave quite like a DC motor that has an equivalent torque-producing back e.m.f. This type of control has been commercially implemented in AC motors using customized digital signal processing (DSP) chips.

A Transfer-Function Model for an Induction Motor

The true dynamic behavior of an induction motor is generally nonlinear and time-varying. But for small variations about an operating point, linear relations can be written. On this basis, a transfer-function model can be established for an induction motor. The procedure described in this section uses the steady-state speed-torque relationship for an induction motor to determine the transfer-function model. The basic assumption here is that this steady-state relationship represents the dynamic behavior of the motor for small changes about an operating point (steady-state) with reasonable accuracy. Suppose that a motor rotor that has moment of inertia J_m and mechanical damping (mainly from the bearings) constant b_m is subjected to a variation in motor torque δT_m and an associated change in rotor speed $\delta \omega_m$, as shown in figure 7.37. In general, these changes may arise from a change in the load torque, denoted by δT_L, and a change in supply voltage, denoted by δv_f. Newton's second law gives

$$\delta T_m - \delta T_L = J_m \, \delta \dot{\omega}_m + b_m \, \delta \omega_m \tag{7.68}$$

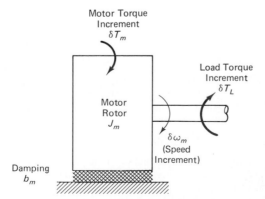

Figure 7.37. Derivation of a transfer-function model for an induction motor.

Now, using a linear steady-state relationship to represent the motor torque variation as a function of $\delta\omega_m$ and supply voltage variation δv_f, we have

$$\delta T_m = -b_e\,\delta\omega_m + k_v\,\delta v_f \tag{7.69}$$

By substituting equation 7.69 in 7.68 and using the Laplace variable s, we obtain

$$\delta\omega_m = \frac{k_v}{[J_m s + b_m + b_e]}\,\delta v_f - \frac{1}{[J_m s + b_m + b_e]}\,\delta T_L \tag{7.70}$$

In the transfer-function equation 7.70, note that $\delta\omega_m$ is the output, δv_f is the control input, and δT_L is an unknown (disturbance) input. The motor transfer function $\delta\omega_m/\delta v_f$ is given by

$$G_m(s) = \frac{k_v}{[J_m s + b_m + b_e]} \tag{7.71}$$

The motor time constant τ is

$$\tau = \frac{J_m}{b_m + b_e} \tag{7.72}$$

Now it remains to identify the parameters b_e (analogous to electrical damping in a DC motor) and k_v (a gain parameter). To accomplish this, we use equation 7.64, which can be written in the form

$$T_m = k(S)v_f^2 \tag{7.73}$$

where

$$k(S) = \frac{uS}{1 + (S/S_b)^2} \tag{7.74}$$

Now, in view of the fact that

$$\delta T_m = \frac{\partial T_m}{\partial\omega_m}\,\delta\omega_m + \frac{\partial T_m}{\partial v_f}\,\delta v_f$$

we have

$$b_e = -\frac{\partial T_m}{\partial\omega_m} \quad \text{and} \quad k_v = \frac{\partial T_m}{\partial v_f}$$

But

$$\frac{\partial T_m}{\partial\omega_m} = \frac{\partial T_m}{\partial S}\frac{\partial S}{\partial\omega_m} = -\frac{1}{\omega_f}\frac{\partial T_m}{\partial S}$$

Thus,

$$b_e = \frac{1}{\omega_f}\frac{\partial T_m}{\partial S} \tag{7.75}$$

where ω_f is the synchronous speed of the motor. Now, by differentiating equation 7.74 with respect to S, we have

$$\frac{\partial k}{\partial S} = a\frac{1 - (S/S_b)^2}{[1 + (S/S_b)^2]^2} \tag{7.76}$$

Hence,

$$b_e = \frac{av_f^2}{\omega_f}\frac{1 - (S/S_b)^2}{[1 + (S/S_b^2]^2} \tag{7.77}$$

Next, by differentiating equation 7.73 with respect to v_f, we have

$$\frac{\partial T_m}{\partial v_f} = 2k(S)v_f$$

Accordingly, we get

$$k_v = \frac{2aSv_f}{1 + (S/S_b)^2} \tag{7.78}$$

where S_b is the fractional slip at the breakdown (maximum) torque and a is a motor torque parameter defined by equation 7.65. If we wish to include the effects of the electrical time constant τ_e of the motor, we may include the factor $\tau_e s + 1$ in the denominator (characteristic polynomial) on the right-hand side of equation 7.70. Since τ_e is usually an order of magnitude smaller than τ as given by equation 7.72, no significant improvement in accuracy is obtained through this modification. Finally, note that the constants b_e and k_v can be obtained graphically using experimentally determined speed-torque curves for an induction motor for several values of the line voltage v_f. (See example 7.6 for a DC motor and example 7.13 for an induction motor.)

Example 7.13

A two-phase induction motor can serve as an AC servomotor. The field windings are identical and are placed in the stator with a geometric separation of 90°, as shown in figure 7.38a. One of the phases is excited by a fixed reference AC voltage ($v_{ref} \cos \omega_p t$). The other phase is 90° out of phase from the reference phase; it is the control phase, with voltage amplitude v_c. The motor is controlled by varying v_c.

1. With the usual notation, obtain an expression for the motor torque T_m in terms of the rotor speed ω_m and the input voltage v_c.
2. Indicate how a transfer function model could be obtained for this AC servo
   (a) Graphically, using the characteristic curves of the motor
   (b) Analytically, using the relationship obtained in part 1

Solution   Note that since $v_c \neq v_f$, the two phases are not balanced. Hence, the resultant magnetic field vector in this two-phase induction motor does not rotate at a constant speed ω_p; as a result, the relations derived previously cannot be applied directly. The first step, then, is to decompose the field vector into two components that rotate at constant speeds. This is accomplished in figure 7.38b. The field component 1 is equivalent to that of an induction motor supplied with a line voltage of ($v_{ref} + v_c)/2$, and it rotates in the clockwise direction at speed ω_p. The field component 2 is equivalent to that generated with a line voltage of ($v_{ref} - v_c)/2$, and it rotates in the counterclockwise direction. Suppose that the motor rotates in the clockwise direction at speed ω_p.

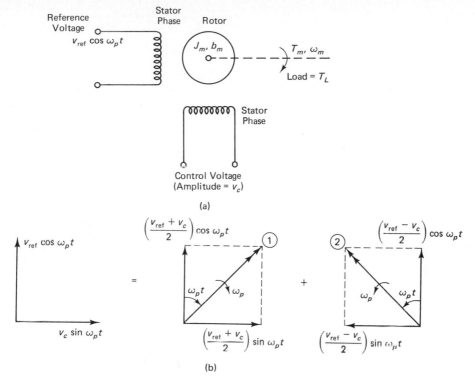

Figure 7.38. (a) A two-phase induction motor functioning as an AC servomotor. (b) Equivalent representation of the magnetic field vector in the stator.

The slip for the equivalent system 1 is

$$S = \frac{\omega_p - \omega_m}{\omega_p}$$

and the slip for the equivalent system 2 is

$$S' = \frac{\omega_p + \omega_m}{\omega_p} = 2 - S$$

in opposite directions. Now, using the relationship for an induction motor with a balanced multiphase supply (equation 7.73), we have

$$T_m = k(S)\left[\frac{v_{\text{ref}} + v_c}{2}\right]^2 - k(2 - S)\left[\frac{v_{\text{ref}} - v_c}{2}\right]^2 \qquad (7.79)$$

We have assumed in this derivation that the electrical and magnetic circuits are linear, so the principle of superposition holds. The function $k(S)$ is given by the standard equation 7.74, with a and S_b defined by equations 7.65 and 7.66, respectively. In this example, there is only one pole pair per phase ($n = 1$). Hence, the synchronous speed ω_f is equal to the line frequency ω_p. The motor speed ω_m is related to S through the usual equation 7.54.

To obtain the transfer-function relation for operation about an operating point, we use the differential relation

$$\delta T_m = \frac{\partial T_m}{\partial \omega_m} \, \delta \omega_m + \frac{\partial T_m}{\partial v_c} \, \delta v_c$$

$$= -b_e \, \delta \omega_m + k_v \, \delta v_c$$

As derived in the last section, the transfer relation is

$$\delta \omega_m = \frac{k_v}{[J_m s + b_m + b_e]} \, \delta v_c - \frac{1}{[J_m s + b_m + b_e]} \, \delta T_L$$

where T_L denotes the load torque. It remains to show how to determine the parameters b_e and k_v both graphically and analytically.

For the graphic method, we need a set of speed-torque curves for the motor for several values of v_c in the operating range. Note that experimental measurements of motor torque contain mechanical damping torque in the bearings. The actual electromagnetic torque of the motor is larger than the measured torque at steady state, the difference being the frictional torque. As a result, adjustments have to be made to the measured torque curve in order to get the true speed-motor torque curves. If this is done, the parameters b_e and k_v can be determined graphically as indicated in figure 7.39. Each curve is a constant v_c curve. Hence, the magnitude of its slope gives b_e. Note that $\partial T_m/\partial v_c$ is evaluated at constant ω_m. Hence, the constant k_v has to be determined on a vertical line ($\omega_m =$ constant). If two curves, one for the operating value of v_c and the other for a unit increment in v_c, are available, as shown in figure 7.39, the value of k_v is simply the vertical separation of the two curves at the operating point. If the increments in v_c are small, but not unity, the vertical separation of the two curves has to be divided by this increment (δv_c) in order to determine k_v, and the result will be more accurate.

To determine b_e and k_v analytically, we must differentiate T_m in equation 7.79 with respect to ω_m and v_c, respectively. Specifically,

$$b_e = -\frac{\partial T_m}{\partial \omega_m} = \frac{1}{\omega_p} \left[\frac{v_{\text{ref}} + v_c}{2} \right]^2 \frac{\partial k(S)}{\partial S} - \frac{1}{\omega_p} \left[\frac{v_{\text{ref}} - v_c}{2} \right]^2 \frac{\partial k(2 - S)}{\partial S} \tag{7.80}$$

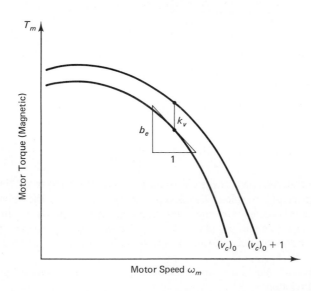

Figure 7.39. Graphic determination of transfer-function parameters for an induction motor.

Continuous-Drive Actuators   Chap. 7

where $[\partial k(S)/\partial S]$ is given by equation 7.76. To determine $[\partial k(2-S)/\partial S]$, we note that

$$\frac{\partial k(2-S)}{\partial S} = \frac{\partial k(2-S)}{\partial(2-S)}\frac{d(2-S)}{dS} = -\frac{\partial k(2-S)}{\partial(2-S)} = -\frac{\partial k(S)}{\partial S}\bigg|_{s=2-s} \quad (7.81)$$

In other words, $[\partial k(2-S)/\partial S]$ is obtained by first replacing S by $2-S$ in the right-hand side of equation 7.76 and then changing the sign. Finally,

$$k_v = \frac{\partial T_m}{\partial v_c} = \tfrac{1}{2}k(S)[v_{\text{ref}} + v_c] + \tfrac{1}{2}k(2-S)[v_{\text{ref}} - v_c] \quad (7.82)$$

Applications of induction motors include traction devices (e.g., ground transit vehicles), machine tools (e.g., lathes and milling machines), heavy-duty factory equipment (e.g., steel rolling mills), and equipment in large buildings (e.g., elevator drives, compressors, and fans).

Single-Phase AC Motors

The multiphase (polyphase) AC motors described previously are normally employed in moderate- to high-power applications (e.g., more than 5 hp). In low-power applications (e.g., motors used in household appliances such as refrigerators, food processors, and hair dryers), single-phase AC motors are commonly used, for they have the advantages of simplicity and low cost.

The stator of a single phase motor has only one set of drive windings, excited by a single-phase AC supply. If the rotor is running close to the frequency of the line AC, this single phase can maintain the motor torque, operating as an induction motor. But a single phase is obviously not capable of starting the motor. To overcome this problem, a second coil that is out of phase from the first coil is used during the starting period and is turned off automatically once the operating speed is attained. The phase difference is obtained either through a difference in inductance for a given resistance in the two coils or by including a capacitor in the second coil circuit.

7.10 SYNCHRONOUS MOTORS

Phase-locked servos and stepper motors can be considered synchronous motors because they run in synchronism with an external command signal (a pulse train) under normal operating conditions. The rotor of a synchronous AC motor rotates in synchronism with a rotating field generated by the stator windings. The generation of the rotating field is just like that in an induction motor. In contrast to an induction motor, however, the synchronous motor has rotor windings that are energized by an external DC source. The rotor poles obtained in this manner will lock themselves with the rotating field generated by the stator and will rotate at the same speed (synchronous speed). For this reason, synchronous motors are particularly suited for constant-speed applications under variable-load conditions. Synchronous motors with permanent-magnet (e.g., samarium-cobalt) rotors are also commercially available.

A schematic representation of the stator-rotor pair of a synchronous motor is shown in figure 7.40. The DC voltage required to energize the rotor windings may come from an independent DC supply, from an external rectified AC supply, or from a DC generator driven by the synchronous motor itself.

One major drawback of the synchronous AC motor is that an auxiliary "starter" is required to bring the rotor speed close to the synchronous speed. The reason for this is that in synchronous motors, the starting torque is virtually zero. To understand this, consider the starting conditions. The rotor is at rest and the stator field is rotating (at the synchronous speed). Consequently, there is 100 percent slip $(S = 1)$. When, for example, an N pole of the rotating field in the stator is approaching an S pole in the rotor, the magnetic force will tend to turn the rotor in the opposite direction of the rotating field. When the same N pole of the rotating field has just passed the rotor S pole, the magnetic force will tend to pull the rotor in the same direction as the rotating field. These opposite interactions balance out, producing a zero net torque on the rotor. One method of starting a synchronous motor is by using a small DC motor. Once the synchronous motor reaches the synchronous speed, the DC motor is operated as a DC generator to supply power to the rotor windings. Alternatively, a small induction motor could be used to start the synchronous motor. A more desirable arrangement that employs this principle is to include several sets of induction-motor-type rotor windings (cage-type or wound-type) in the synchronous motor rotor itself. In all these cases, the supply to the rotor windings of the synchronous motor is disconnected during the starting conditions and is turned on only when the synchronous speed is reached.

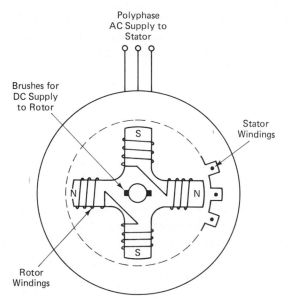

Figure 7.40. Schematic diagram of a stator-rotor configuration of a synchronous motor.

Synchronous Motor Control

Under normal operating conditions, the speed of a synchronous motor is completely determined by the frequency of the AC supply to the stator windings (see equation 7.53), because the motor speed is equal to the speed of the rotating field (ω_f). Hence, speed control can be achieved by the variable-frequency control method described in section 7.8 for induction motors. In some applications of AC motors (both induction and synchronous types), clutch devices that link the motor to the driven load are used to achieve variable-speed control (e.g., using an eddy current clutch system that produces a variable coupling force through eddy currents generated in the clutch). These dissipative techniques are quite wasteful. Furthermore, heat removal methods would be needed to avoid thermal problems. Hence, they are not recommended for high-power applications where motor efficiency is a prime consideration.

Note that unless a permanent-magnet rotor is used, a synchronous motor would require a slip ring and brush mechanism to supply the DC voltage to its rotor windings. This is a drawback that is not present in induction motors.

The steady-state speed-torque curve of a synchronous motor is a straight line parallel to the torque axis. But with proper control (e.g., frequency control), an AC motor can function as a servomotor. Conventionally, a servomotor has a linear torque-speed relationship. Applications of synchronous AC motors include steel rolling mills, rotary cement kilns, hoists, process compressors, recirculation pumps in hydroelectric power plants, and, more recently, servomotors. Synchronous motors are particularly suitable in high-speed, high-power applications where DC motors might not be appropriate. A synchronous motor can operate with a larger air gap between the motor and stator in comparison with an induction motor. This is an advantage for synchronous motors from the mechanical design point of view (e.g., bearing tolerances and rotor deflections due to thermal, static, and dynamic loads). Furthermore, rotor losses are smaller for synchronous motors than for induction motors.

Rectilinear Motors

The motors we have discussed so far are rotatory electromechanical actuators. It is possible to obtain a rectilinear motion from these devices by employing an auxiliary kinematic transducer, such as a cam-follower mechanism, a rack-and-pinion mechanism, or a lead screw. These devices inherently have problems of friction and backlash, however. For improved performance, direct rectilinear electromechanical actuators would be desirable. These devices operate according to the same principle as their rotatory counterparts, except that flat stators and rectilinearly moving elements are employed. Either DC or AC can be used to operate these actuators. Applications include traction devices, liquid-metal pumps, conveyor belt drives, and servovalve actuators.

The solenoid is a common rectilinear actuator that consists of a coil and a plunger. When the coil is activated by a DC signal, it pulls the plunger. The plunger is attached to the load and is resisted by a light spring and a damping element. Solenoids are on/off (push/pull) actuators and can be classified as digital actuators. They are inexpensive devices. Common applications of solenoids include valve actuators, switches, relays, and other two-position systems.

7.11 HYDRAULIC ACTUATORS

The ferromagnetic material in an electric motor saturates at some level of magnetic flux density. This limits the torque/mass ratio obtainable from an electric motor. Hydraulic actuators use the hydraulic power of a pressurized liquid. Since high pressures (on the order of 5,000 psi) can be used, hydraulic actuators are capable of providing very high forces (and torques) at very high power levels. The force limit of a hydraulic actuator can be an order of magnitude larger than that of an electromagnetic actuator. This results in higher torque/mass ratios than those available from electric motors, particularly at high levels of torque and power. This is a principal advantage of hydraulic actuators. Note that the actuator mass considered here is the mass of the final actuating element, not including auxiliary devices such as those needed to pressurize the fluid. Another advantage of a hydraulic actuator is that it is quite stiff when viewed from the side of the load. This is because a hydraulic medium is mechanically stiffer than an electromagnetic medium. Consequently, the control gains required in a high-power hydraulic control system would be significantly less than the gains required in a comparable electromagnetic (motor) control system. Note that the stiffness of an actuator may be measured by the slope of the speed-torque (force) curve.

Components of a Hydraulic Control System

A schematic diagram of a basic hydraulic control system is shown in figure 7.41. The hydraulic fluid (oil) is pressurized using a pump that is driven by an AC motor. Note that the motor converts electrical power into mechanical power, and the pump converts this into fluid power. In terms of through and across variable pairs, these power conversions can be expressed as

$$(i, v) \xrightarrow{\eta_m} (T, \omega) \xrightarrow{\eta_h} (Q, P)$$

in the usual notation. The conversion efficiency η_m of a motor is typically very high (over 90 percent), whereas the efficiency η_h of a hydraulic pump is not as good (about 60 percent), mainly because of dissipation, leakage, and compressibility effects. Depending on the pump capacity, flow rates in the range of 1,000 to 50,000 gallons per minute and pressures from 500 to 5,000 psi can be obtained. A hydraulic valve regulates the fluid into the actuator, controlling both the flow rate (including direction) and the pressure. This valve uses response signals (motion) from the load, in feedback, to achieve the desired response—hence the name *servovalve*. Usually,

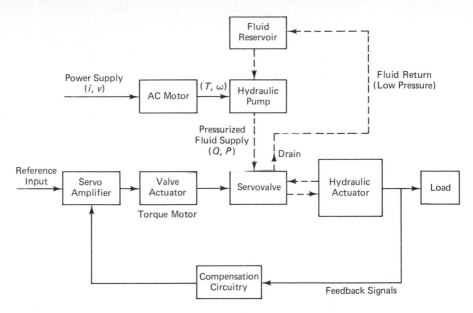

Figure 7.41. Schematic diagram of a hydraulic control system.

the servovalve is driven by an electric *valve actuator,* such as a *torque motor.* This is powered by the output from a *servo amplifier,* which receives a reference input command (corresponding to the desired position of the load) as well as a measured response of the load (in feedback). Compensation circuitry may be used in both feedback and forward paths to modify the signals so as to obtain the desired control action. The hydraulic actuator (typically a piston-cylinder device for rectilinear motions or a hydraulic motor for rotatory motions) converts fluid power back into mechanical power, which is available to perform useful tasks (i.e., to drive a load). Note that some power in the fluid is lost at this stage. The low-pressure fluid at the drain of the hydraulic servovalve is returned to the reservoir and is available to the pump.

One might argue that since the power that is required to drive the load is mechanical, it would be much more efficient to use a motor directly to drive that load. There are good reasons for using hydraulic power, however. For example, AC motors are usually difficult to control, particularly under variable-load conditions. Their efficiency can drop rapidly when the speed deviates from the rated speed, particularly when voltage control is used. They need gear mechanisms for low-speed operation, with associated problems such as backlash, friction, vibration, and mechanical loading effects. Special coupling devices are also needed. Hydraulic devices usually filter out high-frequency noise, which is not the case with AC motors.

Thus, hydraulic systems are ideal for high-power, high-force control applications. In high-power applications, a single high-capacity pump or several pumps can be employed to pressurize the fluid. Furthermore, in low-power applications, several servovalve actuator systems can be operated to perform different control tasks in a distributed control environment, using the same pressurized fluid supply. In this sense, hydraulic systems are very flexible. Hydraulic systems provide excellent speed-force (torque) capability, variable over a wide range of speeds without significantly affecting the power-conversion efficiency because the excess high-pressure fluid is diverted to the return line. Consequently, hydraulic actuators are far more controllable than AC motors.

Hydraulic actuators also have an advantage over electromagnetic actuators from the point of view of heat transfer characteristics. Specifically, the hydraulic fluid promptly carries away any heat that is generated locally and releases it through a heat exchanger at a location away from the actuator. However, special cooling mechanisms are needed in heavy-duty electric motors. Applications of hydraulic control systems include vehicle steering and braking systems, active suspension systems, heavy-duty presses, heavy-duty shakers used in dynamic testing, rolling mills, industrial mechanical manipulators such as hoists, machine tools, industrial robots, and actuators for aircraft control surfaces (ailerons, rudder, and elevators).

Hydraulic Pumps and Motors

The objective of a hydraulic pump is to provide pressurized oil to a hydraulic actuator. Three common types of hydraulic pumps are

1. The vane pump
2. The gear pump
3. The axial piston pump

The pump type used in a hydraulic control system is not very significant, except for the pump capacity, in terms of the control functions of the system. But since hydraulic motors can be interpreted as pumps operating in the reverse direction, we shall make some effort here to understand the operation of these three types of pumps.

A sliding-type *vane pump* is shown schematically in figure 7.42. The vanes slide on the housing. The contact between vanes and housing is maintained by springs in the slots of the rotor. The rotor is eccentrically mounted inside the housing. The fluid is drawn in at the inlet port because of the increasing volume between vane pairs as they rotate. The volume of oil trapped between two vanes is then compressed because of the decrease in the volume of the vane compartment, as shown. Note that a pressure rise will result from pushing the liquid volume to the high-pressure side and not allowing it to return to the low-pressure side of the pump, even when there is no significant change in the volume of the liquid when it moves from the low-pressure side to the high-pressure side. The typical operating pressure (at the outlet port) of these devices is about 1,500 psi. The pressure can be varied by adjusting the rotor eccentricity. A disadvantage of any rotating device with eccentricity is the centrifugal forces that are generated even while rotating at constant speed. Dynamic balancing is needed to reduce this problem.

The operation of an *external gear hydraulic pump* is illustrated in figure 7.43. The two identical gears are externally meshed. The inlet port is facing the gear unmeshing (retracting) region. Fluid is drawn in because of the resulting increase in volume between the two sets of gear teeth. This volume is trapped and transported

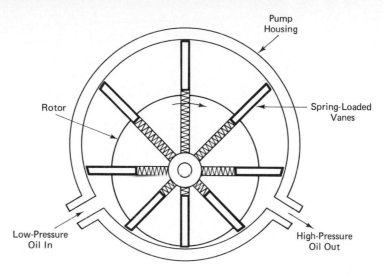

Figure 7.42. A hydraulic vane pump.

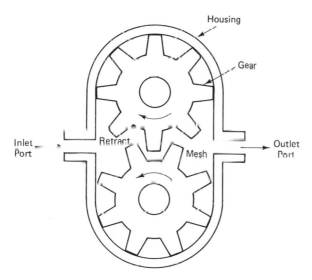

Figure 7.43. A hydraulic gear pump.

around by both gear wheels into the gear meshing region. Here it undergoes a compression, as in the vane pump, by pushing the fluid into the high-pressure side, thus increasing the pressure. Only moderate to low pressures can be realized by gear pumps (1,000 psi maximum), because the volume changes that take place in the unmeshing and meshing regions are small (unlike in vane pumps) and because fluid leakage between teeth and housing can be significant. Gear pumps are robust and low-cost devices, however, and they are probably the most commonly used hydraulic pumps.

A schematic diagram of an *axial piston hydraulic pump* is shown in figure 7.44. The chamber barrel is rigidly attached to the drive shaft. The two pistons also rotate with the chamber barrel, but since the end shoes of the pistons slide inside a

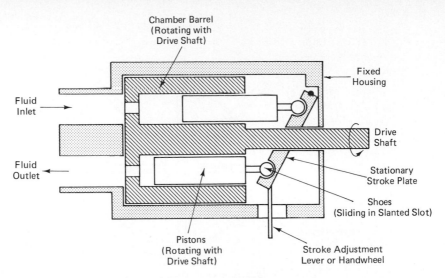

Fluid Inlet

Fluid Outlet

Chamber Barrel
(Rotating with
Drive Shaft)

Fixed
Housing

Drive
Shaft

Stationary
Stroke Plate

Shoes
(Sliding in Slanted Slot)

Pistons
(Rotating with
Drive Shaft)

Stroke Adjustment
Lever or Handwheel

Figure 7.44. An axial piston hydraulic pump.

slanted (skew) slot that is stationary, the pistons also undergo a reciprocating motion in the axial direction simultaneously with the rotatory motion. As a chamber opening reaches the inlet port of the pump housing, fluid is drawn in because of the increasing volume between the piston and the chamber. This fluid is trapped and transported to the outlet port while undergoing compression as a result of the decreasing volume inside the chamber due to the axial motion of the piston. Fluid pressure increases in this process. High outlet pressures (over 2,000 psi) can be achieved using piston pumps. As shown in figure 7.44, the piston stroke can be increased by increasing the inclination of the stroke plate (slot). This, in turn, increases the pressure ratio of the pump. A lever mechanism is usually available to adjust the piston stroke. Piston pumps are relatively expensive.

Hydraulic pump efficiency is given by the ratio of the output fluid power to the motor mechanical power; thus,

$$\eta_P = \frac{PQ}{\omega T} \tag{7.83}$$

where P = pressure increase in the fluid
Q = fluid flow rate
ω = rotating speed of the pump
T = drive torque to the pump

Hydraulic Servovalves

Valves can perform three basic functions:

1. Change flow direction
2. Change flow rate
3. Change fluid pressure

The valves that accomplish the first two functions are termed *flow-control valves*. The valves that regulate the fluid pressure are termed *pressure-control valves*. A simple relief valve controls pressure, whereas a gate valve and a globe valve are on/off flow-control valves. Valves are classified by the number of flow paths present under operating conditions. For example, a four-way valve has four ways in which flow can enter and leave the valve.

Spool valves are used extensively in hydraulic servo systems. A schematic diagram of a four-way spool valve is shown in figure 7.45a. Input displacement (U) applied to the spool rod, using an actuator (torque motor), will regulate the flow rate (Q) to the main hydraulic actuator as well as the corresponding pressure difference (P) available to the actuator. If the land (spool) length is larger than the port width (figure 7.45b), it is an *overlapped land*. This introduces a dead zone in the neighborhood of the central position of the spool, resulting in decreased sensitivity and increased stability problems. Since it is virtually impossible to exactly match the land

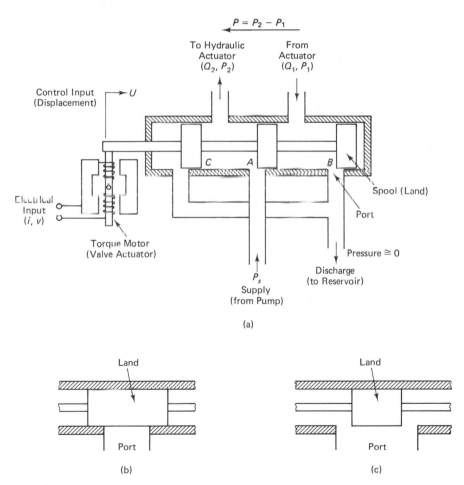

Figure 7.45. (a) A hydraulic servovalve (four-way spool valve). (b) An overlapped land. (c) An underlapped land.

size with the port width, the *underlapped land* configuration (figure 7.45c) is commonly employed. In this case, there is a leakage flow, even in the fully closed position, which decreases efficiency and increases the steady-state error of the hydraulic control system. For accurate operation of the valve, the leakage should not be excessive. The direct flow at various ports of the valve and the leakage flows between the spools (lands) and the valve housing should be included in a realistic analysis of a valve. For small displacements δU about an operating point, the following linearized equations can be written.

Since the flow rate Q_2 into the actuator increases as U increases and decreases as P_2 increases, we have

$$\delta Q_2 = k_q\,\delta U - k_c'\,\delta P_2 \tag{7.84}$$

Similarly, since the flow rate Q_1 from the actuator increases with both U and P_1, we have

$$\delta Q_1 = k_q\,\delta U + k_c'\,\delta P_1 \tag{7.85}$$

The gains k_q and k_c' will be defined later.

In fact, if we disregard the compressibility of the fluid, $\delta Q_1 = \delta Q_2$, assuming that the hydraulic piston (actuator) is double-acting, with equal piston areas on the two sides of the actuator piston. We consider the general case where $\delta Q_1 \ne \delta Q_2$. Note, however, that the inlet port and the outlet port are assumed to have identical characteristics. By adding equations 7.84 and 7.85 and defining an average flow rate

$$Q = \frac{Q_1 + Q_2}{2} \tag{7.86}$$

and an equivalent flow-pressure coefficient

$$k_c = \frac{k_c'}{2} \tag{7.87}$$

we get

$$\delta Q = k_q\,\delta U - k_c\,\delta P \tag{7.88}$$

where the *flow gain* is

$$k_q = \left(\frac{\partial Q}{\partial U}\right)_P \tag{7.89}$$

and the *flow-pressure coefficient* is

$$k_c = -\left(\frac{\partial Q}{\partial P}\right)_U \tag{7.90}$$

Note that the *pressure sensitivity* is

$$k_p = \left(\frac{\partial P}{\partial U}\right)_Q = \frac{k_q}{k_c} \tag{7.91}$$

To obtain equation 7.91, we use the well-known result from calculus:

$$\delta Q = \left(\frac{\partial Q}{\partial U}\right)_P \delta U + \left(\frac{\partial Q}{\partial P}\right)_U \delta P$$

Since $\delta P/\delta U \rightarrow \partial P/\partial U$ as $\delta Q \rightarrow 0$, we have

$$\left(\frac{\partial P}{\partial U}\right)_Q = -\left(\frac{\partial Q}{\partial U}\right)_P \Big/ \left(\frac{\partial Q}{\partial P}\right)_U \tag{7.92}$$

Equation 7.91 follows from equation 7.92.

A valve can be actuated by several methods, including manual operation, the use of mechanical linkages connected to the drive load, and the use of electromechanical actuators such as solenoids and torque motors (or force motors). Solenoids are suitable for on/off control applications, and torque motors are used in continuous control. For precise control applications, electromechanical actuation of the valve (with feedback) is preferred.

Large valve displacements can saturate a valve because of the nonlinear nature of the flow relations at the valve ports. To overcome this saturation problem when controlling heavy loads, several valve stages may be used. In this case, the spool motion of the first stage is the input motion. It actuates the spool of the second stage, which acts as a hydraulic amplifier. The fluid supply to the main hydraulic actuator that drives the load is regulated by the final stage of a multistage valve.

Steady-State Valve Characteristics

Although the linearized valve equation 7.88 is used in the analysis of hydraulic control systems, it should be noted that the flow equations of a valve are quite nonlinear. Consequently, the valve constants k_q and k_c change with the operating point. Valve constants can be determined either by experimental measurements or by using an accurate nonlinear model. In this section, we establish a reasonably accurate nonlinear relationship relating the (average) flow rate Q through the main hydraulic actuator and the pressure difference (load pressure) P provided to the hydraulic actuator.

Assume identical rectangular ports at the supply and discharge in figure 7.45a. When the valve lands are in the neutral (central) position, we take $U = 0$. We assume that the lands match the ports perfectly (i.e., no dead zone or leakage flows due to clearances). The positive direction of U is taken as shown in figure 7.45a. For this positive configuration, the flow directions are also indicated in the figure. The flow equations at ports A and B are

$$Q_2 = Ubc_d \sqrt{\frac{2(P_s - P_2)}{\rho}} \tag{7.93}$$

$$Q_1 = Ubc_d \sqrt{\frac{2P_1}{\rho}} \tag{7.94}$$

where b = land width
c_d = discharge coefficient at each port
ρ = density of the hydraulic fluid
P_s = supply pressure of the hydraulic fluid

Note that the pressure at the discharge is taken to be zero in equation 7.94. For steady-state operation, we use

$$Q_1 = Q_2 = Q \qquad (7.95)$$

Now, squaring equations 7.93 and 7.94 and adding, we get

$$2Q^2 = 2(Ubc_d)^2 \frac{(P_s - P)}{\rho}$$

where the pressure difference supplied to the hydraulic actuator is denoted by

$$P = P_2 - P_1 \qquad (7.96)$$

Consequently,

$$Q = Ubc_d \sqrt{\frac{P_s - P}{\rho}} \qquad \text{for } U > 0 \qquad (7.97)$$

When $U < 0$, the flow direction reverses; furthermore, port A is now associated with P_1 (not P_2) and port C is associated with P_2. It follows that equation 7.97 still holds, except that $P_2 - P_1$ is replaced by $P_1 - P_2$. Hence,

$$Q = Ubc_d \sqrt{\frac{P_s + P}{\rho}} \qquad \text{for } U < 0 \qquad (7.98)$$

Combining equations 7.97 and 7.98, we have

$$Q = Ubc_d \sqrt{\frac{P_s - P \, \text{sgn}(U)}{\rho}} \qquad (7.99)$$

This can be written in the nondimensional form

$$\frac{Q}{Q_{\text{max}}} = \frac{U}{U_{\text{max}}} \sqrt{1 - \frac{P}{P_s} \, \text{sgn}\left(\frac{U}{U_{\text{max}}}\right)} \qquad (7.100)$$

where $U_{\text{max}} = $ maximum valve opening (> 0) and

$$Q_{\text{max}} = U_{\text{max}} bc_d \sqrt{\frac{P_s}{\rho}} \qquad (7.101)$$

Equation 7.100 is plotted in figure 7.46. Note that as with the speed-torque curve for a motor, it is possible to obtain the valve constants k_q and k_c, defined by equations 7.89 and 7.90, from the curves given in figure 7.46 for various operating points. For better accuracy, experimentally determined valve characteristic curves should be used.

Hydraulic Primary Actuators

Rotatory hydraulic actuators (hydraulic motors) operate much like the hydraulic pumps discussed earlier, except that the direction flow is reversed and the mechanical power is delivered by the shaft, rather than taken in. High-pressure fluid enters

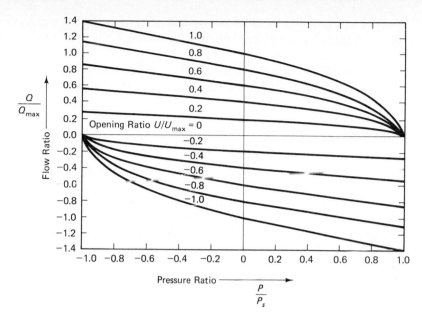

Figure 7.46. Steady-state characteristics for a four-way spool valve.

the actuator. As it passes through the hydraulic motor, the fluid power is used up in turning the rotor, and the pressure is dropped. The low-pressure fluid leaves the motor en route to the reservoir. One of the more efficient rotary hydraulic actuators is the axial piston motor, quite similar in construction to the axial piston pump shown in figure 7.44.

The most common type of rectilinear hydraulic actuator, however, is the hydraulic ram (piston-cylinder actuator). A schematic diagram of such a device is shown in figure 7.47. This is a *double-acting actuator* because the fluid pressure acts on both sides of the piston. If the fluid pressure is present only on one side of the piston, it is termed a *single-acting actuator*. Single-acting piston-cylinder (ram) actuators are also in common use for their simplicity and the simplicity of the other

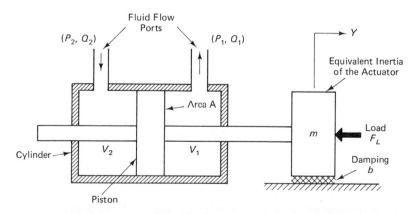

Figure 7.47. Double-acting piston-cylinder hydraulic actuator.

control components, such as servovalves, that are needed, although they have the disadvantage of asymmetry. The fluid flow at the ports of a hydraulic actuator is regulated by a servovalve.

To obtain the equations for the actuator shown in figure 7.47, we must realize that the flow rate Q into a chamber depends primarily on two factors;

1. Increase in chamber volume
2. Increase in pressure (compressibility effect of the fluid)

When a piston of area A moves through a distance Y, the flow rate due to the increase in chamber volume is $\pm A\dot{Y}$. Now, with an increase in pressure δP, the volume of a given fluid mass would decrease by the amount $[-(\partial V/\partial P)\delta P]$. As a result, an equal volume of new fluid would enter the chamber. The corresponding rate of flow is $[-(\partial V/\partial P)(dP/dt)]$. Since the *bulk modulus* (isothermal, or at constant temperature) is given by

$$\beta = -V\frac{\partial P}{\partial V} \tag{7.102}$$

the rate of flow due to the rate of pressure change is given by $[(V/\beta)(dP/dt)]$. Using these facts, the fluid conservation equations for the two sides of the actuator chamber in figure 7.47 can be written as

$$Q_2 = A\frac{dY}{dt} + \frac{V_2}{\beta}\frac{dP_2}{dt} \tag{7.103}$$

$$Q_1 = A\frac{dY}{dt} - \frac{V_1}{\beta}\frac{dP_1}{dt} \tag{7.104}$$

Note that for a realistic analysis, leakage flow rate terms (for leakage between piston and cylinder and between piston rod and cylinder) should be included in equations 7.103 and 7.104. For a linear analysis, these leakage flow rates can be taken as proportional to the pressure difference across the leakage path. Note, further, that V_1 and V_2 can be expressed in terms of Y, as follows:

$$V_1 + V_2 = V_o \tag{7.105}$$

$$V_1 - V_2 = V_o' + 2AY \tag{7.106}$$

where V_o and V_o' are constant volumes that depend on the cylinder capacity and on the piston position when $Y = 0$, respectively. Now, for incremental changes about the operating point $V_1 = V_2 = V$, equations 7.103 and 7.104 can be written as

$$\delta Q_2 = A\frac{d\delta Y}{dt} + \frac{V}{\beta}\frac{d\delta P_2}{dt} \tag{7.107}$$

$$\delta Q_1 = A\frac{d\delta Y}{dt} - \frac{V}{\beta}\frac{d\delta P_1}{dt} \tag{7.108}$$

Note that the absolute equations 7.103 and 7.104 are already linear for constant V. But since the valve equation is nonlinear, and since V changes, we should use the incremental equations 7.107 and 7.108, instead of the absolute equations, in linear models. Adding equations 7.107 and 7.108 and dividing by 2, we get the hydraulic actuator equation

$$\delta Q = A \frac{d\delta Y}{dt} + \frac{V}{2\beta} \frac{d\delta P}{dt} \qquad (7.109)$$

where

$$Q = \frac{Q_1 + Q_2}{2} = \text{average flow into the actuator}$$

$$P = P_2 - P_1 = \text{pressure difference at the actuator piston}$$

The Load Equation

So far, we have obtained the linearized valve equation 7.88 and the linearized actuator actuation 7.109. It remains to determine the load equation, which depends on the nature of the load that is driven by the hydraulic actuator. We may represent it by a load force F_L, as shown in figure 7.47. Note that F_L is a dynamic term that may represent such effects as flexibility, inertia, and the dissipative effects of the load. In additon, the inertia of the moving parts of the actuator is modeled as a mass m, and the energy dissipation effects associated with these moving parts are represented by an equivalent viscous damping constant b. Accordingly, Newton's second law gives

$$m \frac{d^2 Y}{dt^2} + b \frac{dY}{dt} = A(P_2 - P_1) - F_L \qquad (7.110)$$

This equation is also linear already. Again, since the valve equation is nonlinear, we should consider incremental motions δY about an operating point. Consequently, we have

$$m \frac{d^2 \delta Y}{dt^2} + b \frac{d\delta Y}{dt} = A \, \delta P - \delta F_L \qquad (7.111)$$

where, as before

$$P = P_2 - P_1$$

If the active areas on the two sides of the piston are not equal, a net imbalance force would exist. This could lead to unstable response under some conditions.

7.12 HYDRAULIC CONTROL SYSTEMS

The main components of a hydraulic control system are

1. A servovalve
2. A hydraulic actuator
3. A load

4. Feedback control elements

We have obtained linear equations for the first three components as equations 7.88, 7.109, and 7.111. We shall rewrite these equations, denoting the incremental variables about an operating point by lowercase letters.

$$\textit{Valve:} \quad q = k_q u - k_c p \tag{7.112}$$

$$\textit{Hydraulic actuator:} \quad q = A\frac{dy}{dt} + \frac{V}{2\beta}\frac{dp}{dt} \tag{7.113}$$

$$\textit{Load:} \quad m\frac{d^2 y}{dt^2} + b\frac{dy}{dt} = Ap - f_L \tag{7.114}$$

Note that the feedback elements will depend on the specific feedback control method employed. We shall return to this aspect of a hydraulic control system later. Equations 7.112 through 7.114 can be represented by the block diagram shown in figure 7.48a. This is an open-loop control system because no external feedback elements have been used. Note, however, the presence of a "natural" *pressure feedback* path and a "natural" *velocity feedback* path that are inherent to the dynamics of the open-loop system. The block diagram can be reduced to the equivalent form shown in

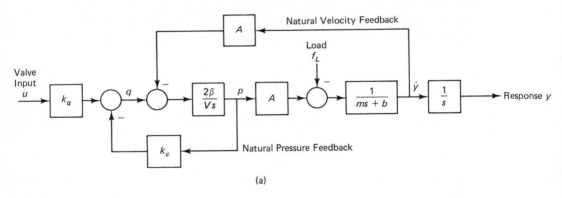

(a)

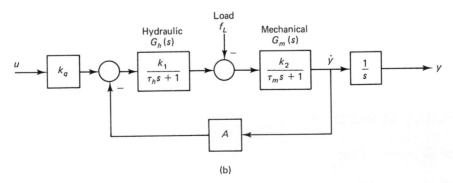

(b)

Figure 7.48. (a) Block diagram for an open-loop hydraulic control system. (b) An equivalent block diagram.

Continuous-Drive Actuators   Chap. 7

figure 7.48b. To obtain this, combine the first two summing junctions and then obtain the equivalent transfer function for the pressure feedback loop. This equivalent transfer function can be obtained using the relationship for equivalent transfer function:

$$G_h = \frac{G}{1 + GH} \qquad (7.115)$$

where G = forward transfer function
H = feedback transfer function

In this case,

$$G = \frac{2\beta}{Vs}$$

and

$$H = k_c$$

Hence,

$$G_h = \frac{k_1}{\tau_h s + 1} \qquad (7.116)$$

where the *pressure gain* parameter is

$$k_1 = \frac{1}{k_c} \qquad (7.117)$$

and the hydraulic time constant is

$$\tau_h = \frac{V}{2\beta k_c} \qquad (7.118)$$

Note that the pressure gain k_1 is a measure of the load pressure p generated for a given flow rate q into the hydraulic actuator. The smaller the pressure coefficient k_c, the larger the pressure gain will be, as is clear from equation 7.90. The hydraulic time constant increases with the volume of the actuator fluid chamber and decreases with the bulk modulus of the hydraulic fluid. This is to be expected, because the hydraulic time constant depends on the compressibility of the hydraulic fluid.

The mechanical transfer function of the hydralulic actuator is represented by

$$G_m = \frac{k_2}{\tau_m s + 1} \qquad (7.119)$$

where the mechanical time constant is given by

$$\tau_m = \frac{m}{b} \qquad (7.120)$$

and $k_2 = 1/b$. Typically, the mechanical time constant is the dominant time constant, since it is usually larger than the hydraulic time constant.

Example 7.14

A model for the automatic gage control (AGC) system of a steel rolling mill is shown in figure 7.49. The rolls are pressed using a single-acting hydraulic actuator with valve displacement u. The rolls are displaced through y, thereby pressing the steel that is being rolled. The rolling force F is completely known from the steel parameters for a given y.

1. Identify the inputs and the controlled variable in this control system.
2. In terms of the variables and system parameters indicated in figure 7.49, write dynamic equations for the system, including valve nonlinearities.
3. What is the order of the system? Identify the response variables.
4. Draw a block diagram for the system, clearly indicating the hydraulic actuator with valve, the mill structure, inputs, and the controlled variable.
5. What variables would you measure (and feed back through suitable controllers) in order to improve the performance of the control system?

Solution   *Part 1:* Valve displacement u and rolling force F are inputs. Roll displacement y is the controlled variable.

Part 2: The structural-dynamic equations are

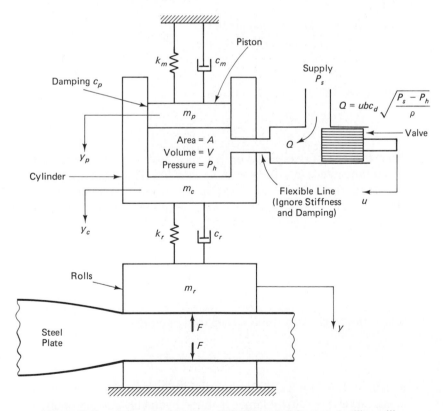

Figure 7.49.   Automatic gage control (AGC) system for a steel rolling mill.

$$m_p \ddot{y}_p = -k_m y_p - c_m \dot{y}_p - c_p(\dot{y}_p - \dot{y}_c) - AP_h \qquad \text{(i)}$$

$$m_c \ddot{y}_c = -k_r(y_c - y) - c_r(\dot{y}_c - \dot{y}) - c_p(\dot{y}_c - \dot{y}_p) + AP_h \qquad \text{(ii)}$$

$$m_r \ddot{y} = -k_r(y - y_c) - c_r(\dot{y} - \dot{y}_c) - F \qquad \text{(iii)}$$

Note that the static forces balance and the displacements are measured from the corresponding equilibrium configuration, so that gravity terms do not enter the equations.

The *hydraulic actuator equation* is derived as follows: For the valve, with the usual notation, the flow rate is given by

$$Q = buc_d \sqrt{\frac{P_s - P_h}{\rho}}$$

For the piston-cylinder,

$$Q = A(\dot{y}_c - \dot{y}_p) + \frac{V}{\beta} \frac{dP_h}{dt}$$

Hence

$$\frac{V}{\beta} \frac{dP_h}{dt} = A(\dot{y}_c - \dot{y}_p) + buc_d \sqrt{\frac{P_s - P_h}{\rho}} \qquad \text{(iv)}$$

Part 3: There are three second-order differential equations (i), (ii), (iii) and one first-order differential equation (iv). Hence, the system is seventh-order. The response variables are the displacements y_p, y_c, y and the pressure P_h.

Part 4: A block diagram for the hydraulic control system of the steel rolling mill is shown in figure 7.50.

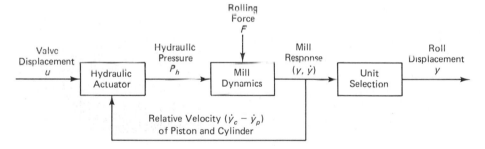

Figure 7.50. Block diagram for the hydraulic control system of a steel rolling mill.

Part 5: The hydraulic pressure P_h and the roll displacement y are the two response variables that can be conveniently measured and used in feedback control. The rolling force F may be measured and fed forward, but this is somewhat difficult in practice.

Example 7.15

A single-stage pressure control valve is shown in figure 7.51. The purpose of the valve is to keep the load pressure P_L constant. Volume rates of flow, pressures, and the volumes of fluid subjected to those pressures are indicated in the figure. The mass of the spool and appurtenances is m, the damping constant of the damping force acting on the moving parts is b, and the effective bulk modulus of oil is β. The accumulator volume

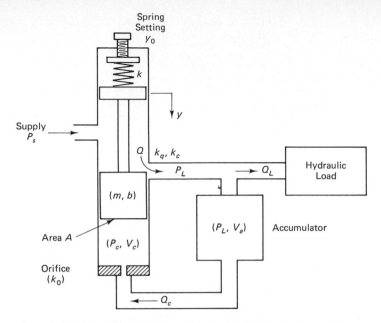

Figure 7.51. A single-stage pressure control valve.

is V_a. The flow into the valve chamber (volume V_c) is through an orifice. This flow may be taken as proportional to the pressure drop across the orifice, the constant of proportionality being k_o. A compressive spring of stiffness k restricts the spool motion. The initial spring force is set by adjusting the initial compression y_o of the spring.

1. Identify the reference input, the primary output, and a disturbance input for the valve system.
2. By making linearization assumptions and introducing any additional parameters that might be necessary, write equations to describe the system dynamics.
3. Set up a block diagram for the system, showing various transfer functions.

Solution *Part 1:*

    Input setting $= y_o$

    Primary response (controlled variable) $= P_L$

    Disturbance input $= Q_L$

Part 2: Suppose that the valve displacement y is measured from the static equilibrium position of the system. The equation of motion for the valve spool device is

$$m\ddot{y} = -b\dot{y} - k(y - y_0) + A(P_s - P_c) \tag{i}$$

The flow through the chamber orifice is given by

$$Q_c = k_0(P_L - P_c) = -A\frac{dy}{dt} + \frac{V_c}{\beta}\frac{dP_c}{dt} \tag{ii}$$

The outflow Q from the spool port increases with y and decreases with the pressure drop $(P_L - P_s)$. Hence the linearized flow equation is

$$Q = k_q y - k_c(P_L - P_s)$$

Note that k_q and k_c are positive constants defined previously by equations 7.89 and 7.90.

The accumulator equation is

$$Q - Q_c - Q_L = \frac{V_a}{\beta} \frac{dP_L}{dt}$$

Substituting for Q and Q_c, we have

$$k_q y - k_c(P_L - P_s) - k_0(P_L - P_c) - Q_L = \frac{V_a}{\beta} \frac{dP_L}{dt}$$

or

$$k_q y - (k_c + k_0) P_L + (k_c P_s + k_0 P_c) - Q_L = \frac{V_a}{\beta} \frac{dP_L}{dt} \tag{iii}$$

The equations of motion are (i), (ii) and (iii).

Part 3: Using equations (i) through (iii), the block diagram shown in figure 7.52 can be obtained. Attention should be paid to the feedback path of load pressure P_L. This feedback is responsible for the pressure control characteristic of the valve.

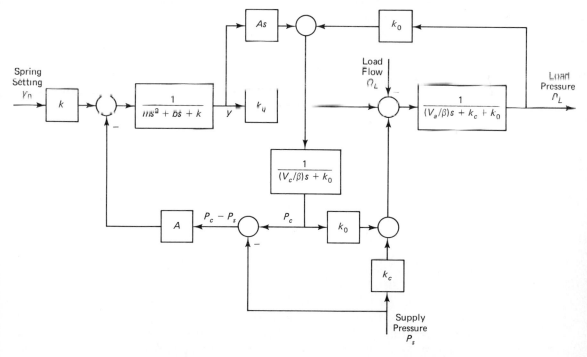

Figure 7.52. Block diagram for the single-stage pressure control valve.

Feedback Control

In figure 7.48a, we have identified two "natural" feedback paths that are inherent in the dynamics of the open-loop hydraulic control system. In figure 7.48b, we have shown the time constants associated with these natural feedback modules. Specifically, we observe the following:

1. A *pressure feedback path* and an associated *hydraulic time constant* τ_h
2. A *velocity feedback path* and an associated *mechanical time constant* τ_m

The hydraulic time constant is determined by the compressibility of the fluid. The larger the bulk modulus of the fluid, the smaller the compressibility, providing a smaller hydraulic time constant. Furthermore, τ_h increases with the volume of the fluid in the actuator chamber; hence, this time constant is related to the capacitance of the fluid as well.

The mechanical time constant has its origin in the energy dissipation (damping) in the moving parts of the actuator. Furthermore, as expected, the actuator becomes more sluggish as the inertia of the moving parts increases, resulting in an increased mechanical time constant.

These natural feedback paths usually provide a stabilizing effect to a hydraulic control system, but it is often not adequate for satisfactory operation of the system. Furthermore, the speed of response, wihich is usually a requirement that conflicts with stability, has to be adequate for proper performance. Consequently, it is necessary to include feedback control into the system by measuring response variables and modifying the system inputs, using the measured responses according to some control law. There are numerous methods of feedback control. Many of the conventional methods implement a combination of the following three basic control actions:

1. Proportional control (P)
2. Derivative control (D)
3. Integral control (I)

In proportional control, the measured responses are used directly in the control action. In derivative control, the measured responses are differentiated before they are used in the control action. Similarly, in integral control, the measured responses are integrated and used in the control action. Modification of the measured responses to obtain the control signal is done in many ways, including by electronic, digital, and mechanical means. For example, an analog hardware unit (termed a compensator or controller) that consists of electronic circuitry may be employed for this purpose. Alternatively, the measured signals, if they are analog, could be digitized and subsequently modified in a required manner through digital processing (multiplication, differentiation, integration, addition, etc.). This is the method used in digital control; either hardware control or software control may be used. Mechanical components may also be used to obtain a required control action. A feedback (closed-loop) hydraulic conrol system is shown by the block diagram in figure 7.53. Note that a feedback control law may be written as

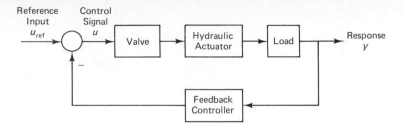

Figure 7.53. A closed-loop hydraulic control system.

$$u = u_{\text{ref}} - f(y) \qquad (7.121)$$

where $f(y)$ denotes the modifications made to the measured output y in order to form the control (error) signal u. The reference input u_{ref} is specified.

Example 7.16

A mechanical linkage is employed as the feedback device for a servovalve of a hydraulic actuator. The arrangement is illustrated in figure 7.54a. The reference input is u_{ref}, the input to the servovalve is u, and the displacement (response) of the actuator piston is y. A coupling element is used to join one end of the linkage to the piston rod. The displacement at this location of the linkage is x.

Show that rigid coupling gives proportional feedback action (figure 7.54b). Now, if a viscous damper (damping constant b) is used as the coupling element and if a spring (stiffness k) is used to externally restrain the coupling end of the linkage (figure 7.54c), show that the resulting feedback action is lead compensation. Next, if the damper and the spring are interchanged (figure 7.54d), what is the resulting feedback control action?

Solution For all three cases of coupling, the relationship between u_{ref}, u, and x is the same. To derive this, we introduce the variable θ to denote the clockwise rotation of the linkage. With the linkage dimensions h_1 and h_2 defined as shown in figure 7.54a, we have

$$u = u_{\text{ref}} + h_1 \theta$$

$$x = u_{\text{ref}} - h_2 \theta$$

Now, by eliminating θ, we get

$$u = (r + 1)u_{\text{ref}} - rx \qquad \text{(i)}$$

where

$$r = h_1/h_2 \qquad \text{(ii)}$$

For rigid coupling (figure 7.54b),

$$y = x$$

Hence, from equation (i), we have

$$u = (r + 1)u_{\text{ref}} - ry \qquad (7.122)$$

Clearly, this is a proportional feedback control law.

Next, for the coupling arrangement shown in figure 7.54c, by equating forces in the spring and the damper, we have

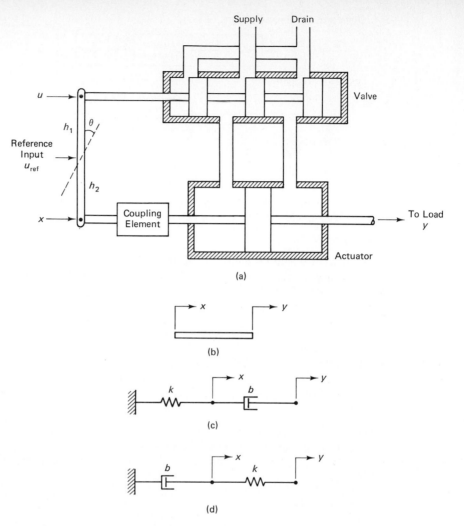

Supply   Drain

Valve

u

h_1 θ

Reference
Input
u_{ref}

h_2

x

Coupling
Element

To Load
y

Actuator

(a)

x y

(b)

x b y

k

(c)

x k y

b

(d)

Figure 7.54. (a) A servovalve and actuator with a feedback linkage. (b) Rigid coupling (proportional feedback). (c) Damper-spring coupling (lead compensator). (d) Spring-damper coupling (lag compensator).

$$kx = b(\dot{y} - \dot{x}) \qquad \text{(iii)}$$

Introducing the Laplace variable, we have the transfer-function relationship corresponding to equation (iii):

$$x = \frac{bs}{bs + k} y \qquad \text{(iv)}$$

By substituting equation (iv) in (i), we get

$$u = (r + 1)u_{\text{ref}} - \frac{rbs}{bs + k} y \qquad (7.123)$$

Note that the feedback transfer function

$$G_c(s) = \frac{rbs}{bs + k} \tag{7.124}$$

is a lead compensator, because the numerator provides a pure derivative action.
Finally, for the coupling arrangement shown in figure 7.54d, we have

$$b\dot{x} = k(y - x) \tag{v}$$

The corresponding transfer-function relationship is

$$x = \frac{k}{bs + k} y \tag{vi}$$

By substituting equation (vi) in (i), we get the transfer-function relationship for the feedback controller:

$$u = (r + 1)u_{\text{ref}} - \frac{rk}{bs + k} y \tag{7.125}$$

Note that the feedback transfer function in this case is

$$G_c(s) = \frac{rk}{bs + k} \tag{7.126}$$

This is clearly a lag compensator, because the denominator dynamics of the transfer function provide the lag action and the numerator has no dynamics (independent of s).

Constant-Flow Systems

So far, we have discussed only *valve-controlled* hydraulic actuators. There are two types of valve-controlled systems:

1. Constant-pressure systems
2. Constant-flow systems

We have considered only the constant-pressure system, in which the supply pressure P_s to the servovalve is maintained constant. Since there are four flow paths for a four-way spool valve, an analogy can be drawn with a Wheatstone bridge circuit operating under a constant supply voltage. Each arm of the bridge corresponds to a flow path. The flow through the actuator is represented by a load resistance connected across the bridge. In a *constant-flow system*, the supply flow is kept constant at varying pressure levels. This system is analogous to a constant-current Wheatstone bridge. Constant-flow operation requires a constant-flow pump, which is more economical than a variable-flow pump. Despite this, constant-pressure systems are more commonly used in practical applications.
Valve-controlled hydraulic actuators are the most common type in industrial applications. They are particularly useful when more than one actuator is powered by the same hydraulic supply.

Pump-Controlled Hydraulic Actuators

Pump-controlled hydraulic drives are suitable when only one actuator is needed to drive a process. A typical configuration of a pump-controlled hydraulic drive system is shown in figure 7.55. A variable-flow pump is driven by an electric motor (typically, an AC motor). The pump feeds a hydraulic motor, which in turn drives the load. Control is provided by the flow control of the pump. This may be accomplished in several ways—for example, by controlling the pump stroke (see figure 7.44) or by controlling the pump speed using a frequency-controlled AC motor. Typical hydraulic drives of this type can provide positioning errors less than 1° at torques in the range 200 to 2,000 lbf.in.

Hydraulic Accumulators

Since hydraulic fluids are quite incompressible, one way to increase the hydraulic time constant is to use an *accumulator*. An accumulator is a tank that can hold excessive fluid during pressure surges and release this fluid to the system when the pressure slacks. In this manner, pressure fluctuations can be filtered out from the hydraulic system. There are two common types of hydraulic accumulators:

1. Gas-charged accumulators
2. Spring-loaded accumulators

In a gas-charged accumulator, the top half of the tank is filled with air. When high-pressure liquid enters the tank, the air compresses, making room for the incoming liquid. In a spring-loaded accumulator, a movable piston, restrained from the top of the tank by a spring, is used in place of air. The operations of the two types are quite similar.

Pneumatic Control Systems

Pneumatic control systems operate in a manner similar to hydraulic control systems, and pneumatic pumps, servovalves, and actuators are quite similar in design to their hydraulic counterparts. The basic differences include the following:

1. The working "fluid" is air, which is far more compressible than hydraulic oils. Hence, thermal effects and compressibility should be included in any meaningful analysis.

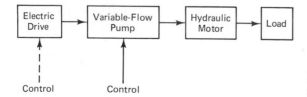

Figure 7.55. Configuration of a pump-controlled hydraulic drive system.

2. The outlet of the actuator and the inlet of the pump are open to the atmosphere (no reservoir tank is needed).

By connecting the pump (hydraulic or pneumatic) to an accumulator, the flow into the servovalve can be stabilized and the excess energy can be stored for later use. This minimizes undesirable pressure pulses, vibration, and fatigue loading. Hydraulic systems are usually employed in heavy-duty control tasks, whereas pneumatic systems are particularly suitable for medium to low-duty tasks. Both hydraulic and pneumatic control loops might be present in the same control system. For example, in a manufacturing cell, hydraulic control can be used for parts transfer and machining operations, and pneumatic control can be used for fine manipulation and minor adjustments in tools or work pieces. In general, pneumatic devices are less costly, primarily because the cost of the working fluid is not a major factor.

We will not extend our analysis of hydraulic systems to include air as the working fluid. The reader may consult a book on pneumatic control for information on pneumatic actuators and valves.

Flapper Valves

Flapper valves, which are relatively inexpensive and operate at low-power levels, are commonly used in pneumatic control systems. This does not rule them out for hydraulic control applications, however. A schematic diagram of a single-jet flapper valve used in a piston-cylinder actuator is shown in figure 7.56. If the nozzle is completely blocked by the flapper, the two pressures P_1 and P_2 will be equal, balancing

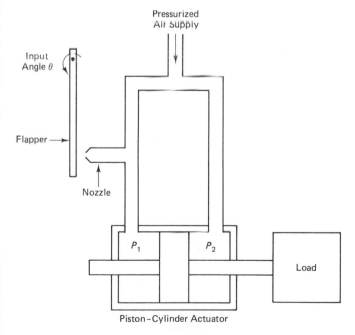

Figure 7.56. A pneumatic flapper valve system.

the piston. As the clearance between the flapper and the nozzle increases, the pressure P_1 drops, thus creating an imbalance force on the piston of the actuator. For small displacements, a linear relationship between the flapper clearance and the imbalance force can be assumed.

Note that the operation of a flapper valve requires fluid leakage at the nozzle. This does not create problems in a pneumatic system. In a hydraulic system, however, this not only wastes power but also wastes hydraulic oil unless a collecting tank and a return line to the oil reservoir are employed. For more stable operation, double-jet flapper valves should be employed. In this case, the flapper is mounted symmetrically between two jets. The pressure drop is still highly sensitive to flapper motion, potentially leading to instability. To reduce instability problems, pressure feedback, using a bellows unit, can be employed.

Example 7.17

Draw a schematic diagram to illustrate the incorporation of pressure feedback, using a bellows, in a flapper-valve pneumatic control system. Describe the operation of this feedback control scheme, giving the advantages and disadvantages of this method of control.

Solution   One possible arrangement for external pressure feedback in a flapper valve is shown in figure 7.57. Its operation can be explained as follows: If pressure P_1 drops, the bellows will contract, thereby moving the flapper closer to the nozzle, thus increasing P_1. Hence, the bellows acts as a *mechanical feedback* device that tends to regulate pressure disturbances.

The advantages of such a device are

1. It is a simple, robust, low-cost mechanical device.
2. It provides mechanical feedback control of pressure variations.

The disadvantages are

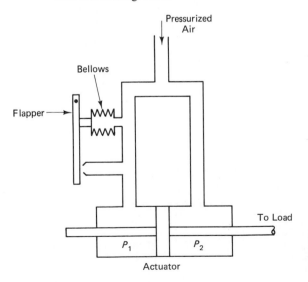

Figure 7.57.   External pressure feedback for a flapper valve, using a bellows.

1. It can result in a low-bandwidth system if the inertia of the bellows is large.

2. It introduces a time delay, which can have a destabilizing effect, particularly at high frequencies.

Example 7.18

What is a multistage servovalve? Describe its operation. What are the advantages of using several valve stages?

Solution A multistage servovalve uses several servovalves in series to drive a hydraulic actuator. The output of the first stage becomes the input to the second stage. A common combination is a hydraulic flapper valve and a hydraulic spool valve, operating in series. A multistage servovalve is analogous to a multistage amplifier.

The advantages of multistage servovalves are

1. A single-stage servovalve will saturate under large displacements (loads). This is overcome by using several stages, with each stage being operated in its linear region. Hence, a large operating range (load variations) is possible without introducing excessive nonlinearities, particularly saturation.

2. Each stage will filter out high-frequency noise, giving a lower overall noise-to-signal ratio

The disadvantages are

1. They cost more and are more complex than single-stage servovalves.

2. Because of series connection of several stages, failure of one stage will bring about system failure (a reliability problem).

3. Multiple stages result in decreased overall bandwidth (lower speed of response).

Hydraulic Circuits

A typical hydraulic control system consists of several components—such as pumps, motors, valves, piston-cylinder actuators, and accumulators—that are interconnected through piping. It is convenient to represent each component with a standard graphic symbol, so that the overall system can be represented by a circuit diagram, with the symbols for various components joined by lines to denote flow paths. Circuit representations of some of the many hydraulic components are shown in figure 7.58. A few explanatory comments would be appropriate. The inward solid pointers in the motor symbols indicate that a hydraulic motor receives hydraulic energy. Similarly, the pointers in the pump symbols show that a hydraulic pump gives out hydraulic energy. In general, the arrows inside a symbol show fluid flow paths. The external spring and arrow in the relief valve symbol shows that the unit is adjustable. There are three basic types of hydraulic line symbols. A solid line indicates a primary hydraulic flow. A broken line with long dashes is a *pilot line,* which indicates the control of a component. For example, the broken line in the relief valve symbol indicates that the valve is controlled by pressure. A broken line with short dashes represents a drain line or leakage flow. In the spool valve symbols, P denotes the supply port (pressure P_s) and T denotes the discharge port to the reservoir (zero pres-

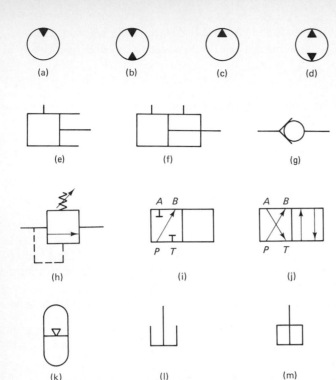

Figure 7.58. Typical graphic symbols used in hydraulic circuit diagrams: (a) motor; (b) reversible motor; (c) pump; (d) reversible pump; (e) single-acting cylinder; (f) double-acting cylinder; (g) ball-and-seat check valve; (h) relief valve (adjustable and pressure-operated); (i) two-way spool valve; (j) four-way spool valve; (k) accumulator; (l) vented reservoir; (m) pressurized reservoir.

sure). Finally, note that ports *A* and *B* of a four-way spool valve are connected to the two ports of a double-acting hydraulic cylinder (see figure 7.45a).

PROBLEMS

7.1. Prepare a table to compare and contrast the following types of motors:
    (a) Conventional DC motor
    (b) Torque motor
    (c) Stepper motor
    (d) Induction motor
    (e) AC synchronous motor
In your table, include terms such as power capability, speed controllability, speed regulation, linearity, operating bandwidth, starting torque, power supply requirements, commutation requirements, and power dissipation. Discuss a practical method for reversing the direction of rotation in each of these types of motors.

7.2. What factors generally govern
    (a) The electrical time constant
    (b) The mechanical time constant
of a motor? Compare typical values for these parameters and discuss how they affect the motor response.

7.3. In equivalent circuits for DC motors, iron losses (e.g., eddy current losses) in the stator are usually neglected. A way to include these effects is shown in figure P7.3. Iron

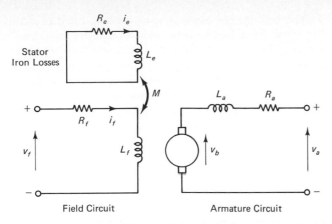

Stator Iron Losses

R_e i_e

L_e

M

Field Circuit

R_f i_f

v_f

L_f

Armature Circuit

L_a R_a

v_b

v_a

Figure P7.3. Equivalent circuit for a separately excited DC motor with iron losses in the stator.

losses in the stator poles are represented by a circuit with resistance R_e and self-inductance L_e. The mutual inductance between the field circuit and the iron loss circuit is denoted by M. Note that

$$M = k \sqrt{L_f L_e}$$

where L_f is the self-inductance in the field circuit and k denotes the coupling constant. With perfect coupling (no flux leakage between the two circuits), we have $k = 1$. But usually, k is less than 1. The circuit equations are

$$v_f = R_f i_f + L_f \frac{di_f}{dt} - M \frac{di_e}{dt}$$

$$0 = R_e i_e + L_e \frac{di_e}{dt} - M \frac{di_f}{dt}$$

The parameters and variables are defined in figure P7.3. Obtain the transfer function for i_f/v_f. Discuss the case $k = 1$ in reference to this transfer function. In particular, show that the transfer function has a "phase lag" effect.

7.4. Explain the operation of a brushless DC motor. How does it compare with the principle of operation of a stepper motor?

7.5. The following parameter values are given for a DC motor:

$$R_a = 5 \, \Omega, \qquad R_f = 20 \, \Omega, \qquad k = 1 \, \text{N.m/A}^2$$

Note that

$$T_m = k i_f i_a$$

Assume that the supply voltage is 115 V. Plot the steady-state torque-speed curves for
(a) A shunt-wound motor
(b) A series-wound motor
(c) A compound-wound motor with $R_{f1} = R_{f2} = 10 \, \Omega$
Using these curves, compare the steady state performance of the three types of motors.

7.6. What is the electrical damping constant in a DC motor? Determine expressions for this constant for the three types of DC motors mentioned in problem 7.5. In which case is this a constant value? Explain how the electrical damping constant could be experimentally determined. How is the dominant time constant of a DC motor influenced by the electrical damping constant? Discuss ways to decrease the motor time constant.

7.7. Explain why the transfer-function representation (speed/input voltage) for a separately excited DC motor (field-controlled or armature-controlled) is more accurate than that for shunt-wound, series-wound, or compound-wound DC motors.

7.8. Explain the differences between full-wave circuits and half-wave circuits in thyristor (SCR) control of DC motors. For the SCR drive circuit shown in figure 7.19a, sketch the armature current time history.

7.9. In the chopper circuit shown in figure 7.19a, suppose that $L_o = 100$ mH and $v_{ref} = 200$ V. If the worst-case amplitude of the armature current is to be limited to 1 A, determine the minimum chopper frequency.

7.10. For a DC motor, the rated torque and rated speed are known and are denoted by T_o and ω_o, respectively. The rotor inertia is J. Determine an expression for the dominant time constant of the motor.

7.11. Compare DC motors with AC motors in general terms. In particular, consider mechanical robustness, cost, size, maintainability, and speed control capability.

7.12. Compare frequency control with voltage control in induction motor control, giving advantages and disadvantages. The steady-state slip-torque relationship of an induction motor is given by

$$T_m = \frac{aSv_f^2}{[1 + (S/S_b)^2]}$$

with the parameter values $a = 1 \times 10^{-3}$ N.m/V^2 and $S_b = 0.25$. If the line voltage $v_f = 241$ V, calculate the breakdown torque. If the motor has two pole pairs per phase and if the line frequency is 60 Hz, what is the synchronous speed (in rpm)? What is the speed corresponding to the breakdown torque? If the motor drives an external load that is modeled as a viscous damper of damping constant $b = 0.03$ N.m/rad/s, determine the operating point of the system. Now, if the supply voltage is dropped to 163 V through voltage control, what is the new operating point? Is this a stable operating point?

7.13. Chopper circuits are used to "chop" a DC voltage so that a DC pulse signal results. This type of signal is used in DC motor control because the pulse width for a given pulse frequency determines the mean voltage of the pulse signal. Inverter circuits are used to generate an AC voltage from a DC voltage. The switching (triggering) frequency of the inverter determines the frequency of the resulting AC signal. This (inverter circuit) method is used in frequency control of AC motors. Both types of circuits use thyristor elements for switching. Indicate how an AC signal could be obtained using a chopper and a high-pass filter.

7.14. Using sketches, describe how pulse-width modulation (PWM) effectively varies the average value of the modulated signal. Explain how one could obtain
(a) A positive average
(b) A zero average
(c) A negative average
by pulse-width modulation. Show how PWM is useful in the control of DC motors. List the advantages and disadvantages of pulse-width modulation.

7.15. Show that the root-mean-square (rms) value of a rectangular wave can be changed by phase-shifting it and adding to the original signal. What is its applicability in the control of induction motors?

7.16. The direction of the rotating magnetic field in an induction motor (or any other type of AC motor) can be reversed by changing the supply sequence of the phases to the stator poles (i.e., phase-switching). An induction motor can be decelerated quickly in this manner. This is known as "plugging" an induction motor. The slip versus torque relationship of an induction motor may be expressed as

$$T_m = k(S)v_f^2$$

Show that the same relationship holds under plugged conditions, except that $k(S)$ is replaced by $-k(2 - S)$. Sketch the $k(S)$, $k(2 - S)$, and $-k(2 - S)$ curves from $S = 0$ to $S = 2$. Using these curves, indicate the nature of the torque acting on the rotor during plugging. (*Hint:* $k(S) = (aS)/[1 + (S/S_b)']$).

7.17. What is a servomotor? AC servomotors that can provide torques on the order of 7,000 oz.in. at 3,000 rpm are commercially available. Describe the operation of an AC servomotor that uses a two-phase induction motor. A block diagram for an AC servomotor is shown in figure P7.17. Describe the purpose of each component in the system and explain the operation of the overall system. What are the advantages of using an AC amplifier after the inverter circuit in comparison to using a DC amplifier before the inverter circuit?

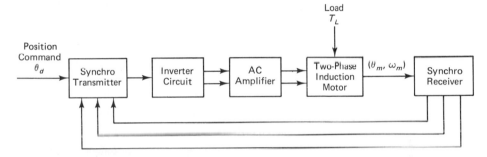

Figure P7.17. AC servomotor, using a two-phase induction motor and synchro.

7.18. Consider the two-phase induction motor discussed in example 7.13. Show that the motor torque T_m is a linear function of the control voltage v_c when $k(2 - S) = k(S)$. How many values of speed (or slip) satisfy this condition? Determine these values.

7.19. Consider the induction motor in problem 7.12. Suppose that the line voltage $v_f = 200$ V and the line frequency is 60 Hz. The motor is rigidly connected to an inertial load. The combined moment of inertia of the rotor and load as $J_{eq} = 5$ kg.m². The combined damping constant is $b_{eq} = 0.1$ N.m/rad/s. If the system starts from rest, determine, by computer simulation, the speed time history $\omega_L(t)$ of the load (and rotor). (*Hint:* Assume that the motor is a torque source, with torque represented by the steady-state speed-torque relationship.)

7.20. What are common techniques for controlling
  (a) DC motors?
  (b) AC motors?
  Compare these methods with respect to speed controllability.

7.21. Describe the operation of a single-phase AC motor. List several applications of this actuator. Is it possible to realize three-phase operation using a single-phase AC supply? Explain your answer.

7.22. In drive systems for robotic manipulators, it is necessary to minimize backlash. Discuss the reasons for this. Conventional techniques for reducing backlash in gear drives include preloading, the use of bronze bearings that automatically compensate for wear, and the use of high-strength steel and other alloys that can be machined accurately and that have minimal wear problems. Discuss the shortcomings of some of the conventional methods of backlash reduction. Discuss the operation of a drive that has virtually no backlash problems.

7.23. In some applications, it is necessary to apply a force without creating a motion. Discuss one such application. Discuss how an induction motor could be used in such an application. What are the possible problems?

7.24. List three types of hydraulic pumps and compare their performance specifications. A position servo system uses a hydraulic servo along with a synchro as the feedback sensor. Draw a schematic diagram and describe the operation of the control system.

7.25. Giving typical applications and performance characteristics (bandwidth, load capacity, controllability, etc.), compare and contrast DC servos, AC servos, hydraulic servos, and pneumatic servos.

7.26. What is a multistage servovalve? Describe its operation. What are the advantages of using several valve stages?

7.27. Discuss the origins of the hydraulic time constant in a hydraulic control system that consists of a four-way spool valve and a double-acting cylinder actuator. Indicate the significance of this time constant. Show that the dimensions of the right-hand-side expression in equation 7.118 are [time].

7.28. Sometimes either a pulse width–modulated (PWM) alternating current (AC) signal or a DC signal with a superimposed constant-frequency AC signal (dither) is used to drive the valve actuator (torque motor) of a hydraulic actuator. What is the main reason for this? Discuss the advantages and disadvantages of this approach.

7.29. Compare and contrast valve-controlled hydraulic systems with pump-controlled hydraulic systems. Using a block diagram, explain the operation of a pump-controlled hydraulic motor. What are its advantages and disadvantages over a frequency-controlled AC servo?

7.30. Explain why accumulators are used in hydraulic systems. Sketch two types of hydraulic accumulators and describe their operation.

7.31. Explain the components of the hydraulic system given by the circuit diagram in figure P7.31. Describe the operation of the overall system.

7.32. If the load on the hydraulic actuator shown in figure 7.47 consists of a rigid mass restrained by a spring, with the other end of the spring connected to a rigid wall, write equations of motion for the system. Draw a block diagram for the complete system consisting of a four-way spool valve, and give the transfer function that corresponds to each block.

7.33. Suppose that the coupling of the feedback linkage shown in figure 7.54c is modified as shown in figure P7.33. What is the transfer function of the controller? Show that this feedback controller is a lead compensator.

7.34. The moment of inertia of the rotor of a motor (or any other rotating machine) can be determined by a run-down test. With this method, the motor is first brought up to an acceptable speed and then quickly turned off. The motor speed versus time curve is obtained during the run-down period that follows. A typical run-down curve is shown in figure P7.34. Note that the motor decelerates because of its resisting torque τ_r during this period. The slope of the run-down curve is determined at a suitable value of speed $(\overline{\omega}_m)$ in figure P7.34. Next, the motor is brought up to this speed $(\overline{\omega}_m)$, and the torque

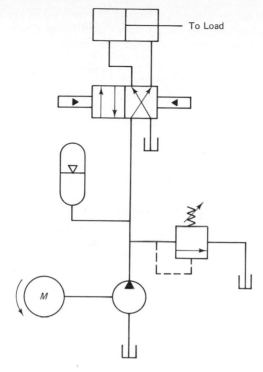

Figure P7.31. Hydraulic circuit diagram.

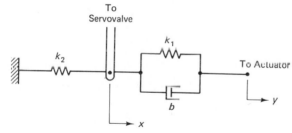

Figure P7.33. A mechanical coupling with lead action for a hydraulic servovalve.

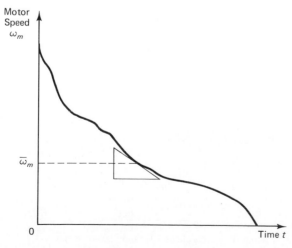

Figure P7.34. Data from a run-down test on an electric motor.

$(\bar{\tau}_r)$ needed to maintain the motor steady at this speed is obtained (either by direct measurement of torque or by computing, using field current measurement and a known value for the torque constant). Explain how the rotor inertia J_m could be determined from this information.

7.35. The sketch in figure P7.35 shows a half-sectional view of a flow control valve that is intended to keep the flow to a hydraulic load constant regardless of variations of the load pressure P_3 (disturbance input).

   (a) Discuss briefly how the valve works physically, remembering that the flow will be constant if the pressure drop across the fixed area orifice is constant.

   (b) Write all equations describing the dynamics. The mass, damping constant, and spring constant of the valve are m, b, and k, respectively. The volume of oil under pressure P_2 is V, and the bulk modulus is β. Make the usual linearizing assumptions.

   (c) Set up a block diagram for the system from which the dynamics and stability could be studied.

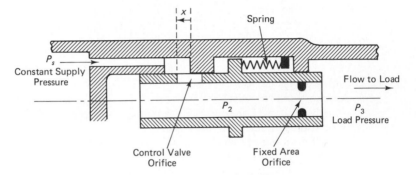

Figure P7.35. A flow control valve.

7.36. A schematic diagram of a pump stroke–regulated hydraulic power supply is shown in figure P7.36. The system uses a three-way spool valve of the type described in example 7.15 (see figure 7.51). This valve controls a spring-loaded piston that, in turn, regulates the pump stroke by adjusting the swash plate angle of the pump. The load pressure P_L is to be regulated. This pressure can be set by setting the preload x_0 of the spring in the pressure control valve (y_0 in figure 7.51). The load flow Q_L enters into the hydraulic system as a disturbance input.

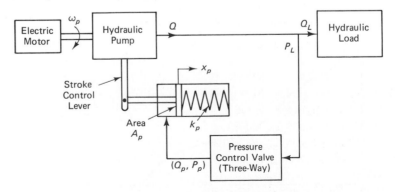

Figure P7.36. A pump stroke-regulated hydraulic power supply.

(a) Describe briefly how the control system works physically.

(b) Write all equations for the system dynamics, assuming that the pump stroke mechanism and the piston inertia can be replaced by an equivalent mass m_p moving through x_p. The corresponding spring and damping constants are k_p and b_p, respectively. The piston area is A_p. The mass, spring constant, and damping constant for the valve are m, k, and b, respectively. The valve area is A_v and the valve spool movement is x_v. The volume of oil under pressure P_L is V_t, and the volume of oil under pressure P_p is V_o (volume of oil in the cylinder chamber). The bulk modulus of the oil is β.

(c) Set up a block diagram for the system from which the behavior of the system could be investigated. Indicate the inputs and outputs.

(d) If Q_p is relatively negligible, indicate which control loops can be omitted from the block diagram. Hence, derive an expression for the transfer function $x_p(s)/x_v(s)$ in terms of the system parameters.

7.37. A schematic diagram of a solenoid-actuated flow control valve is shown in figure P7.37a (see page 424). The valve rod is spring-loaded, with a compressive spring of stiffness k. The mass of the valve rod assembly (all moving parts) is m, and the associated equivalent viscous damping constant is b. The downward movement of the valve rod is denoted by x. The voltage supply to the valve actuator (solenoid) is denoted by v_i. For a given voltage v_i, the solenoid force is a nonlinear function of x. This steady-state characteristic curve is shown in figure P7.37b, along with the spring characteristic. Assuming that the inlet pressure and the outlet pressure of the fluid flow are constant, the flow rate will be determined by the valve position x. Hence, the objective of the valve actuator would be to set x using v_i.

(a) Show that for a given input voltage v_i, the resulting equilibrium position (x) of the valve will always be stable.

(b) Describe how the relationship between v_i and x could be obtained
 (i) Under quasi-static conditions
 (ii) Under dynamic conditions

7.38. Component sizing is an important consideration in the design of a hydraulic control system. You are asked to design a hydraulic system for a radar positioning drive. Specifically, you must

(a) Select a suitable hydraulic motor and suitable gearing to drive the load (radar).

(b) Select a suitable pump for continuous hydraulic power supply.

(c) Determine the gear reduction from motor to load (radar).

(d) Determine the inlet pressure at the hydraulic motor.

(e) Determine the pump outlet pressure.

The following data are given:

Load intertia = 1,000 slug.ft² (lb.ft.s²)

Maximum load speed = 1 rad/s

Maximum load acceleration = 10 rad/s²

Wind torque = 1,000 lbf.ft

Distance from pump to hydraulic motor = 10 ft

Size of pipeline (steel) = 0.5 in. O.D. and 0.035 in. thickness

Maximum supply pressure of fluid = 3,000 psi

Hydraulic power loss in pipeline = 5%

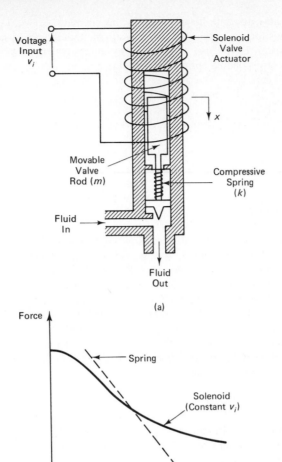

Figure P7.37. (a) A solenoid-actuated flow control valve. (b) Steady-state characteristics of the valve.

Pump leakage = motor leakage = 5%
Motor efficiency = gear box efficiency = 95%

Assume that the hydraulic fluid is MIL-H-5606. Design

1. A pump-controlled system in which the pump directly supplies a controlled flow to hydraulic motor

2. A valve-controlled system in which a valve is used between the pump and the hydraulic motor to supply a controlled flow to the motor (assume a 5 percent valve leakage).

You may use a mechanical engineering handbook to obtain the pump and hydraulic motor specifications (usually the same specs are given for both pumps and motors) and to estimate the pressure loss in steel piping carrying MIL-H-5606 oil.

7.39. In some types of (indirect-drive) robotic manipulators, joint motors are located away from the joints and torques are transmitted to the joints through transmission devices such as gears, chains, cables, and timing belts. In some other types of (direct-drive) manipulators, joint motors are located at the joints themselves, the rotor being on one link and the stator being on the joining link. Discuss the advantages and disadvantages of these two designs.

7.40. What are the advantages and disadvantages of hydraulic (and pneumatic) actuators in comparison with electric motors? A pneumatic rack-and-pinion actuator is an on/off device that is commonly used as a rotary valve actuator. A piston or diaphragm in the actuator is moved by allowing compressed air into the valve chamber. This rectilinear motion is converted into a rotary motion through a rack-and-pinion device in the actuator. Single-acting types with spring return and double-acting types are commercially available. Using sketches, explain the operation of a piston-type single-acting rack-and-pinion actuator with a spring-restrained piston. Could the sensitivity of the device be improved by using two pistons and racks coupled with the same pinion? Explain.

7.41. Consider the problem of servovalve selection for a hydraulic drive. The first step is to choose a suitable hydraulic actuator (ram or motor) that meets the load requirements (see problem 7.38). This establishes the load flow Q_L and the load pressure P_L at the operating speed of the load. The supply pressure P_s is also known. Note that under no-load conditions, the pressure drop across the servovalve is P_s, and under normal operating conditions, it is $P_s - P_L$. In manufacturers' specifications, the rated flow of a servovalve is given at some specified pressure drop (e.g., 1,000 psi). Since the flow is proportional to the square root of the pressure drop, we can determine the required flow rating for the valve (typically, by increasing the computed value of the rated flow by 10 percent to allow for leakage, fluctuations in load, etc.). This rated flow is one factor that governs the choice of a servovalve. The second factor is the valve bandwidth, which should be several times larger than the primary resonant frequency of the load for proper control. Typical information available from servovalve manufacturers includes the frequency corresponding to the 90° phase lag point of the valve response. This frequency may be used as a measure of the valve bandwidth. Select a servovalve for the valve-controlled radar drive given in problem 7.38. The first step is to obtain a catalog with a data sheet from a well-known servovalve manufacturer. Assume that the resonant frequency of the radar system is 10 Hz.

7.42. In brushless motors, commutation is achieved by switching the stator phases at the correct rotor positions (e.g., at the points of intersection of the static torque curves corresponding to the phases, for maximum average static torque). In the text, we noted that the switching points can be determined by measuring the rotor position using an incremental encoder. Since we need only know the switching points (continuous measurement of rotor position is not necessary), and since these points are uniquely determined by the stator magnetic field distribution, we can use Hall effect sensors instead of an encoder to detect the switching points. Specifically, Hall effect sensors are located at switching points on the stator, and a magnet assembly is located on the rotor. As the rotor rotates, a magnetic pole will trigger an appropriate Hall effect sensor, thereby generating a switching signal (pulse) for commutation at the proper rotor position. Since Hall effect sensors have several disadvantages—such as hysteresis, thermal problems, and noise due to stray magnetic fields (see problem 5.14)—it might be better to use fiber optic sensors for brushless commutation. Describe how the fiberoptic method of motor commutation works.

7.43. A magnetically levitated rail vehicle uses the induction motor principle for traction. Magnetic levitation is used for suspension of the vehicle slightly above the emergency guide rails. Explain the operation of the traction system of this vehicle, particularly identifying the stator location and the rotor location. What kinds of sensors would be needed for the traction and levitation control systems? What type of control strategy would you recommend for the vehicle control?

7.44. Write an expression for the back e.m.f. of a DC motor. Show that the armature circuit of a DC motor may be modeled by the equation

$$v_a = i_a R_a + k\phi\omega_m$$

where v_a = armature supply voltage
i_a = armature current
R_a = armature resistance
ϕ = field flux
ω_m = motor speed

and k is a motor constant. Suppose that $v_a = 20$ V DC. At standstill, $i_a = 20$ A. When running at a speed of 500 rpm, the armature current was found to be 15 A. If the speed is increased to 1,000 rpm while maintaining the field flux constant, determine the corresponding armature current.

7.45. Consider a pneumatic speed sensor that consists of a wobble plate and nozzle arranged like a pneumatic flapper valve. The wobble plate is rigidly mounted at the end of a rotating shaft, so that the plane of the plate is inclined to the shaft axis. Using a sketch, explain the principle of operation of the wobble plate pneumatic speed sensor.

7.46. Friction drives (traction drives) that use rollers making frictional contact have been proposed as transmission devices. One possible application is for joint drives in robotic manipulators that presently use gear transmission. A prototype 2 d.o.f. traction drive joint for a space robot is shown in figure P7.46a. One advantage of friction roller drives is the absence of backlash. Another advantage is finer motion resolution in com-

(a)

Figure P7.46. The NASA-ORNL traction drive joint: (a) The prototype (courtesy of NASA Langley Research Center); (b) a schematic representation.

J = Moment of Inertia
b = Damping Constant
K = Stiffness
T = Torque
n = Gear Ratio

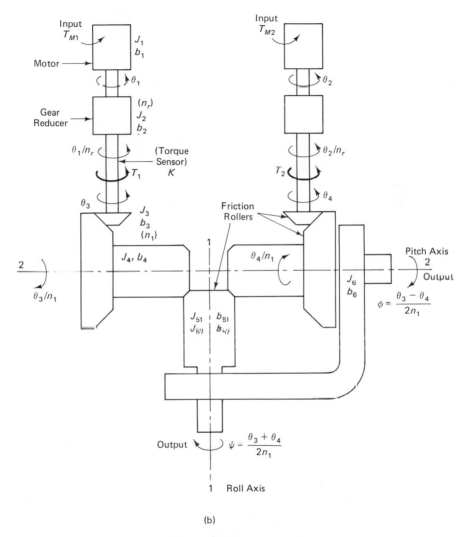

(b)

Figure P7.46. *(continued)*

parison to gear drives. Give two other possible advantages and several disadvantages of friction roller drives. A schematic representation of the NASA traction drive joint is shown in figure 7.46b. Write dynamic equations for this model for evaluating its behavior.

7.47. In the harmonic drive configuration shown in figure 7.22, the outer rigid spline is *fixed* (stationary), the wave generator is the *input* member, and the flexispline is the *output* member. Five other possible combinations of harmonic drive configurations are tabulated here. In each case, obtain an expression for the gear ratio in terms of the standard ratio (for figure 7.22) and comment on the drive operation.

Case	Rigid spline	Wave generator	Flexispline
1	Fixed	Output	Input
2	Output	Input	Fixed
3	Input	Output	Fixed
4	Output	Fixed	Input
5	Input	Fixed	Output

REFERENCES

BLACKBURN, J. F., REETHOF, G., and SHEARER, J. L. (Ed.). *Fluid Power Control*. MIT Press, Cambridge, 1960.

DATTA, S. K. *Power Electronics and Controls*. Reston, Reston, Va., 1985.

DESILVA, C. W. "Motor Controllers." *Measurements and Control Journal* 115: 271–74, February 1986.

FITZGERALD, A. E., KINGSLEY, C., JR., and UMANS, D. D. *Electric Machinery*, 4th ed. McGraw-Hill, New York, 1983.

GIBSON, J. E., and TUTEUR, F. B. *Control System Components*. McGraw-Hill, New York, 1958.

MERRIT, H. E. *Hydraulic Control Systems*. Wiley, New York, 1967.

PALM, W. J., III. *Control Systems Engineering*. Wiley, New York, 1986.

Answers to Numerical Problems

Chapter 1

1.10. f_o, 23.5 dB

Chapter 2

2.20. 7.5%
2.22. 1.32%
2.25. (a) 2 mm; (b) 2.5×10^3 N/m²; (c) 74.0 dB
2.29. (b) 0.99%
2.30. (b) $e_{JM} - \pm 1.1\%$, $e_{J\ell} = \pm 11.0\%$, $e_r = \pm 1.2\%$, $e_{\alpha \ell} = \pm 1.0\%$
2.32. 0.5°

Chapter 3

3.5. 31.2%, 17.6%, −9.6%, compare with −20%
3.6. At $\theta = \theta_{max}$
3.16. 100 seconds
3.19. 98.37%, 99.99%, 170 Hz
3.22. 1,000 Hz
3.33. 9,500 Ω

Chapter 4

4.12. 2.5%
4.14. 144.0, 5.9%
4.24. 0.01 N, 2 mm, 72 dB

Chapter 5

5.5. 0.088°

5.11. 12 bits, 12 tracks, 4,096 sectors

Chapter 6

6.3. 200 steps/revolution, 6 teeth/pole, 50 rotor teeth

6.4. 15°, 7.5°, 15°

6.6. 5 phases, 20 stator poles, 0.72°, 100 rotor teeth

6.8. 2 ms, 50 steps/second

6.20. 100, 50, 25, and 5 steps/second

6.26. **(a)** 30°; **(b)** 15°

6.27. 7.2° pitch, 1.8° step, 200 steps/revolution, 2 poles/phase, 6 teeth/pole or 4 teeth/pole with interpolar gaps

Chapter 7

7.9. 500 Hz

7.12. 7.26 N.m; 1,800 rpm; 1,350 rpm; 5 N.m, 1,620 rpm; 2.5 N.m, 810 rpm

7.44. 10 A

Index